THEODERIC IN ITALY

Theoderic in Italy

JOHN MOORHEAD

CLARENDON PRESS · OXFORD

Oxford University Press, Great Clarendon Street, Oxford OX2 6DP
Oxford New York
Athens Auckland Bangkok Bogota Bombay
Buenos Aires Calcutta Cape Town Dar es Salaam
Delhi Florence Hong Kong Istanbul Karachi
Kuala Lumpur Madras Madrid Melbourne
Mexico City Nairobi Paris Singapore
Taipei Tokyo Toronto

and associated companies in
Berlin Ibadan

Published in the United States by
Oxford University Press Inc., New York

First published by Oxford University Press 1992
Special edition for Sandpiper Books Ltd., 1997

British Library Cataloguing in Publication Data
Data available

Library of Congress Cataloging in Publication Data

ISBN 0-19-814781-3

5 7 9 10 8 6 4

Printed in Great Britain by
Bookcraft Ltd
Midsomer Norton, Somerset

ACKNOWLEDGEMENTS

This book had its genesis in 1987, when the Institute for Advanced Studies in the University of Edinburgh honoured me with the election to a fellowship. It is hard to imagine a more congenial place for reading and thinking, and I am grateful to the Director, Professor Peter Jones, the staff, and those whose fellowships overlapped with mine for making my visit so happy. Norman James kindly read a draft of the book, to its great benefit, and later it profited from the secretarial skills of Mary Kooyman and Suzanne Lewis; I particularly wish to thank the latter for help generously given at a difficult time. I gladly thank Richard Bartlett for discussing aspects of the thesis he is writing on Ennodius. For many kindnesses during the period during which work was proceeding I shall always be grateful to Meryl McLeod. Finally, I must record that for years I have been benefiting from discussions on late antiquity with Robert Markus. I hope he will not find this book unworthy of all I have learned from him.

John Moorhead

Feast of St Ambrose of Milan, 1991

TABLE OF CONTENTS

ABBREVIATIONS

AASS	Acta Sanctorum
ACO	Acta Conciliorum Oecumenicorum, ed. E. Schwartz
CIL	Corpus Inscriptionum Latinarum
CC	Corpus Christianorum
SL	Series Latina
CM	Continuatio Medievalis
CSEL	Corpus Scriptorum Ecclesiasticorum Latinorum
Ennodius,	
pan.	panegyricus dictus Theoderico
V. Epiph.	*Vita Epiphani*
MGH	Monumenta Germaniae Historica
AA	Auctores Antiquissimi
Ep.	Epistulae
Leg.	Leges
SRLI	Scriptores Rerum Langobardicarum et Italicarum
SRM	Scriptores Rerum Merovingicarum
SS	Scriptores
PG	Patrologia Graeca, ed. J. P. Migne
PL	Patrologia Latina, ed. J. P. Migne
PLS	Patrologia Latina Supplementum
PLRE	*The Prosopography of the Later Roman Empire*, i, ed. A. H. M. Jones, J. R. Martindale, and J. Morris (Cambridge, 1971); ii, ed. J. Martindale (Cambridge, 1980)
Procopius,	
anec.	Anecdota
build.	De aedificiis
BG	Bellum Gothicum
BP	Bellum Persicum
BV	Bellum Vandalicum
PW	A. Pauly and G. Wissowa, eds., *Real-Encyclopädie der classischen Altertumswissenschaft*
RIS	Rerum Italicarum Scriptores
SC	Sources chrétiennes
Settimane di studio	Settimane di studio del centro Italiano di studi sull'alto medioevo

Introduction

The study which follows describes the achievements of the ruler of a group of barbarians who came to hold power in Italy from 493 to 526. Theoderic is certainly the best-documented barbarian of his era, his career in Italy having attracted the detailed notice of a number of contemporary authors, and the issues which have concerned me in writing about it are those which have seemed accessible by means of written evidence. This is not to deny the importance of work being done by archaeologists, or the validity of approaches which stress social and economic considerations more than I have, but for this enquiry, centred as it is on one person, the precision which documentary evidence affords, at least in principle, has seemed more useful than the broader perspectives offered by other classes of evidence, whatever their interest in suggesting contexts in which Theoderic's achievement can be located. There are, of course, positive dangers which my approach entails: that of assuming that topics which contemporary authors found worthy of their attention, such as theological and ecclesiastical affairs, were as important in reality as their writings suggest, and, more subtly, that of approaching a barbarian by means of sources written in their great majority by Romans, who, naturally enough, described Theoderic in terms familiar to themselves which may have failed to do justice to the degree to which he remained a barbarian. It therefore follows that the approach I adopt here has involved considerable attention being paid to these authors, and for the convenience of readers it may be as well to introduce briefly the five most important of them.

The most significant body of evidence bearing on Theoderic is the letters of Cassiodorus. Cassiodorus was born, probably in the late 480s, into a family with civil service traditions which held estates at Squillace, the classical Scyllaeum, on the toe of Italy. In the course of a distinguished career, which included a consulship in 514, he held the offices of quaestor under Theoderic (507–11), *magister officiorum* under Theoderic and his son Athalaric (523–7), and praetorian prefect of Italy under Theodahad and Witigis (533–7), producing a remarkable amount of official correspondence on behalf of the Gothic administration which he edited and published under the title *Variae*, alluding to the various styles in which the letters were composed. The first five books of the *Variae* contain 235 letters written on behalf of Theoderic while Cassiodorus was quaestor and *magister officiorum*, and are of primary importance in understanding the king's policies during these years, as well as affairs in the state which he ruled. In general, the letters are characterized by a chancery style typical of late antiquity and by an interest in the recondite which seems to have distinguished Cassiodorus, features which indicate that we would be unwise to accept the sentiments in them as being necessarily those of the figures in whose names they were written. Nevertheless, we shall later suggest that on at least some occasions Theoderic's personal views were expressed. Cassiodorus also composed some works of lesser importance in understanding the career of Theoderic: a chronicle, written on the occasion of the consulship of Theoderic's son-in-law Eutharic in 519, a history which is now lost but was drawn on by another writer discussed below, and some orations. State patronage stood behind all these works.

A major difficulty attends the *Variae*, however. While Cassiodorus was preparing his material for publication after he relinquished the office of praetorian prefect in 537, an army which the emperor Justinian had dispatched to overthrow the Gothic state had already occupied much of Italy, and one cannot help wondering whether Cassiodorus tailored his material in view of the current realities. Whichever way the war turned out, one of its losers had been king Theodahad, murdered by the Goths in December 536 after they had chosen Witigis to replace him, and it need not be accidental that Cassiodorus published letters critical of this figure (*var*. 4.39, 5.12, and, by implication, 12.20). On the other hand, the Carthaginian writer Liberatus reports that Theodahad wrote to Pope Agapetus and the senate threatening to murder the senators, their wives, and their sons and daughters if the emperor were not asked to withdraw the army he had sent to Italy (*brev*. 21); if such correspondence occurred, there is no trace of it in the *Variae*. Cassiodorus published letters praising people involved in the fall of Boethius, which occurred towards the end of Theoderic's reign (esp. *var*. 5.3f., 5.40f.), yet included in his collection letters which praised Boethius (esp. *var.* 1.45; see too 1.10, 2.40). On the whole, it is difficult to detect any clear line of bias running through the *Variae*, and we may enjoy the thought that in this collection of letters Cassiodorus, a man whose career had not been free of expediency, presents a selection of material which has not been slanted.

Another author who flourished in Italy in the time of Theoderic is less well known. Ennodius was born in the south of Gaul, perhaps at Arles, in 473 or 474, but found his way to Italy before the advent of Theoderic, and served among the clergy of the churches of Pavia and Milan before being elevated in about 513 to the see of Pavia, which he occupied until the time of his death in 522. He enjoyed close relations with the Church of Rome, being a member of a circle loyal to Pope Symmachus during a schism in that Church and a friend of that pope's successor, Hormisdas, and cultivated a number of Roman aristocrats. During the period 495/6–513 Ennodius produced a body of writing of exceptional interest for the reign of Theoderic, which included a panegyric in honour of the king, apparently produced in 507; 297 letters, the great majority to clergy and laity resident at Rome or Ravenna, written over the period 501–13; and a Life of St Epiphanius, who was bishop of Pavia at the beginning of Theoderic's reign. Ennodius' works have never received the attention they deserve, doubtless in part because of the difficulties presented by his latinity, with the result that his significance both as a historical source and a literary figure has been underestimated, and while working with Ennodius for the purpose of this study I have increasingly come to realize how much interest there would be in a study devoted to him in his own right.

The most useful narrative source for the reign of Theoderic is the second part of a work which is often called the *Excerpta Valesiana*, named from its first editor, Henri de Valois, who published his edition in Paris in 1636. The first part of this document deals with the reign of Constantine; the second, the *pars posterior*, by a different author, deals with Theoderic and his antecedents. It is this second author whom I style 'Anonymous Valesianus', while acknowledging that he is to be distinguished from the author of the first part, who could equally be so described. The style and content of the second section suggest that it is the work of one author who found himself telling a story which became less pleasant towards its end, and the order in which the paragraphs appear in the manuscripts seems to me to be that which the author assigned to them, although this is not to say that the sequence in which they occur is necessarily the order in which the events described took place. I take this author to have been writing not long after the death of Theoderic and to stand in some relation to both Italian annalistic material and Eastern sources.

The narrative of the *pars posterior* begins at para. 36 of the full text, and deals with the coming of Nepos to Italy in 474, and by para. 70 the author has advanced to the year 500. Thereafter he discusses various good works of Theoderic in general terms (71–3), the succession to the emperor Anastasius, who died in 518 (74–8), and the alleged illiteracy of Theoderic (79), before resuming his narrative with the consulship of Eutharic in 519 (80) and continuing as far as the death of Theoderic. It therefore follows that this author describes the period before Theoderic's coming to power in Italy, his early years there, and the last years of his reign, but with a hiatus in the middle which extends for approximately half his reign. This is particularly unfortunate given that no work of Ennodius can be dated to later than 513 and that Cassiodorus' tenure as *magister officiorum* only commenced in 523. The decade 510–20 thus emerges as poorly covered by our chief sources.

Two less full narrative accounts of the reign of Theoderic were written in Constantinople at about the middle of the sixth century. In a work which has been published under the title *De origine actibusque Getarum*, Jordanes, a Catholic Goth, purports to describe the history of the Goths from an original migration 'from the island of Scandza' until the time of the marriage of the Ostrogothic princess Matasuentha to Germanus, a nephew of the emperor Justin, which occurred in 551 or 552. Jordanes represents himself as abbreviating a work of Cassiodorus 'on the origin and deeds of the Goths' (*praef.* 1); this will have been the twelve-volume work on the origin of the Goths which Cassiodorus twice describes himself as having written (*Var. praef.* 11, 9.25.4). Unfortunately Cassiodorus' work has not survived, and both the degree to which Jordanes was indebted to it and the circumstances in which it became available to him remain conjectural. The discussion of Theoderic provided by Jordanes is chiefly concerned with military and diplomatic affairs, which he invariably interprets in a way favourable to him.

Procopius, the author of a voluminous work on some of Justinian's wars, accompanied the emperor's general Belisarius to

Sicily in 535 and seems to have remained with him in Italy until 540. The account of Theoderic he provides in the opening pages of the *Gothic War* must therefore have been based on material gathered in Italy something like a decade after the death of the king. Although Procopius' works are all in Greek, he was a good latinist, capable of quoting the Sibylline oracles (*BG* 1.24.28–31). He clearly had extensive dealings with the inhabitants of Italy while he was there, and may well also have spoken with some of the Italian refugees who flocked to Constantinople during the 540s. His familiarity with Italian affairs can be deduced from the large number of the native inhabitants of Italy whom he names in the course of his narrative, whereas very few of the inhabitants of Africa are named in his *Vandal War*. On the whole his attitude towards the Ostrogothic state is cool, although he describes Theoderic in positive terms.

The bibliography at the end of this work lists all the primary sources I have consulted, except for standard classical and patristic texts cited by way of background and for comparison. It cannot claim to be so full for secondary works, which are multitudinous and become more voluminous annually. Reading one's way into the literature concerning Theoderic is an exercise in humility: one comes to appreciate more and more the powerful nature of much of the scholarship in the field, while yet becoming increasingly aware of the extent to which historians reflect the concerns of the times in which they write. The relationship between the Germanic and Latin parts of Western Europe, the conduct proper to those who find themselves living in territory occupied by another power, the role of the Catholic Church in Europe, the survival of minority cultures: these are issues of recent or contemporary urgency which all find resonance in the story of Theoderic. Modern scholars who have written about him, including some of those I admire the most, have frequently been influenced by political or cultural questions which have seemed important to their generation or themselves. I cannot hope to have escaped this danger; indeed, given that there are so many excellent guides, my responsibility for errors of interpretation, not to mention fact, is the greater.

I

The Way to Italy

ODOVACER

The fall of the Roman Empire in the West was a protracted affair. The first half of the fifth century saw the invasions of Italy by the Visigoths which culminated in the sack of Rome in 410 and their subsequent imperially approved settlement in south-west Gaul and Spain, the effective abandonment of Britain, the loss of Africa to the Vandals, the settlement of the Sueves in Spain, and the advance of the Burgundians in the East of Gaul, which combined to produce a situation in which imperial power was increasingly confined to Italy and dependent on the goodwill of barbarians for its exercise. Hence, when Attila the Hun launched an attack on the north of Gaul in 451 at the head of a host which included Ostrogoths, Gepids, Rugians, Sciri, Heruls, and Thoringians, he was confronted by a coalition led by the Roman *magister militum* Aetius and the Visigothic king Theoderic. At a battle conventionally located at the Catalaunian Fields Attila was checked; an expedition he launched against Italy in the following year proved fruitless; and in 453 the great king of the Huns died.[1] His demise was followed by important events among both his erstwhile opponents and supporters.

In 454 Attila's opponent Aetius was murdered in Rome, together with his friend the praetorian prefect Boethius, apparently at the instigation of a wealthy noble, Petronius Maximus. Looking back at the event from the early sixth century the Byzantine chronicler Marcellinus *comes* commented that with Aetius, Roman sovereignty had fallen in the West, something which had not been restored at the time when he wrote;[2] and however we choose to assess the various stages in the crumbling of Roman power in the

[1] A. H. M. Jones, *The Later Roman Empire* (Oxford, 1964), 182ff. provides a general account of these and the following events.

[2] Marcellinus, *chron. s.a.* 454.2; cf. Sidonius Apollinaris, *carm.* 7.357ff. and Procopius, *BV* 1.4.28.

West there can be no doubt that 454 marked an important stage in its decline. In the following year the emperor Valentinian III was murdered, quite possibly at the instigation of Petronius Maximus, who quickly succeeded him as emperor. But a few months later the Vandal king Geiseric attacked Rome. Petronius fled but was captured and murdered by the people of the city; his body was thrown into the Tiber and the way was left open for the Vandals to spend a fortnight sacking the city. Among the booty they took home were Valentinian's wife Eudoxia and his daughters Eudocia and Placidia. Before long, *de facto* power in Italy lay in the hands of the barbarian *magister militum* and patrician Ricimer; following his death in 472 the Burgundian king Gundobad saw fit to intervene. By the time Nepos became emperor in 474 Gundobad had withdrawn, and the following year the Pannonian Orestes forced the emperor Nepos to withdraw to Dalmatia, and made his son Romulus 'Augustulus' emperor. In practical terms the office of emperor had become superfluous, and in 476 it was done away with.

Our most detailed sources for the events of 476 were written in the sixth century, mid-way through the century in the cases of Jordanes and Procopius, and are difficult to evaluate. But it seems that the barbarian troops in Italy, who by now may well have constituted the entire army, made demands of Orestes concerning their maintenance.[3] Orestes unwisely turned them down, whereupon the troops approached one of their own number, Odovacer, who agreed to do as they wished if they appointed him ruler. On 23 August Odovacer was made king; five days later Orestes was captured and put to death near Placentia. Odovacer then advanced on Ravenna, near which city Orestes' brother Paul was killed on 4 September.[4] The young emperor Romulus was still alive, but Odovacer declined to murder him. He was given an annual income of 6,000 solidi and sent to live with his relations at the castellum Lucullanum in Campania.[5] So ended the line of Roman emperors in the West.

The significance of the deposition of Romulus by Odovacer has been widely discussed. Procopius was aware that Rome had ceased to be subject to the Romans in 476, and Marcellinus offered a famous interpretation: 'The Western empire of the Roman race,

[3] See the discussion below, at the beginning of Ch. 2.

[4] Main sources: anon. Vales. 37; Jordanes, *get.* 242 and *rom.* 344; Procopius, *BG* 1.1.4–6. More sources and dates: *PLRE* 2.792.

[5] Anon. Vales. 38; Jordanes, *get.* 242 and *rom.* 344; Marcellinus, *chron. s.a.* 476.

which Octavian Augustus first began to hold 709 years after the founding of the city, perished with this Augustulus in the 522nd year that his predecessors had been emperors of the realm; since then kings of the Goths have held Rome.'[6] But Procopius and Marcellinus were Byzantine authors, and we may doubt whether the people of Rome in 476 were aware that an epochal event had occurred; indeed, as we have seen, Marcellinus himself felt that the disagreeable situation which had begun to obtain in 454, no less than that which commenced in 476, had yet to be rectified. Barbarian strongmen were scarcely a novelty in the West, as the examples of Stilicho at the end of the fourth and the beginning of the fifth centuries and later of Ricimer indicate, and according to a story current in Rome at the beginning of the sixth century the senate itself had invited the barbarian Aspar to become emperor; if we accept that this occurred, it is probably to be dated to 457.[7] One wonders how important the events of 476 were in the eyes of those who lived through them in Italy. The absence of an emperor was not particularly startling, as there had been periods of over 12 months in 456–7 and again in 465–7 when there had been no emperor, and in any case the early coinage of Odovacer has been interpreted as implying an uneasy acceptance of the claims of Nepos, who only died in 480;[8] certainly the emperor Zeno, when confronted in about 477 with simultaneous embassies from Odovacer seeking his approval to rule in the West without an emperor and from Nepos seeking his aid in regaining his throne, counselled Odovacer to submit to the authority of Nepos and welcome him when he returned.[9] Further, Odovacer may not have been as exotic at the head of Italy as has sometimes been assumed, for if we accept, as has recently been suggested, that he was a brother of the *magister militum* Armatus, he would have been the nephew of Basiliscus, who briefly usurped the imperial throne in 475–6, and the empress Verina, wife of the emperor Leo, hence easily assimilable in the

[6] Procopius: *BG* 1.14.14. Marcellinus: *chron.* 476.2, followed by Jordanes, *get.* 243 and *rom.* 345. Consult on this passage in particular B. Croke, 'A.D. 476: The Manufacture of a Turning Point', *Chiron*, 13 (1983), 81–119, and below, p. 159.

[7] *MGH AA* 12.425.

[8] J. C. P. Kent, 'Julius Nepos and the Fall of the Western Empire', in *Corolla memoriae Erich Swoboda dedicata* (Böhlau, 1966), 146–50.

[9] Malchus, *frag.* 10 (Blockley, 14). Note too that Odovacer's forwarding to Zeno booty taken in his later war with the Rugians can be construed as implying some recognition of the emperor's sovereignty: M. McCormick, 'Odoacer, Emperor Zeno and the Rugian Victory Legation', *Byzantion*, 47 (1977), 212–22.

ranks of the military aristocracy of the empire.[10] Indeed, there would seem to have been no reason for Odovacer not making himself emperor, and his reticence in this matter may support an interpretation of the events of 476 as falling far short of the epochal in their significance.

We therefore have no reason to see the advent of Odovacer as marking an important change in Italy, and in this respect, above all others, the reign of Odovacer anticipated the Ostrogothic period which was to follow it. The two most powerful elements in Italian society, the Catholic Church and the landowning aristocracy, emerged from the demise of the empire in the West substantially unscathed. Odovacer obviously went out of his way to conciliate influential sectors of Roman opinion; for example, from 480 to 490 a Western consul was appointed annually, which implies a measure of imperial acceptance of his position for that period, and it was the first of these consuls, Basilius, who, acting on Odovacer's behalf, was to be influential in securing the election of Pope Felix III, one of the few aristocratic popes of the period, in 483. Similarly, the seats of the senators in the Flavian amphitheatre were restored,[11] so that in the sixth century Odovacer was remembered as having been a man of good will, despite his having favoured the Arian heresy.[12] It is true that in the time of Theoderic some Romans described Odovacer in the blackest terms, but later in this chapter we shall suggest that their condemnations were politically interested. Italian contemporaries can have vested the events of 476 with little significance, and as the years passed they would have found little to complain of in Odovacer's administration.

Yet however much it is necessary to demythologize the year 476, it is clear that it occurred within a period of great change for Italy. Odovacer's coming to power was followed by a series of territorial adjustments: he annexed Dalmatia, following the death of Nepos; gained Sicily at the expense of a tribute to be paid to the Vandals;

[10] S. Krautschick, 'Zwei Aspekte des Jahres 476', *Historia*, 35 (1986), 344–71 at 349, on the basis of John of Antioch, *frag*. 209.1 (Blockley, 372) read as *'Οδόακρος πατρὸς δε 'Ιδικῶος, καί ἀδελφὸς 'Ονούλφου καί 'Αρμτίου*, but see already G. B. Picotti, 'Il "patricius" nell'ultima età imperiale e nei primi regni barbarici d'Italia', *Archivio storico italiano*, 7th ser. 9 (1928), 3–80 at 78. Such a reading would force us to confront the possibility of a connection between the revolts of Basiliscus in 475 and Odovacer in 476.

[11] A. Chastagnol, *Le Sénat romain sous le règne d'Odoacre* (Bonn, 1966).

[12] Anon. Vales. 48.

and was forced to cede Provence to the Visigoth Euric.[13] It is possible to interpret these events as the culmination of a decades-long trend which saw Italy shed control over provinces ever closer to itself and finally emerge as an isolated political unit.[14] A similar process was under way in economic life. By the end of the second century the hegemony of Italy in the Roman Empire had been replaced by that of Africa, but recent work by archaeologists is making it clear that as the fifth century progressed African exports were in turn declining, and there is no strong evidence for Italy importing more goods from elsewhere.[15] The Italy over which Odovacer and later Theoderic presided was more isolated, politically and economically, than it had been for centuries.

It is therefore not to be wondered at that Odovacer's fall was the result of developments outside Italy. His doom was sealed by an obscure series of events in the east and north. In 484 the *magister militum* Illus, whose relations with Zeno had been strained for some years, rebelled against the emperor, and among those whose help he sought was Odovacer.[16] It is not clear whether Odovacer took steps to intervene, but it may not be accidental, given the ability to interfere in Western affairs which Eastern emperors retained, that within a few years he found himself involved in a war with the Rugians, who had been moving into Noricum, which could roughly be described as lying to the south of the Danube between Passau and Vienna. Trouble broke out when Fredericus, the son of the Rugian king Feletheus (also known as Feba and Fevva) killed the king's brother Ferderuchus (also known as Fredericus), and war with Odovacer followed. Our sources fail to make it clear which party was the aggressor, but the Rugians were soundly defeated. Feletheus and his wife were captured and taken to Italy, while Fredericus fled. Later it came to Odovacer's attention that Fredericus had returned, whereupon he sent against him his brother Onoulf with a large force. Fredericus again fled, but Odovacer felt that the task of defending Noricum had become too great, and

[13] Dalmatia: Cassiodorus, *chron. sa* 481; *Auct. haun. ordo prior et ordo post, sa* 482. Africa: Victor of Vita, *hist. pers.* 1.14, with Ennodius, *pan.* 70. Provence: Procopius, *BG* 1.12.20.

[14] A classical statement in L. M. Hartmann, *Geschichte Italiens im Mittelalter*, 1. *Das Italienische Königreich* (Leipzig, 1897), 35.

[15] See the comments of A. Caradini, C. Panella, and F. Villedieu, in A. Giardina, ed., *Società romana e impero tardoantico*, 3. *Le merci gli insediamenti* (Rome, 1986), esp. 10, 160, 452.

[16] John of Antioch, *frag.* 214.2.

Onoulf announced that all the Romans were to migrate to Italy. This they did, the operation being overseen by Count Pierius.[17] Among the baggage of the Romans were the relics of the holy man Severinus, who had died in 482. They were taken to a fort named Feleter or Felether, but later, on the initiative of the widow Barbaria and with the blessing of Pope Gelasius, hence in the period 492–6, they were placed in the castellum of Lucullanum, near Naples,[18] presumably to be identified with the place to which Romulus retired in 476. Odovacer may have felt that with the evacuation of Noricum his dealings with the Rugians had come to an end, but such thoughts would have been sadly inaccurate, for when Fredericus fled for the second time it was to Theoderic the Ostrogoth, who at the time was based at Novae, the modern Svishtov, in Moesia.[19] Shortly afterwards Theoderic marched against Odovacer.

THEODERIC AND THE GOTHS

Little is known of the early years of Theoderic. He was the son of Theodemer who, together with his brothers Valamer and Videmer, the former of whom enjoyed seniority, had led the Ostrogothic forces which formed part of the confederation of nations fighting on behalf of Attila in his confrontation with the Romans and Visigoths in 451, the Ostrogoths having been for some time under the sway of the Huns.[20] His mother was Erelieva, a concubine of Theodemer and a Catholic who had taken in baptism the name Eusebia.[21] She has been variously described as an Ostrogothic princess, for which

[17] Eugippius, *V. Sev.* 44.3–5; cf. John of Antioch, *frag.* 214.7. Western sources disagree as to whether the aggressor was Odovacer (Eugippius, *V. Sev.* 44.4) or the king of the Rugians (*Auct. haun. ordo prior, s.a.* 487).

[18] Eugippius, *V. Sev.* 44.5–7, 46.

[19] Ibid. 44.4. Note that according to Procopius the Rugians were a Gothic nation (ἔθνος Γοτθικόν: *BG* 3.2.1); it is presumably to this period that the alliance described in this passage should be dated.

[20] Son of Theodemer: Jordanes, *get.* 80, 269 and *rom.* 347; Cassiodorus, *var.* 11.1.19, whose testimony is to be preferred to that of Byzantine authors (cited *PLRE* 2.1077) and anon. Vales. 42, 58 that he was the son of Valamer. This is not the last time we will find the Anonymous Valesianus writing from an Eastern perspective. Three brothers and Attila: Jordanes, *get.* 199f. On the general background of the people generally referred to as the Ostrogoths, H. Wolfram, *History of the Goths* (Berkeley, Calif., 1988) is in a class by itself; see too T. S. Burns, *A History of the Ostrogoths* (Bloomington, Ind., 1984).

[21] Anon. Vales. 58.

there is no evidence, and as possibly of Roman origin, but one of our sources unambiguously calls her *gothica*, and from her taking a new name at her baptism we may deduce she was a convert to Catholicism, presumably from paganism.[22] She certainly remained a Catholic, and it is tempting to believe that the influence which Pope Gelasius later felt she could bring to bear on her son in the interests of the Church already existed when Theoderic was young, but we have no evidence of this.[23] The birth of the son of Theodemer and Erelieva must be set in the context of the rapid decline of the power of the Huns following the death of Attila in 453. The chronology is obscure, but it seems that after the Gepids had defeated the Huns and killed Attila's eldest son at the battle of the Nedao, an engagement at which the Ostrogoths were conspicuous by their failure to support the victorious side, the three brothers, who commanded the allegiance of at least a large number of the Ostrogoths, came to an agreement with the Romans and settled with their followers in Pannonia, south of the Danube and to the east of Noricum.[24] This turn of events brought no joy to the Huns, and more of Attila's sons attacked Valamer. They were defeated, and Valamer sent a messenger with the joyful news to Theodemer. On the very day he arrived, Theoderic was born.[25]

As with many great people, the circumstances of Theoderic's birth are obscure. It is dated by our sources only with reference to this battle, but the battle itself cannot be assigned a precise date: Attila died in 453, quite possibly early in the year,[26] but the chronology of the following events is difficult to deduce from the narrative of Jordanes. Nevertheless, it is possible to throw some light on the question of when Theoderic was born by working back-

[22] Ostrogothic princess: M. Schönfeld, *Wörterbuch der Altgermanischen Personen- und Völkernamen* (Heidelberg, 1911), 75. Possible Roman origin: Wolfram, *History*, 262, but see W. Ensslin, *Theoderich der Grosse*, 2nd edn. (Munich, 1959), 10.

[23] Gelasius to Erelieva, *Epistulae Theodericanae*, 4f. (*MGH AA* 12.390f.); cf. Ennodius, *pan*. 42, 'sancta mater'.

[24] Jordanes, *get*. 262–4, 268, with Cassiodorus, *var*. 3.23.2 on Pannonia as an earlier dwelling place of the Goths. A. Kiss, 'Ein Versuch die Funde und das Siedlungsgebiet der Ostgoten in Pannonien zwischen 456 und 471 zu bestimmen', *Acta archaeologica*, 31 (1979), 329–39 has brought archaeological data into play, suggesting a region between the Danube and the Mecsek mountains for Theodemer's zone.

[25] Jordanes, *get*. 268f.

[26] 453 is supplied by Prosper Tiro, *chron*. 1370; that Attila died early in the year may be implied by Hydatius, according to whom he retired to his own lands after his expedition to Italy in 452 and soon (*mox*) died: *cont. chron. Hier* 154.

wards from later events which can be securely dated. Theoderic celebrated his tricennalia in 500.[27] It is not clear what event was being commemorated, but it is hardly likely to have been anything prior to his return to the Ostrogoths from being a hostage at Constantinople and the victory he won shortly afterwards against the Sarmatian king Babai. When Theoderic defeated Babai he was 18 years of age,[28] hence he would have been born *c.*452/3. This is not impossible to reconcile with Jordanes' placing it some time after Attila's death in 453, and a date of 453 seems to create fewer difficulties than earlier or later dates.[29]

Despite their having been settled in Pannonia by the emperor, relations between that part of the Ostrogothic people led by the three brothers and Constantinople were not good. The Goths set to work devastating Illyricum early in the reign of Leo, and as a condition of the peace which was thereafter concluded with this emperor, Theoderic, who was just entering on his eighth year, was sent as a hostage to Constantinople. Jordanes believed that he was an attractive youth and so had deserved the favour which the emperor bestowed upon him.[30] He was to spend a decade in the royal city, and such experiences as he had there must have been significant in the development of the way in which he looked at the world. Jordanes was under no illusions as to the impact a trip to Constantinople could have on a barbarian; he attributed to the Visigoth Athanaric, who visited the city in 381, wonder at its suitability for ships, its walls, and the various races that flowed together there: 'Who can doubt that the emperor is a god (God?) on earth, and that whoever lifts a hand against him will be guilty of his own life.'[31] Again, the Georgian Nabarnugios (Peter), sent to Constantinople as a hostage at the age of 12 in 421, fell under the influence of pious men and became an ascetic, and it is worth recalling that Sidonius Apollinaris, admittedly seeking to be paradoxical, described Ravenna, by his time the seat of Roman government in the West, as

[27] Anon. Vales. 67 with Cassiodorus, *chron. s.a.* 500. [28] Jordanes, *get.* 282.

[29] Wolfram, *History*, 262 suggests 451 and makes the Hunnish defeat of which news came to Theodemer on the day Theoderic was born the battle of the Catalaunian Fields. Yet, according to Jordanes, Theodemer was present at that battle (*get.* 199), and so early a date flies in the face of this author's chronology more than I would care to. Ensslin originally dated Theoderic's birth to 456, but subsequently retracted this (O. J. Maenchen-Helfen, *The World of the Huns* (Berkeley, Calif., 1973), 157; cf. Ensslin, *Theoderich*, 10). *PLRE* 2.1078 supplies no evidence in support of its 'probably in 454'.

[30] Jordanes, *get.* 271. [31] Ibid. 143.

a place where eunuchs took to arms and *foederati* took to letters.[32] Our sources seem to take it for granted that Theoderic profited intellectually from his stay in Constantinople, and whether they assert this on the basis of informed knowledge or merely a sense that this had probably occurred we may assume that he did, in fact, become acquainted with profane wisdom.[33] While it would certainly be possible to make too much of the extent to which Theoderic appropriated contemporary and past Greek culture while in Constantinople,[34] we may accept that the Byzantines were fully capable of inculcating upon a young barbarian a world-view favourable to themselves, and so that Theoderic received some intellectual formation along classical lines, such culture as he acquired being in Greek rather than Latin.

After a decade in Constantinople Theoderic returned to his people. By this time Valamer had been killed by a group of Sciri and leadership had passed to Theodemer, who took vengeance in an expedition which saw the defeat of the Sueves and the Alamanni. Returning home to Pannonia he met Theoderic, who had been sent back by the emperor Leo with lavish gifts.[35] But Theoderic was quick to make it clear that he proposed to be his own man for, 18 years old as he was, without his father's knowledge he gathered together a force which our only source estimates at 6,000 men, crossed the Danube, and attacked Babai, king of the Sarmatians, who had recently defeated a Roman force. The young Goth was victorious. Babai was killed, and Theoderic carried off his household and property to his father. He went on to capture Singidunum (Belgrade) but, declining to return it to the Romans, subjected it to

[32] Nabarnugios: D. M. Lang, trans., *Lives and Legends of the Georgian Saints* (London, 1976), 59ff.; cf. Sidonius Apollinaris, *ep*. 1.8.2. One cannot help but note, however, that the infamous Vandal Huneric had been a hostage at Rome: Procopius, *BV* 1.4.13.

[33] John of Nikiu, trans. Charles, 88.48. This text must be used with care, however: Charles's translation is of an Ethiopian translation of a lost Arabic translation of a lost Greek (and Coptic?) original, and it is impossible to tell what degree of distortion has entered the Ethiopian text, itself apparently corrupt. But see too Ennodius, *pan*. 11, 'educavit te in gremio civilitatis Graecia'; Theophanes, *chron*. AM 5977, *οὐδὲ λόγων ἀμοίρου* (he had the best teachers), and John Malalas, *chron*. 383.

[34] W. Capelle, *Die Germanen der Völkerwanderung* (Stuttgart, 1940), 353 has him learning to read Homer and proceeds to deny that he was a barbarian. The impact of these years is stressed in particular by German scholars, e.g. A. Nagl, 'Theoderic 1der Grosse', PW 2/10.1746.58ff. and Ensslin, *Theoderich*, 14ff.; for a reaction against this general tendency, although not the specific case, see P. Riché, *Education and Culture in the Barbarian West* (Columbia, s.c, 1976), 52.

[35] Jordanes, *get*. 276–81.

his own sway.[36] Such exploits were anticipatory of later events in Theoderic's career, but for the time being they had little impact on the situation of the Goths, who, lacking food and clothing, determined to leave Pannonia. Lots were cast, and Videmer led some of the people to the West. He died on the march, but his son, also named Videmer, received gifts when in Italy from the emperor Glycerius; Glycerius' period as augustus, 473–4, provides the only date for this sequence of events. From Italy the Goths moved on to Gaul, where they formed one body with the Visigoths.[37] Theodemer, on the other hand, in conjunction with Theoderic moved eastward and advanced as far as Thessalonica, whereupon the Goths were granted seven towns in Macedonia. Peace was restored, but shortly afterwards Theodemer died, having first called together the Goths and designated Theoderic the heir to his kingdom.[38] By 476 the new king had led his following to lower Moesia.[39]

Theoderic's accession was followed by difficult years, dominated by his people's shortage of land. His position was complicated by the existence nearby of another Ostrogothic leader, his relative Theoderic Strabo ('the squinter') the son of Triarius, who commanded the allegiance of a considerable following;[40] given the high degree of fluidity characteristic of barbarian peoples in the settlement period, the later dominance which Theoderic the son of Theodemer exercised over the Ostrogoths was something which had to be achieved. Further, his relationship with Constantinople was awkward. As emperors frequently played off groups of barbarians against each other, and as the hold on the throne of the emperor Zeno, who succeeded Leo in 474, was often precarious, the stage was set for complex wheeling and dealing.[41] The task of

[36] Ibid. 282. Babai had been a member of the army recently defeated by Theodemer.

[37] Ibid. 283 f., with *rom*. 347.

[38] Jordanes, *get*. 285–8.

[39] Anon. Vales. 42.

[40] See for Theoderic Strabo, *PLRE* 2.1073 ff., with stemma 39 (p. 1331) for his relationship with Theoderic, and note as well references to Bigelis, 'Getarum rex' (Jordanes, *rom*. 336) and the imperial ally Sidimund, apparently an Ostrogoth (Malchus, *frag*. 18 (Blockley, *frag*. 20, pp. 438–40)). Hence the justice of Mommsen's comment that Theoderic became not a Volkskönig but a Gaukönig: T. Mommsen, *Gesammelte Schriften*, 6 (Berlin, 1910), 478.

[41] Some of the points made here are illustrated by a passage of Malchus which describes Zeno deciding to support Theoderic Strabo against Theoderic the son of Theodemer (*falso* Valamir) because the former was gaining strength, and the later threat of the latter's followers to defect to his rival: Malchus, *frag*. 14 f. (Blockley, 18.1 f.).

the historian who would understand these years is further complicated by the virtual silence of Jordanes and the circumstance that our sources, while extremely detailed at certain points, are fragmentary.[42] However, it is clear that Zeno asked for the help of Theoderic at the time of the usurpation of Basiliscus, who was himself allied to Theoderic Strabo, in 475–6, and this seems to have been forthcoming, for Theoderic was shortly afterwards made patrician, *magister militum praesentalis*, and Zeno's son-at-arms, as well as being designated the emperor's 'friend'.[43] But in about 478 Theoderic Strabo prevailed upon his namesake to abandon Zeno's cause, and the two Theoderics sent a joint embassy to the emperor, the son of Theodemer's demands including land and enough grain to tide his people over till the next harvest. Zeno responded by offering him lavish payments and the hand of the daughter of Olybrius, who had briefly been emperor in the West in 472, or of another woman of the aristocracy, if he made war on Theoderic Strabo.[44] Nothing came of these proposals, and Zeno subsequently allied himself with Theoderic Strabo, while the son of Theodemer devoted himself to plundering imperial territory, even after the death of his rival and namesake in 481; on one occasion he took Dyrrhachium (Durazzo, now Durrës), a mere 140 kilometres by sea from southern Italy, and suggested to the emperor that he might go to Dardania to restore the emperor Nepos.[45] But in 483 he came to terms with Zeno. Theoderic was given gifts, made *magister militum*, and held the office of consul in 484, while a statue of him on horseback was erected in front of the palace and his people were given lands in Dacia and Moesia; at the instigation of Zeno, Theoderic, in an act anticipatory of his later assassination of Odovacer, murdered the son of Theoderic Strabo, his own cousin

[42] The arguments of M. Errington, 'Malchos von Philadelphia, Kaiser Zenon und die zwei Theoderiche', *Museum Helveticum*, 40 (1983), 82–110, seem to have been met by R. C. Blockley, 'On the Ordering of the Fragments of Malchus' History', *Liverpool Classical Monthly*, 9 (1984), 152f.

[43] Request for help: anon. Vales. 42. Basiliscus and Theoderic Strabo: Malchus, *frag.* 11 (Blockley, 15). References to the distinctions received by Theoderic are conveniently supplied in *PLRE* 2.1079.

[44] Malchus, *frag.* 15–17 (Blockley, 18.2–4).

[45] See in particular Marcellinus, *chron. s.a.* 479, 482; Malchus, *frag.* 18 (Blockley, 20); John of Antioch, *frag.* 211.4, 213. Death of Theoderic Strabo: John of Antioch, *frag.* 211.5; Jordanes, *rom.* 346; Marcellinus, *chron. s.a.* 481. Capture of Dyrrhachium: John of Antioch, *frag.* 211.4; Malchus, *frag.* 18 (Blockley, 20); *Paschale campanum, s.a.* 478 (*MGH AA* 9.310).

Recitach.[46] In the year of his consulship he was dispatched to do battle with the rebel patrician Illus, who had sought the aid of Odovacer. The sources indicate that Zeno may have doubted Theoderic's trustworthiness; he was recalled and thereafter never helped Zeno.[47] In 486 he devastated Thrace,[48] and in the following year he went so far as to proceed to Constantinople with a large force, cutting an aqueduct of the city and only withdrawing when Zeno had paid him much money and sent his assailant's sister from the court.[49] It was probably on his return to his headquarters at Novae that Odovacer's enemy Fredericus met him. In 488 he agreed with Zeno that he would go to Italy and seek to overcome Odovacer. Accordingly, he departed for the West later that year.

WAR IN ITALY

At this point it will be worth our while to pause and interrogate our sources more closely, for they preserve very different traditions as to the origin of the Ostrogothic expedition to Italy. Some see it as having been initiated by the emperor Zeno. This perspective is very clear in the narrative of Procopius, who asserts, immediately after a brief account of the troubles Theoderic brought upon Constantinople, that the emperor Zeno, who understood how to settle to his advantage any situation in which he found himself, advised Theoderic to proceed to Italy, attack Odovacer, and win for himself and the Goths the western dominion. Jordanes, in his *Romana*, and Anonymous Valesianus are of similar import.[50] Other sources, however, disagree. Ennodius, in his panegyric written in honour of Theoderic in about 507, excludes any imperial involvement in the coming of the Goths to Italy; for him, it was simply a question of

[46] Jordanes, *get.* 289; Marcellinus, *chron. sa* 483; John of Antioch, *frag.* 214.3.

[47] John of Antioch, *frag.* 214.3–6, with Evagrius, *hist. eccl.* 3.27 and Theophanes, *chron. am* 5977.

[48] John of Antioch, *frag.* 214.7; Zachariah, *hist.* 6.6.

[49] John of Antioch, *frag.* 214.8; John Malalas, *chron.* 383; Marcellinus, *chron. sa* 487; Michael the Syrian, *chron.* 9.6; Procopius, *BG* 1.1.9; Theophanes, *chron. am* 5977. The unnamed sister, mentioned only by John of Antioch, may have been Amalafrida, but certainty is impossible.

[50] Procopius, *BG* 1.1.10, trans. Dewing, where I take the αὐτῷ to refer to Theoderic and not Zeno; cf. Theophanes, *chron. am* 5977 (Theoderic impelled to go to Italy); Jordanes, *rom.* 348 and Anon. Vales. 49 (*mittens eum ad Italiam*). See in general J. Moorhead, 'Theoderic, Zeno and Odovacer', *Byzantinische Zeitschrift*, 77 (1984), 261–6.

Theoderic's seeking revenge for the murder of some of his relatives by Odovacer. In his *Getica*, Jordanes held that Theoderic had been leading a life of luxury in Constantinople, when he heard that the situation of his people in Illyricum was not entirely satisfactory. He sought from Zeno permission to lead them to Rome, which the emperor was reluctant to give, but being unwilling to cause Theoderic unhappiness he agreed.[51] Evagrius, writing in the late sixth century, was aware of two traditions: according to one, which is credited to Eustathius of Epiphaneia, who died in 503, Theoderic went to Rome because he became aware that Zeno was plotting against him, but the other asserted that Theoderic went to Rome on Zeno's instruction.[52] Such divergent traditions impose caution on the historian. But it can be said that the account provided by Jordanes in his *Getica* must be false, for, as we have seen, prior to his departure for Italy Theoderic had not been living happily in Constantinople but had attacked the city. As is well known, Jordanes claimed that his *Getica* was little more than an abbreviation of the subsequently lost history of the Goths in 12 volumes written by Cassiodorus, and we may well see the propagandist hand of Cassiodorus behind this account, so much at variance with that supplied by Jordanes himself in the *Romana*.[53] Ennodius' account is of a piece with a tendency frequently encountered in his works to minimize the role of the empire in the West and, indeed, to assimilate Theoderic into the category of emperors.[54] Once these two sources, which both stand near the court of Theoderic, are removed, it becomes less difficult to credit Zeno with the initiative in the dispatch of Theoderic to Italy. The tactic of encouraging various barbarian groups to fight each other was, as Theoderic was already well aware from his own experience, commonly employed by em-

[51] Ennodius, *pan.* 25; Jordanes, *get.* 289–92. See too the brief treatments of Marcellinus, *chron. s.a.* 489, John of Nikiu, *chron.* 115.

[52] Evagrius, *hist. eccl.* 3.27.

[53] Jordanes, *get.* 1. His account of how the Goths came to be in Italy, which is legitimist yet attributes the initiative to Theoderic, may be compared with his account of how the Goths had earlier come to be in Pannonia: 'maluerunt a Romano regno terras petere quam cum discrimine suo invadere alienas, accipientesque Pannoniam . . .' (*get.* 264).

[54] This will be discussed in the following chapter, but note in particular *V. Epiph.* 109: Theoderic came to Italy 'dispositione caelestis imperii'. The phrase almost seems designed to exclude any role which might have been attributed to an earthly empire; see further M. Reydellet, *La Royauté dans la littérature latine de Sidoine Apollinaire à Isidore de Seville* (Paris, 1981), 153f.

perors,[55] and the history of the fifth century contains various examples of their allowing barbarian peoples to settle in defined areas; the settling of the Visigoths in Gaul would have furnished a relevant precedent. In 484 Zeno had deployed Theoderic against Illus, who had sought the support of Odovacer even if he had not enjoyed it; in 487 Theoderic moved against Constantinople; in 488 Odovacer's enemy Fredericus arrived at Novae. What would have been more natural than for Zeno to dispose of Theoderic by sending him against Odovacer? It is possible that Theoderic was in some way related to the Rugian royal house,[56] and the dire straits in which Zeno found himself in 487 would have been enough to lead him to take the initiative in the journey of the Goths to Italy. The great seventh-century chronicle attributed to Fredegarius tells a story according to which the people of Rome and Italy asked the emperor Leo to make Theoderic a patrician to deal with Odovacer.[57] The tale is wildly inaccurate, for Leo died in 473 and we have no reason to believe that the people of Rome and Italy were anything but satisfied with Odovacer's rule after he came to power in 476. But there is no need to question its implication that the emperor stood behind the move of the Goths to Italy. It is characteristic of the failure which so often attended Zeno's policies that the vacuum created by the removal of the Ostrogoths was soon filled by the arrival of the Bulgars.

It was probably towards the end of 488 that Theoderic led his followers towards Italy.[58] They constituted an unwieldy band of men, women, and children, their household goods carried in wagons, and were accompanied by such livestock as they possessed.[59] Our most detailed source goes so far as to say that a whole world migrated with Theoderic; although the number of people who accompanied him cannot be calculated, a figure in the order of 100,000 seems plausible, but is not necessarily even approximate.[60]

[55] Cf. Theoderic Strabo's evaluation of Roman intentions: 'For they, remaining at peace, wish the Goths to wear each other down. Whichever of us falls, they will be the winners with none of the effort' (Malchus, *frag.* 11 (Blockley, 15, whose translation I follow)).

[56] Note Ennodius, *pan.* 25, 'in propinquorum tuorum necem'.

[57] Fredegarius, *chron.* 2.5.7 (ed. *MGH SRM* 2.79).

[58] The time may be deduced from Ennodius, *pan.* 27 on the harshness of winter, shortly after the departure of the Goths.

[59] Ennodius, *pan.* 26; Procopius, *BG* 1.1.12.

[60] 'Migrante tecum ad Ausoniam mundo': Ennodius, *pan.* 26. The numbers involved are discussed below at the beginning of Ch. 3.

Any temptation to equate the host led by Theoderic with the Ostrogothic people should be resisted. Just as there had been Ostrogoths independent of the authority of Theoderic's father and uncles in the period following the death of Attila, so now there were some who preferred to stay in the East.[61] On the other hand, Theoderic was accompanied by people who were not Ostrogoths. Fredericus, who arrived at Novae in 488, was to play an inglorious role during the war between Theoderic and Odovacer, and he, together with the Rugians who caused trouble in Pavia during the war, presumably made their way to Italy in the company of Theoderic;[62] however, despite being 'absorbed into the Gothic nation', the Rugians refused to intermarry with the Goths and were successful in preserving their own identity in Italy.[63] Huns and Romans may also have been among the heterogeneous host which followed Theoderic.[64]

Their march took them along the Danube through imperial territory to Sirmium,[65] but as they progressed further to the west they entered lands controlled by the Gepids, old enemies of the Ostrogoths according to Jordanes.[66] They took shelter behind the river Ulca, probably to be identified with the modern Vulca, a tributary of the Danube which flows into it upstream from

[61] Jordanes, *get.* 292 is ambiguous (Theoderic led 'omnem gentem Gothorum, qui tamen ei prebuerunt consensum'), but see Procopius, *Buildings*, 3.7.13 for Goths who remained in the Crimea. In 503 the Ostrogoths Godidisclus and Bessas fought for the Byzantines against the Persians (Procopius, *BP* 1.8.3), and Bessas was later involved in the Byzantine conquest of Italy from his own people (*PLRE* 2.226–8). See too, perhaps, Jordanes, *get.* 266.

[62] Note Ennodius, *pan.* 26: 'tunc a te conmonitis longe lateque viribus innumeros diffusa per populos gens una contrahitur', and Jordanes, *rom.* 349 for Theoderic going to Italy as 'rex gentium'. Fredericus at Novae: Eugippius, *V. Sev.* 44.4. Role in war: Ennodius, *pan.* 55. Rugians in Pavia: Ennodius, *V. Epiph.* 118f.

[63] Procopius, *BG* 3.2.1–3.

[64] Agathias, *Hist.* 2.13.3 refers to one Ragnaris who fought for the Goths at the time of Narses, claiming that he was not a Goth but a member of the Hunnic tribe of the Bitgors, apparently seeking to correct Procopius' assertion that he was a Goth (*BG* 4.26.4, 4.34.9); the name, however, is Gothic: Schönfeld, *Wörterbuch*, 184. Artemidorus, an ambassador of Zeno to Theoderic in 479 (Malchus, *frag.* 18 (Blockley, *frag.* 20)), could be spoken of in a letter of 509/10 as having served Theoderic for a long time (Cassiodorus, *var.* 1.44.2), and may have come to Italy with him. The classic discussion of the tendency of migrating groups to grow like avalanches is R. Wenskus, *Stammesbildung und Verfassung* (Cologne, 1961), 439–45; specifically on Theoderic, 483f.

[65] Jordanes, *get.* 292, to be followed in preference to Procopius, *BG* 1.1.13, perhaps confusing Theoderic's route in 488–9 with his seizure of Dyrrhachium in 479.

[66] Jordanes, *get.* 94–100.

Belgrade, and battle was joined, perhaps in February 489.[67] Initially the Goths fared poorly, but, if we are to believe the details provided in the panegyric of Ennodius, Theoderic rallied his troops with a rousing speech, demanded a cup of wine for the sake of augury, took up his reins, and devastated the enemy in the manner of a river raging in the midst of fields, or a lion in the midst of a herd. The coming of night allowed some of the Gepids to escape, but many were killed, among them King Trapstila.[68] We may be well advised not to take all these details literally, but from Ennodius' description of how the storehouses of the Gepids were found to be crammed with goods which not only satisfied the needs of the Goths but finally occasioned some ennui we may deduce the importance of this victory to Theoderic's travel-worn host.[69] As the year progressed fortune continued to smile on Theoderic, for he defeated the Sarmatians, the people against whom he had won his first victory in about 472,[70] and proceeded westwards.

Odovacer must have known for some time of Theoderic's advance, and his sending of a victory legation to Zeno after his defeat of the Rugians in 487 implies a desire to conciliate him, while the acceptance in the East of Odovacer's nominee as consul in 490 may indicate that his star had not set as far as Constantinople was concerned.[71] He is described as having called forth all the nations against Theoderic, so many kings coming to fight with him that their soldiers could scarcely be supported.[72] The identity of these kings is unknown, but any help they may have given was not evident when Theoderic appeared at the river Isonzo to the east of Aquileia on 28 August, for Odovacer, perhaps alarmed by the size of Theoderic's forces, retreated, possibly before battle had been joined. He made his way to Verona where, on 27 September, he prepared a fortified camp. Verona was probably a predictable site, for, located as it was at the junction of the viae Claudia Augusta, Gallica, and Post-

[67] For this and what follows, Ennodius, *pan.* 28–34. Place and date: H. Löwe, 'Theoderichs Gepidensieg im Winter 488/489', in *Historische Forschungen und Probleme: Festschrift Peter Rassow* (Wiesbaden, 1961), 1–16.

[68] Paul the Deacon, *Hist. rom.* 15.15. Paul is only cited in these notes when he reports details not known from other sources.

[69] Ennodius, *pan.* 34.

[70] Ibid. 35.

[71] Victory legation: McCormick, 'Odoacer'. I accept that Odovacer and not Theoderic nominated Faustus, with J. Sundwall, *Abhandlungen zur Geschichte des ausgehenden Römertums* (Helsinki, 1919), 187, against E. Loncao, *Fondazione del regno di Odoacre* (Scansano, 1907), 42.

[72] Ennodius, *pan.* 36.

humia, it was a key centre for the defence of Italy, and was subsequently to become important to Theoderic for this reason.[73] But Odovacer was quickly followed, and Ennodius describes Theoderic on the night before the battle looking at the fires of his enemies, which shone like stars. But he knew no fear, and the next morning, when his mother and sister, tossed between hope and fear, came to see him, he supplied reassurance: it was a true man (*vir*) to whom his mother had given birth, and on that day he was going to show himself a man; the glories won by his ancestors would not perish through him! He asked the women to bring his best clothes, such as would make him more easily recognized, and on a field by the River Adige battle was joined. Both sides sustained heavy losses, but Odovacer was finally obliged to quit the field, leaving victory to Theoderic. The field was covered with bodies; some 18 years later Ennodius complained that hungry cattle were destroying evidence of the victory provided by the bones that still lay there.[74] Odovacer fled, almost certainly to Ravenna.[75] Part of his army remained at Milan under the command of Tufa, whom Odovacer had appointed *magister militum* on 1 April, but as Theoderic approached Tufa surrendered with a large part of his army, and the town of Pavia apparently followed suit.[76] Tufa's defection was a major gain for Theoderic, who sent him against Odovacer at the head of an army which included some Gothic nobles. He laid siege to Faenza, but when Odovacer arrived from Ravenna Tufa again changed sides. He handed over to Odovacer some of Theoderic's counts who had accompanied him; they were put in irons and taken to Ravenna.[77] Theoderic suddenly

[73] *Auct. haun. sa* 490, the most detailed source; anon. Vales. 50; Cassiodorus, *chron. sa* 489; Ennodius, *pan*. 37f.; Jordanes, *get*. 292f. Only Jordanes and the Anonymous suggest actual fighting by the Isonzo, and as the latter employs a weak formula to describe it (*pugnans victus*; cf. *ceciderunt ab utraque parte*, used of three later engagements: 50, 53, 54) we should probably conclude that Odovacer retreated without giving battle; so L. Simeoni, 'Note Teodericiane', *Accademia delle scienze dell' Istituto di Bologna: Classe di scienze morali memorie*, 4th ser. 8 (1945–8), 149–98 at 160–2, see too M. A. Wes, *Das Ende des Kaisertums im Westen des römischen Reichs* (The Hague, 1967), 64.

[74] Ennodius, *pan*. 39–47 (assuming a date of 507 for this work, with Sundwall, *Abhandlungen*, 42f.), with anon. Vales. 50; *Auct. haun. sa* 490; Cassiodorus, *chron. sa* 499; Jordanes, *get*. 293.

[75] Anon. Vales. 50, to be preferred to the colourful story of Paul the Deacon, *hist. rom*. 15.16, who reports that Odovacer fled to Rome but found the gates of the city shut against him, whereupon he destroyed what he could and made his way back to Ravenna. Paul has been followed, unwisely, by G. Romano and A. Solmi, *Le dominazioni barbariche in Italia (395–888)* (Milan, 1940), 152.

[76] Anon. Vales. 51; Ennodius, *V. Epiph*. 109, 111.

[77] Anon. Vales. 51f.; cf. Ennodius, *V. Epiph*. 111.

found his position worse, and as winter was approaching he moved his people into Pavia, where the cramped conditions must have been productive of ill tempers.[78]

By the end of 489 Italy had been effectively divided into various zones, one of them controlled by Theoderic and another by Odovacer. The former sought legitimacy from the emperor. It was probably in 490 that he sent Festus *caput senatus* to Constantinople to seek recognition from Zeno, and Theoderic issued coins in Zeno's name from Milan, obviously before Zeno's death, which occurred on 9 April 491, became known in Italy; shortly afterwards he issued coins in Rome as well.[79] Odovacer, for his part, seems to have distanced himself from Constantinople by nominating his son Thela caesar, a constitutionally impossible act which implied that Odovacer was an augustus.[80] Towards the end of his reign Odovacer was minting coins in which the image of Zeno was replaced by representations of Rome and Ravenna,[81] and it may be significant that one eastern consular list, the *Fasti Heracliani*, does not register the Western consuls for 489 and 490, although other lists do name them.[82] But Italy was not merely divided into areas under the control of two rulers who defined themselves differently with respect to Constantinople, for Tufa's reconciliation with Odovacer was short-lived. Before long he was operating independently, apparently in co-operation with Theoderic's now erstwhile Rugian ally Fredericus, whose father, we may remember, had been captured by Odovacer in 487. Finally they fell out, and at a battle fought somewhere between Trent and Verona in 493 Tufa was defeated, and while Fredericus' fate is unknown, the Rugians may have continued

[78] Ennodius, *V. Epiph.* 111 ff.

[79] Mission of Festus: anon. Vales. 53. Coins in Milan: W. Hahn, *Moneta imperii Byzantini*, 1 (Vienna, 1973), table x, no. 20. Coins in Rome: ibid. tables x and xi; M. F. Hendy, *Studies in the Byzantine Monetary Economy* (Cambridge, 1985), 490.

[80] John of Antioch, *frag*. 214a; for comparison, some 20 years earlier the emperor Leo had proclaimed Patricius, the son of the *magister militum* Aspar, caesar (references in *PLRE* 2.842, 'Iulius Patricius 15'). We also know of two *magistri militum* who served Odovacer during the war against Theoderic: Tufa, appointed on 1 Apr. 489 (*PLRE* 2.1131) and Libila, who was killed in July 491 (*PLRE* 2.681). On the name of Odovacer's son, consult P. Classen, 'Der erste Römerzug in der Weltgeschichte', in H. Beumann, ed., *Historische Forschungen für Walter Schlesinger* (Cologne, 1974), 325–47 at 340.

[81] Hahn, *Moneta*, 79, although the reason for the date 'um 486' is not clear; cf. W. Wroth, *Catalogue of the Coins of the Vandals, Ostrogoths and Lombards in the British Museum* (London, 1911), p. xxx.

[82] *Fasti Heracliani* at *MGH AA* 13.406; but cf. *chron. pasc.*, ed. Dindorf, 606; Marcellinus, *chron. sa* 489, 490.

to cause trouble in Pavia subsequently.[83] If this was not enough, other powers intervened. By the summer of 490 an unknown number of Visigoths had arrived to support Theoderic,[84] in 491 the Vandals are reported to have stopped attacking Sicily,[85] and at the same time the Burgundian king Gundobad, an old hand at Italian affairs whose earlier presence in Rome was still remembered in the next decade, advanced into Liguria, only retiring when he had taken over 6,000 captives.[86] He may have been one of the many kings whose support Odovacer was said to have obtained, but Gundobad, who had murdered the emperor Anthemius in 472 and been involved in the subsequent elevation of Glycerius, was probably merely fishing in troubled waters.[87] One has the distinct impression of vultures hovering over Italy.

The following summer saw Odovacer take the field again. Proceeding from Cremona he advanced to Milan, and on 11 August 490 a battle was fought by the river Addua (now the Adde), to the east of the city. It was Theoderic's fourth battle by a river since leaving Moesia, and his third against Odovacer. The result was the same: aided by the Visigoths Theoderic defeated his foe, killing in the process Odovacer's *comes domesticorum* Pierius. Odovacer again retreated to Ravenna, but this time Theoderic followed him, and laid siege to the city from the wood called the Pineta, some 5 kilometres east of the city.[88] It was the place where Odovacer had murdered Paul, the brother of Orestes, in 476; a beautiful passage in Dante's *Purgatorio* describes the forest as it stood early in the fourteenth century, and parts of it still stand.[89] Ravenna was a notoriously difficult city to attack, it being precisely for this reason that the emperor Honorius had moved there in 402/3.[90] In late

[83] Ennodius, *pan.* 55 (for the phrase 'nata est inter sceleratos . . . discordia', cf. *pan.* 25, 'nata est felicis inter vos [Theoderic and Odovacer] causa discordiae'); *auct. haun. s.a.* 493.2, *fasti Vind. priores, s.a.* 493. The most straightforward way of reading Ennodius, *V. Epiph.* 117 and 119 is to have the Goths in Pavia for three years and the Rugians for two subsequent years.

[84] Anon. Vales. 53. [85] Cassiodorus, *chron. s.a.* 491; cf. perhaps *var.* 1.3.3f.

[86] Ennodius, *V. Epiph.* 172. For Gundobad's earlier involvement in Italy, see *PLRE* 2.524: for his presence in Rome, Cassiodorus, *var.* 1.46.2.

[87] *PLRE* 2.524; cf. A. Lumpe, 'Ennodiana', *Byzantinische Forschungen*, 1 (1966), 200–10 at 206f. The reconstruction proposed by O. Perrin, *Les Burgondes* (Neuchâtel, 1968), 441 is fanciful.

[88] Anon. Vales. 53; Jordanes, *get.* 293; Cassiodorus, *chron. s.a.* 490.

[89] Murder of Paul: anon. Vales. 37. Dante describes the Pineta at *Purgatorio* 28.1–21.

[90] J. Matthews, *Western Aristocracies and Imperial Court* (Oxford, 1975), 274.

antiquity branches of the river Po still ran on both the north and south sides of the town and there were swamps to the east, so that it could reasonably be compared to an island; its watery location made it ideal for the cultivation of asparagus.[91] Theoderic contented himself with taking the other cities in which Odovacer maintained a military presence, until finally only Caesena and Ravenna held out,[92] and waited for his enemy to act. On the night of 9–10 July 491 Odovacer advanced with a force of Heruls towards Theoderic's encampment, but to no avail: he was forced to retire behind the walls of Ravenna, and Tufa's successor as *magister militum*, Libila, was killed in flight near the river Bedens.[93] On 29 August 492 Theoderic launched a successful attack on the port of Ariminum (now Rimini), as a result of which he was able to tighten the blockade of Ravenna.[94] The price of grain soared, famine ensued, and in February 493 Bishop John came forth to negotiate a settlement: he opened the gates and came out with crosses, thuribles, and gospels, and prostrated himself before Theoderic.[95]

Negotiating with invaders was a frequent task for bishops in the fifth century, and there is no need to impute treason to John,[96] for it is easy to believe that Odovacer was becoming desperate as the siege wore on. Our sources make it clear that Odovacer took the initiative in the negotiations,[97] but the terms to which Theoderic agreed were unexpected: he and Odovacer were to share the rule over the Romans.[98] Given that Theoderic had ostensibly come to Italy to dislodge Odovacer and had been provided with no mandate to share power with him, and given further that Odovacer's elevation of his son and coinage issues smacked of rebellion against the

[91] Jordanes, *get.* 148f.; cf. Procopius, *BG* 1.1.18; Sidonius Apollinaris, *ep.* 1.5.5; Zosimus, *hist. nov.* 5.30.2 and, briefly, F. W. Deichmann, *Ravenna Haupstadt des spätantiken Abendlandes* (Wiesbaden, 1958–76), 1.2f. Cultivation of asparagus: Martial, 13.21; Pliny, 19.54. See now N. Christie and S. Gibson, 'The City Walls of Ravenna', *Papers of the British School at Rome*, 56 (1988), 156–97, at 156–9.

[92] Procopius, *BG* 1.1.14f.

[93] Anon. Vales. 54; *Auct. haun. s.a.* 491 and *fasti Vind. priores, s.a.* 491; Cassiodorus, *chron. sa* 491; cf. Ennodius, *pan.* 53.

[94] *Auct. haun. sa* 493 and *fasti Vind. priores, sa* 493 (in both cases the entry occurs under what is presumably the wrong year); Agnellus, *codex*, 39 (ed. Testi-Rasponi, 107).

[95] Agnellus, *codex*, 39 (ed. Testi-Rasponi, 108).

[96] As does Deichmann, *Ravenna*, 1.6.

[97] Anon. Vales. 54, 'coactus Odoacar'; cf. *auct. haun. sa* 493 and Jordanes, *get.* 294f., against Procopius, *BG* 1.1.24.

[98] John of Antioch, *frag.* 214a; cf. Procopius, *BG* 1.1.24.

imperial authority which Theoderic, in theory, represented, the arrangement represented a compromise for the Goth, and it is not surprising that Italian sources fail to mention it. How it would have worked in practice is not clear and may never have been made explicit; perhaps Odovacer and Theoderic would have ruled their respective followings, presumably fairly distinct in terms of race, but how was their authority over the Romans to have been exercised?[99] In any case, Odovacer's son Thela was handed over to Theoderic, and on 5 March Bishop John opened the gates of the city and invited 'the new king who came from the East' to enter.[100] But the agreement between the two rulers was not to last. On 15 March Theoderic invited his colleague to a banquet, and there produced his sword. 'Where is God?' Odovacer was later believed to have cried. Theoderic's reply as he took his colleague's life was cryptic: 'This is what you did to mine.' We may note in passing that these details are only known from Eastern sources; doubtless people writing under the Ostrogoths felt it prudent not to be explicit.[101] As we have seen, Ennodius was later to suggest that the origin of the war between Theoderic and Odovacer was the murder of some of Theoderic's relatives (*pan*. 25), and it is tempting to see Theoderic as having taken vengeance for Odovacer's slayings, it being possible that the Rugian royal family was in some way connected to Theoderic.[102] On the same day Theoderic gave orders that all of Odovacer's soldiers were to be killed, wherever they could be found, together with all his family; his brother Onoulf sought sanctuary in a church but was killed, his wife Sunigilda was imprisoned and subsequently starved to death, while his son Thela fled to Gaul but was killed when he returned to Italy.[103] That all of Odovacer's troops were killed is distinctly unlikely, for, apart from the logistic difficulties such an operation would have involved, a king Sinduald who fought the Goths under Narses in Justinian's Gothic war is described by a later source as one of the *stirps* of the Heruls

[99] See Mommsen, *Gesammelte Schriften*, 6.335.

[100] Anon. Vales. 54; *auct. haun. s.a.* 493; Agnellus, *codex*, 39 (ed. Testi-Rasponi, 109).

[101] John of Antioch, *frag*. 214a; cf. Procopius, *BG* 1.1.25. According to John, Theoderic commented after the murder that the villain did not seem to have any bones.

[102] Consult on blood vengeance H. Rosenfeld, 'Ost- und Westgoten', *Die Welt als Geschichte*, 17 (1957), 245–58 at 255, and R. Wenskus, *Stammesbildung*, 11, 23.

[103] Anon. Vales. 56; John of Antioch, *frag*. 214a; cf. Ennodius, *pan*. 51.

Odovacer had brought to Italy.[104] Some authors of the sixth century were aware of reports that Odovacer had been plotting against Theoderic, which would have allowed the latter's murder of his rival to be excused as a pre-emptive measure,[105] but given that people writing during the reigns of Theoderic and his successors would have had a vested interest in describing Theoderic's conduct favourably, and that works based on such traditions would have reproduced their bias, whereas no one after 493 stood to gain by writing favourably of Odovacer, it seems reasonable to conclude that Theoderic was at fault in the murder of his colleague.

It had been a hard war. Bishop John of Ravenna reported to Pope Gelasius that war and famine were dangerously depleting the ranks of the clergy, and Gelasius prescribed to the bishops of Lucania, Brutium, and Sicily, not areas one would think likely to have been seriously affected, that monks could be made *lector, notarius*, and *defensor* as needed, and thereafter promoted to the ranks of acolyte, subdeacon, deacon, and priest at three-monthly intervals, which entailed a considerable weakening of earlier papal legislation and a lowering of the standards which had previously been current in Italy.[106] The writings of Gelasius contain numerous references to barbarian incursions and the continued tempest of wars, some of which could be held to suggest that trouble continued after the murder of Odovacer,[107] and because of 'various difficulties' he was not able to announce his accession to Bishop Aeonius of Arles until 23 August 494, even though he had become pope on 1 March 492. The period during which Gelasius was unable to communicate partially overlapped with the war between Odovacer and Theoderic, which may provide an explanation for part of his delay.[108] So great was the flood of people who came to Rome during Gelasius'

[104] Paul the Deacon, *hist. lang.* 2.3.

[105] Most forcibly Cassiodorus, *chron. s.a.* 493 and anon. Vales. 55; see as well Jordanes, *rom.* 349 (Odovacer murdered 'ac si suspectus') and Procopius, *BG* 1.1.25 (but the report is prefaced ὥς φασιν).

[106] Gelasius, *ep.* 14.1f., with which cf. Ennodius, *V. Epiph.* 8, 18, 26, 34; *const. Silv.* 11 (*PL* 8.838; on the origin of this document see below, Ch. 4); A. H. M. Jones, *Later Roman Empire*, 912f.

[107] Gelasius, *ep.* 6.1, 7.1 (to the bishops of Picenum, an area of heavy Gothic settlement); *frag.* 9, 35, and *Lettre contre les Lupercales*, 13 (SC 65.172). I take this last work to have been written by Gelasius, against Y.-M. Duval, 'Les Lupercales de Constantinople aux Lupercales de Rome', *Revue des études latines*, 55 (1977), 222–70 at 246–50.

[108] *MGH Ep.* 3.33 (no. 22 = Thiel *ep.* 19), on which see W. Ullmann, *Gelasius I (492–496)* (Stuttgart, 1981), 217.

pontificate, perhaps seeking refuge, that famine threatened; the cities of Liguria, we are told, were ruined and the people dying from swords and hunger; and it has plausibly been suggested that hoards of coins which have been uncovered at Reggio Emilia, Zeccone, and Braone were laid down during the war.[109] It is easy to see how rumours could have spread in 493 concerning the birth of Antichrist.[110]

Such auspices were not propitious for Theoderic's rule in Italy, and invite consideration of the regard the Romans had for Odovacer. It has been widely assumed that he was unpopular, and that any support he may have enjoyed quickly evaporated when Theoderic appeared in 489,[111] and a certain amount of evidence can be held to suggest that Odovacer was disliked. Procopius asserts that some Goths who negotiated with Belisarius during the Gothic war claimed that Odovacer had 'treated the land outrageously' for ten years (*BG* 2.6.21), but it is highly unlikely that they said anything of the kind, for in the West Odovacer's rule was accepted as having run for thirteen or fourteen years.[112] Procopius elsewhere credits Odovacer with a rule of ten years (*BG* 1.1.8), possibly counting from the death of Nepos in 480, although on another occasion he seems to misunderstand when this occurred (*BV* 1.7.15), and so he presumably followed the standard practice of putting in the mouths of speakers words, however tendentious, he felt would have been appropriate. It is worthwhile emphasizing this, for some of our sources go out of their way to denigrate Odovacer. Cassiodorus described him as a crooked prince who governed in worthless times when avarice was not considered a crime.[113] For Ennodius he was a

[109] Rome: *lib. pont.* 255.5f., with Gelasius, *frag.* 35. Cities of Liguria: Ennodius, *V. Epiph.* 121; *opusc.* 5.20 (ed. Vogel, 303.6f.). Coin hoards: V. Bierbrauer, *Die ostgotischen Grab- und Schatzfunde in Italien* (Spoleto, 1975), 215. Specifically on Reggio Emilia: M. Degani, *Il tesoro romano-barbarico di Reggio Emilia* (Florence, 1959), 36, 105–7, but see most recently S. Gelichi, in Giardina, ed., *Società romana*, 634f. On Braone: G. P. Bognetti, in *Storia di Brescia*, 1 (Brescia, 1963), 398f. On Zeccone: A. Peroni, *Oreficerie e metalli lavorati tardoantichi e altomedievali del territorio di Pavia* (Spoleto, 1967), 104.

[110] *Pasch. camp. s.a.* 493.

[111] See e.g. M. Brion, *Théoderic roi des Ostrogoths 454–526* (Paris, 1935), 184–7; J. M. Wallace-Hadrill, *The Barbarian West 400–1000*, 3rd edn. (London, 1967), 33; C. Wickham, *Early Medieval Italy* (London, 1981), 20. John Malalas, *chron.* 383 could be cited in support of this position, but the inaccuracies of this author do not inspire confidence.

[112] Anon. Vales. 48, quoting Eugippius, *V. Sev.* 32.2 (13 or 14 safe years); Jordanes, *get.* 243 (13 years).

[113] Cassiodorus, *var.* 8.17.2, 5.41.5, 3.12.3; cf. Boethius, *cons. phil.* 1.4.10 for *avaritia* as a characteristic of barbarians.

plunderer who practised daily depredations and whose revolt against Orestes had been inspired by the Devil; in an autobiographical passage he commented that Theoderic's coming to Italy had been long desired.[114] Jordanes, possibly reproducing the sentiments of Cassiodorus, represents Theoderic as reminding Zeno that Rome, the head and mistress of the world, was being driven to and fro under the tyranny of this king of the Thorcilingi and Rugians, who was oppressing the senate with a tyrannical yoke and part of the state under the servitude of captivity; his reign, Jordanes commented elsewhere, was one of terror, so that when Theoderic arrived all Italy was quick to call him lord.[115] But, oddly enough, in other passages Jordanes suggests that Odovacer and the Ostrogothic monarchs were equally 'kings of the Goths' (*rom.* 345, *get.* 243), as does Marcellinus *comes* (*chron. s.a.* 476), and it must be said that less interested sources offer a more positive perspective on Theoderic's predecessor. Before Odovacer went to Italy Severinus of Noricum predicted that he would become renowned; later, the king having invited him to make a request, Severinus petitioned for the recall from exile of one Ambrosius, and Odovacer obeyed his command with joy.[116] Anonymous Valesianus describes him as a man of good will, a phrase he elsewhere applies to Theoderic (48; cf. 59), and another chronicler commented on his wisdom (*auct. haun. s.a.* 476). His relations with the senate were good, as demonstrated by the work carried out on the Flavian amphitheatre, and by the appointment of a Western consul annually from 480 to 490 in a sequence marked by what may have been a judicious *mélange* of representatives of different families; for comparison, there had only been three western consuls in the 470s.[117] Odovacer was capable of generosity towards his friends, for in 489 Pierius, his *comes domesticorum*, received from him estates in Sicily which would yield an annual revenue of just over 40 solidi, in addition to estates in Sicily

[114] Daily depredations: *pan.* 23. Diabolical inspiration: *V. Epiph.* 95 (yet Theoderic came to power because of God, ibid. 109). Coming desired: *opusc.* 5.20 (ed. Vogel, 303.5ff.). Is there a hint in the allusion to Gen. 22: 17 at *pan.* 29 (not noticed by Ennodius' editor Vogel) that the Goths were seen by Ennodius as a chosen people? See too *pan.* 24 (Odovacer as *tyrannus*), 52 (his *praesumptio*).

[115] Jordanes, *get.* 291 f., 294 with 243.

[116] Eugippius, *V. Sev.* 7.1, 32.1 f.

[117] See in general A. Chastagnol, *Le Sénat romain*, esp. 52–6; L. C. Ruggini, 'Nobilità romana e potere nell'età di Boezio', in L. Obertello, ed., *Congresso internazionale di studi Boeziani Atti* (Rome, 1981), 73–96 at 82 f.; G. B. Picotti, 'Sulle relazioni fra re Odoacre e il senato e la chiesa di Roma', *Rivista storica Italiana*, 5th ser. 4 (1939), 363–86.

and Dalmatia worth 650 solidi per annum which he had already received;[118] as we have seen, he died fighting for Odovacer in August 490. Likewise Liberius, who held an unknown office under Odovacer, remained loyal to him and only entered on his distinguished career of service to the Ostrogoths after Odovacer had been murdered (Cassiodorus, *var.* 2.16.2).

There is nothing here to suggest a speedy adherence to Theoderic on the part of the Italians; indeed, the attitude of the Sicilians at the beginning of Theoderic's rule was such as to have raised the possibility of his taking vengeance had not Cassiodorus' father intervened (*var.* 1.3.3). Nor is there any reason to believe that the Italian episcopate rallied to Theoderic's cause. True, Pope Gelasius took pride in having resisted certain orders of Odovacer, described by him as a 'barbarian heretic',[119] but this does not imply that he would have been likely to carry out orders issued by Theoderic, surely no less of a barbarian heretic. Bishop Epiphanius of Pavia is described as having been impressed by Theoderic, but our only evidence for this is a highly rhetorical passage in the *Vita* of him written by Ennodius, himself an adherent of Theoderic, and as far as one can judge the bishop's task of interceding on behalf of his people continued to be exercised in the reign of Theoderic in precisely the same way as it had been in the time of Odovacer.[120] Laurentius of Milan is said to have resisted the blandishments of Odovacer, but we have no reason to believe that he actively supported Theoderic.[121] We may conclude that some of the Italians supported Odovacer, while others, presumably including Festus, who had travelled to Constantinople on Theoderic's behalf in 490, supported the invader, while others supported neither but awaited developments. After the war Theoderic decreed that only those Romans who had supported him were to enjoy the rights of citizen-

[118] The document is edited and discussed by L. Santifaller, 'Die Urkunde das Königs Odovaker vom Jahre 489', *Mitteilungen des Österreichische Geschichtsforschung*, 60 (1952), 1–30, and Tjäder, *Papyri*, 1.279–93.

[119] *Coll. avel.* 95.63 (*CSEL* 35.391.19–21).

[120] Ennodius, *V. Epiph.* 109f. Epiphanius' qualities as a political and pragmatic saint, at least as described by Ennodius, are well brought out by E. Pietrella, 'La figura del santo-vescovo nella "Vita Epifani" di Ennodio di Pavia', *Augustinianum*, 24 (1984), 213–26, esp. at 221–5.

[121] Ennodius, *dict.* 1.12ff., ed. Vogel, 2f. It will be clear I cannot accept the assessments of G. M. Cook, *The Life of Saint Epiphanius by Ennodius: A Translation with an Introduction and Commentary* (Washington, DC, 1942), 199f., and E. Stein, *Histoire du bas-empire*, 2 (Paris/Bruges, 1949), 56.

ship; others would not be allowed to testify or to dispose of their property. It is clear that more than a small group was affected, for our only source reports that many people found the laws trodden underfoot and all Italy was oppressed. Bishops Epiphanius of Pavia and Laurentius of Milan went to Ravenna to intercede for those affected, whether fulfilling the traditional function of bishops as interceders for the oppressed or with an eye to the properties which churches, as major beneficiaries of wills, stood to lose, whereupon Theoderic issued a general indulgence for all except a few ringleaders, who were no longer to live where they had hitherto.[122] It seems too that Theoderic planned to act against those Odovacer had adlected to the senate, a speech by the senator Symmachus being necessary to dissuade him.[123] In short, while a war of a few years had transformed the Ostrogoths from a group of plundering marauders to the masters of Italy, much remained to be done before Theoderic would commend himself to the Romans he now governed.

[122] Ennodius, *V. Epiph.* 122ff. The *Edictum Theodorici* explicitly allows the making of wills (cap. 28). W. Ensslin, 'Der erste bekannte Erlass des Königs Theoderich', *Rheinisches Museum für Philologie*, 92 (1944), 266–80 and D. Claude, 'Universale und particulare Züge in der Politik Theoderichs', *Francia*, 6 (1978), 19–58 at 46 offer different interpretations of what Theoderic's action implies for his policies with respect to the empire and Italian autonomy, but I am not certain that such matters seemed important to Theoderic in this case.

[123] H. Usener, 'Das Verhältnis des Römischen Senates zur Kirche in der Ostgotenzeit', in *Commentationes philologiae in honorem Theodori Mommseni* (Berlin, 1877), 759–67.

2

The Securing of the State

The state which Theoderic founded in 493 was to become mighty, and the modern student can all too easily interpret the king's position at the beginning of his reign in the light of the successes which were to follow. But care is called for, for even after the murder of Odovacer difficult tasks were still to be carried out. His barbarian followers had to be provided with means of support, his relationships with both the emperor and the other monarchs of the West placed on a sound footing, and good relations established with both the Catholic Church and the senatorial aristocracy of Rome, those resilient bodies which had happily outlived the empire in the West and whose support for him, as we have seen, was by no means inevitable. These were the concerns of Theoderic in the years immediately following the murder of Odovacer.

SETTLING THE GOTHS

As Procopius told the story, Odovacer had come to power in 476 by satisfying the demand of the barbarian followers of Orestes to be given one-third of the land of Italy (*BG* 1.1.4–8). This, it has widely been believed, would have been in keeping with the Roman system of *hospitalitas* expressed in legislation of the emperors Arcadius and Honorius in 398, according to which one-third of houses was to be made available for the use of those on state business, although the term *hospitalitas* is never used of the settling of the followers of Odovacer or Theoderic, and the change from houses to 'all the land of Italy' is a large one, as is the shift from temporary billeting to what must already have seemed likely to become permanent control of territory, but the *Epistula Honorii* of the early fifth century could be held to mark a stage in such a development.[1] When Theoderic

[1] *Cod. theod.* 7.8.5 of 398; E. Demougeot, 'Une lettre de l'empereur Honorius sur *l'hospitium* des soldats', *Revue historique de droit français et étranger*, 4th ser. 34 (1956), 25–49.

came to power in Italy, Procopius continues, 'the Goths distributed among themselves the portions of the lands which Odovacer had given to his own partisans' (*BG* 1.1.28, trans. Dewing). It is hard to take these words literally, and any procedure along such lines would have been more complicated in practice: the followers of Theoderic, whether or not restricted in this case to Goths, were presumably more numerous than Odovacer's adherents, and despite the reports of our sources it is hard to believe that all of the latter's followers were killed, so a simple transfer of property from one group to the other would have been awkward. Such considerations are consonant with a letter of Cassiodorus which indicates that under Theoderic an allotment of 'thirds' (*tertiarum deputatio*) had been carried out through the agency of the praetorian prefect Liberius, a man who is known to have occupied this office from early in Theoderic's reign till 500; so skilfully did he discharge his task that the hearts as well as the possessions of Goths and Romans were joined together, and friendly relations between the peoples grew because of the losses (*var.* 2.16.5). There exists as well a letter written by Ennodius to Liberius in 511, which has generally been taken to refer to the same phenomenon. In it the deacon of Milan observes that his correspondent had enriched countless hordes of Goths by a widespread handing over of estates (*larga praediorum conlatio*) in such a way that the victors desired no more and those who had been overcome felt no loss (*ep.* 9.23.5). According to a literal interpretation of 'thirds' as 'thirds of land', however, not all the 'thirds' would have been resumed, for the state is known to have collected payments called 'tertiae', which would presumably have been paid in lieu of the expropriation of estates (Cassiodorus, *var.* 1.14, 2.17); perhaps resumptions would have been more likely to have occurred in the north, where, as we shall see, most of the Goths were based. Broadly similar systems of dividing the land are widely believed to have been employed in the cases of the Visigoths and Burgundians in Gaul.[2]

That the lands of Italy were divided along these lines was generally accepted by modern scholars until the publication of a study by Walter Goffart, who justly points to what would have been an

[2] See the classic discussions of E. T. Gaupp, *Die germanischen Ansiedlungen und Landtheilungen in den Provinzen des römischen Westreiches* (Breslau, 1844), and F. Lot, 'Du régime de l'hospitalité', *Revue belge de philologie et d'histoire*, 7 (1928), 975–1011.

extraordinary absence of resistance among the Roman aristocracy, a class whose wealth was based on land, to the expropriation of one-third of their estates.[3] The avidity displayed during Theoderic's reign for the office of consul, one which entailed heavy expense, is suggestive of prosperity among the great landowners, rather than of a class which had recently suffered severe losses, and indeed we shall later suggest that the Ostrogothic period was one of prosperity for the great landholders. Arguing against the scholarly consensus, Goffart proposes that the passages of Cassiodorus and Ennodius referred to above should not be taken at face value; rather, he argues that the 'tertiae' allotted to the Goths were, universally, thirds of tax assessments, so that the Goths received income coming from the land rather than the land itself.[4] The thesis is attractive because of its apparent solution to the problem of lack of resistance and the remarkable silence as to any massive expropriations in later sources, such as the Pragmatic Sanction issued by Justinian in 554; it would have been extraordinary for so radical a step to have created so little contemporary resentment and such tiny historiographical ripples.[5] The new interpretation would also clarify the status of some Goths described as 'millenarii' who were summoned to Ravenna from Picenum and Samnium to receive donatives from Theoderic (Cassiodorus, *var*. 5.26f.) and were apparently not commanders of units of one thousand, as the name would suggest, but the holders of tax assessments, for the 'millena' referred to in *var*. 2.37 is a tax.[6]

It therefore seems reasonable, provisionally at any rate, to accept that the 'tertiae' which supported the Goths were tax assessments. As to how the income from taxes was divided amongst the Goths we

[3] W. Goffart, *Barbarians and Romans* (Princeton, 1980), esp. 58, 60, 70–2. It is scarcely enough to attribute Goffart's views to his being an American rather than a European of our day, as did H. Wolfram, 'Zur Ansiedlung reichsangehöriger Föderaten', *Mitteilungen des Instituts für Österreichische Geschichtsforschung*, 91 (1983), 5–35 at 8.

[4] *Barbarians and Romans*, 60–102, discusses the fifth century.

[5] T. S. Burns, 'Ennodius and the Ostrogothic Settlement', *Classical Folia*, 32 (1978), 153–68, seems to me to force the evidence. It is unclear in what sense the 'damna' of Cassiodorus *var*. 2.16.5 are to be taken.

[6] Goffart's interpretation (*Barbarians and Romans*, 80–8), for which he finds precedent in Mommsen (*Gesammelte Schriften*, 6.438 n. 1), may be compared with such texts as Isidore of Seville, *etym*. 9.3.30, Procopius, *BV* 1.5.18 and 2.3.8 on 'chiliarchs', and *Codicis Euricioni fragmenta*, 322 (*MGH* Leg. 1.1.23. 1f.), and the discussion of D. Claude, 'Millenarius und Thiuphadus', *Zeitschrift der Savigny-Stiftung für Rechtsgeschichte, Germanische Abt.*, 88 (1971), 181–90.

have no information. It would seem unlikely that all received the same entitlement, but it is obvious that they would now have enjoyed incomes on a scale far beyond what they had enjoyed during the preceding decades, and that this income could have been used, among other things, for the purchase of land. Most Goths would certainly have found themselves better off than they had been before coming to Italy, although some would have found themselves doing much better than others, while the Roman landowners, contemplating the means by which the Vandals were settled in Africa, must have felt that Theoderic's forces had been settled in a very happy manner which involved little more than a refinement of the manner in which Roman taxation had traditionally brought about a transfer of wealth from the land to the army.[7]

RELATIONS WITH CONSTANTINOPLE

Another area which required Theoderic's attention was the question of his relationship with Constantinople. It is a fair assumption that the state he founded was important to Constantinople, for not only was it already large and strong at the beginning of his reign, but it was directly adjacent to imperial territory. Theoderic obviously retained the power to harm Byzantine interests after his long march to Italy, and even within the West, where imperial diplomacy was certainly active in ways which are now hard to trace, his kingdom must have weighed heavily in the emperor's counsels. Such importance must have been enhanced by the circumstance that he controlled Rome, and we may recall that it was in Constantinople rather than Italy that the deposition of the last emperor in 476 was seen as significant. When the time came to attack the Ostrogothic state, Justinian was prepared to commit resources for two decades

[7] For further discussion of Goffart's thesis consult M. Cesa, 'Hospitalitas o altre "techniques of accomodation"? A proposito di un libro recente', *Archivio storico italiano*, 140 (1982), 539–52, and S. J. B. Barnish, 'Taxation, Land and Barbarian Settlement in the Western Empire', *Papers of the British School at Rome*, 54 (1986), 170–95, arguing for a limited and careful barbarian land-taking in north-east Italy; most recently, it has gained support from M. F. Hendy, 'From Public to Private: The Western Barbarian Coinages as a Mirror of the Disintegration of Late Roman State Structures', *Viator*, 19 (1988), 29–78 at 42f. n. 42, and Wolfram, *History*, 295ff., where the argument in the German original of the book is significantly altered. See further S. Weber, 'Zur Ansiedlung der Germanen nach den Leges Barbarorum', *Zeitschrift für Archäologie*, 19 (1985), 207–11.

to bring about its overthrow. These considerations make the vagueness which surrounds the agreement Zeno contracted with Theoderic the more annoying. The nature of the status Zeno and Theoderic assumed the latter would enjoy if his venture in Italy was successful is not at all clear, and it is quite possible that they had not discussed the matter before Theoderic's departure in 488; the question is complicated by the death of Zeno in 491, it not being clear whether his successor Anastasius subsequently acted in the way Zeno would have done, and by the fact that those of our sources which are most explicit on the issue were all written well into the sixth century, by which time not only would recollections have been hazy, but the whole issue obscured by the propaganda which we may be sure surrounded claims made for and against the legitimacy of the Gothic regime in the time of Justinian.

Our most precise source, Anonymous Valesianus, asserts that Theoderic, in the event of defeating Odovacer, was to rule in Zeno's place until he came (*loco eius, dum adveniret*), and terms the Goth *patricius*; as he assigns the same title to Orestes, and Odovacer is known to have been a patrician, we may be justified in attributing to him the view that Theoderic's initial status in Italy was similar to that of his forebears.[8] But Theoderic was concerned to put his relationship with the emperor on a sound footing as quickly as possible, for it was probably as early as 490 that the senator Festus was sent to Constantinople to negotiate on his behalf. Festus, *caput senatus*, had been consul in 472 and, as the only surviving Western consul to have held office prior to Odovacer's usurpation in 476, could be held to have been a particularly appropriate ambassador, but given Theoderic's failure to gain control over much of Italy by the time Festus was dispatched his request to be allowed to assume royal garb, whatever this meant in practice, was premature; in any case, Zeno's death on 9 April 491 may have foreclosed a decision.[9] Theoderic's second ambassador to Constantinople, Faustus *niger*, was Odovacer's last consular nominee. It is very difficult to assign a date to his embassy, but when Faustus returned

[8] Anon. Vales. 49; see ibid. 36–8 for Orestes, and Malchus, *frag*. 10 (Blockley, 14) for Odovacer as *patricius*; further, the versions preserved in Procopius, *BG* 2.6.14–22, with which cf. *BG* 1.1.10f.; Jordanes, *get*. 291f. (Italy given to Theoderic by Zeno as a gift); Jordanes, *rom*. 348 (Theoderic sent to Italy as Zeno's *cliens*). I doubt whether the truth can be sifted from these accounts.

[9] Anon. Vales. 53, where the mission is dated to 490 and Festus' failure implied; see further Ensslin, *Theoderich*, 68f.

from Constantinople he and his colleague Irenaeus, who is otherwise unknown, brought a complaint from Zeno's successor Anastasius that Pope Gelasius had failed to write to him on becoming pope. As Gelasius' pontificate began 1 March 492 this comports with a date a fair way into 492 for Faustus' mission, and if we choose to see a connection between Faustus' negotiations with Anastasius and the holding of office of Theoderic's first consular nominee in 493 such a date becomes the more plausible.[10] By this time Theoderic's position in Italy was much stronger than it had been in 490, but the second mission to Constantinople was as fruitless as the first. It is possible that Anastasius delayed recognition of Theoderic in the hope that he would put pressure on the pope to enter into communion with the Church of Constantinople, the two sees having been out of communion since the beginning of the Acacian schism in 484,[11] and indeed the outcome of the third embassy of Theoderic to Constantinople, undertaken, probably in 497, by Festus, looks very much like a *quid pro quo*. Nevertheless, as the events of the Laurentian schism in the Roman Church were to show, Festus was something of a religious enthusiast, and we may doubt whether considerations of this kind weighed heavily on Anastasius in 492: Odovacer still controlled Ravenna, and following the example of Zeno when confronted by the rival claims of Julius Nepos and Odovacer in about 477, the emperor may have deemed it prudent not to commit himself.

By the time Festus returned to Italy events had taken a new turn: not only had Theoderic entered Ravenna and killed Odovacer, but the Goths, without waiting for orders from the emperor, had confirmed Theoderic as their king. Jordanes, it is true, provides in his *Getica* a very different account, according to which Zeno advised Theoderic in the third year after he came to Italy to abandon the clothes worn by the Goths and adopt a kingly

[10] Anastasius' complaint: E. Schwartz, *Publizistische Sammlungen zum Acacianischen Schisma* (Munich, 1934), 19.18–20. Consul nominated for 493: Sundwall, *Abhandlungen*, 191 f. See in general Ensslin, *Theoderich*, 74; Claude, 'Universale und partikulare Züge', 46, curiously dates Faustus' mission to the first half of 492. The assertion of Anon. Vales. 57 that news of Zeno's death reached Ravenna before Faustus returned but after Theoderic killed Odovacer is incomprehensible.

[11] Stein, *Histoire*, 112. The authenticity of Gelasius' letter to Faustus, which figures in this argument (Schwartz, *Publizistische Sammlungen*, 16–19), has been affirmed in the face of denial by E. Dekkers, *Clavis patrum Latinorum* (Steenbrugge, 1961), 369.

form of dress, as he was now the ruler of Goths and Romans.[12] This version, however, may easily be dismissed, for not only had Zeno's death occurred well before Theoderic's third year in Italy, but Jordanes in his *Getica* persistently seeks to legitimize Theoderic's role, and this improbable tale may possibly be an indirect reflection of the embarrassment the action of the Goths in 493 was to cause Cassiodorus, from whose lost history of the Goths he presumably took it.[13] It is difficult to establish what the confirmation of Theoderic as king signified. Anonymous Valesianus only terms Theoderic *rex* after this event, persistently styling him patrician beforehand, and so seems to have regarded Theoderic's royal dignity as based on this election.[14] But there can be no doubt Theoderic had succeeded his father Theodemer as *rex* in about 474, so that in some sense he simply remained what he had been.[15] It is possible that the confirmation of 493 was in reality an extension of Theoderic's kingship over non-Goths, whether members of other races who had accompanied him to Italy[16] or the remnants of Odovacer's following, but our one source is explicit that the deed was carried out by Goths. The effect of Theoderic's confirmation on his relations with Constantinople is difficult to judge. True, in 494 there were two western consuls, but their appointment may date from the period before the confirmation,[17] as there was a Western nominated consul for 495. But that for 496, Speciosus, does not seem to have been recognized in the East;[18] and there was no Western consul in 497, although this may merely reflect the lack of a candidate.[19] In any case, Festus was again sent by Theoderic to Constantinople, probably in 497, and if we choose we may connect this with the accession of the philobyzantine Pope Anastasius in 496;[20] he was certainly accompanied by two bishops, Cresconius of Todi and Germanus of Pesaro, who attempted to heal the Acacian schism.[21] Festus' mission was, at least in its political aspect, successful. Peace was made concerning

[12] Anon. Vales. 57; Jordanes, *get.* 295.

[13] Both Anonymous Valesianus and Jordanes imply that Zeno lived for some time after 491. I cannot account for this; Zeno is correctly credited with a reign of 17 years by Jordanes at *rom.* 341.

[14] D. Claude, 'Zu Königserhebung Theoderichs des Grossen', in K. Hauck and H. Mordek, eds., *Geschichtschreibung und Geistigen Leben in Mittelalter: Festschrift für Heinz Löwe* (Cologne, 1978), 1–18 at 4.

[15] Mommsen, *Gesammelte Schriften*, 386.

[16] Cf. Ensslin, *Theoderich*, 74f.

[17] Wes, *Das Ende*, 160.

[18] *PLRE* 2.1025.

[19] Sundwall, *Abhandlungen*, 198, reads too much into the evidence.

[20] So Stein, *Histoire*, 115, and Wolfram, *History*, 284.

[21] *Lib. pont.* 44.

Theoderic's assumption of the kingdom (or 'of the rule'; the expression 'Facta pace . . . de praesumptione regni' is indicative of hitherto bad relations, and the *praesumptio* may have been Theoderic's confirmation in 493), and all the ornaments of the palace which had remained in the city after the attack of the Vandals in 455 and then been transmitted to Constantinople by Odovacer were returned.[22] We may take it that, by the end of the decade, Constantinople had recognized the rule of Theoderic. But it is not clear in what capacity he was recognized.

THEODERIC'S STATUS

We know from various classes of evidence that Theoderic's preferred title was 'king'; the forms 'Theodericus rex' and 'rex Theodericus' occur over and over again.[23] It is, of course, a barbarian title which Theoderic would have assumed in succession to his father, king Theodemer[24] but one cannot help noticing that the formulation was not inevitable. Theoderic could have styled himself 'king of the Goths', and indeed is so termed by various authors,[25] just

[22] Anon. Vales. 64. 'Praesumptio' is discussed by L. Várady, *Epochenwechsel um 476: Odoaker, Theoderich d. gr. und die Unnwandlungen* (Budapest, 1984), 34. On the removal of 'imperialia ornamenta' from Rome by the Vandals, *cod. iust.* 1.27.6f. (*Corpus iuris civilis*, 2).

[23] 'Theodericus *rex*': Roman synod documents of 502 (*MGH AA* 12.419.4 and 'Flavius Theodericus *rex*', ibid. 420.18, 424.2; on the significance of 'Flavius' consult H. Wolfram, *Intitulatio*, 1. *Lateinische Königs- und Fürstentitel bis zum Ende des 8. Jahrhunderts* (Graz, 1967), 50ff., 75), Cassiodorus, *var.* bks. 1–5 *passim*; Fiebiger/Schmidt, *Inschriftensammlung*, no. 181 (*CIL* 11.280). *Rex* Theodericus: medallion of Senigalla; synod of 502 (*MGH AA* 12.426.7); Cassiodorus, *chron.*, *passim* (except *s.a.* 490, 'Theoderichus *rex*'); Fiebiger/Schmidt, *Inschriftensammlung*, e.g. nos. 179 (*CIL* 11.10), 183 (*CIL* 11.310). Theoderic is styled *rex* in Justinian's Pragmatic Sanction of 554 (*corp. iur. civ.* app. 7.8); Totila, the only other Ostrogothic sovereign to be given a title, is *tyrannus* (app. 7.2).

[24] He bore the title ῥήξ, 'for thus the barbarians are accustomed to call their leaders' (Procopius, *BG* 1.1.26), a gloss which inclines me to believe that the Gothic *reiks* rather than the Latin *rex*, representing the Gothic *thiudans*, is meant here; see Wolfram, *Intitulatio*, 40ff., against Reydellet, *Royauté*, 202–5. Ennodius sees him as 'rex genitus' (*pan.* 13; on the significance of this Reydellet, *Royauté*, 165–9). Given the barbarian associations of *rex*, Cassiodorus' 'habui multos reges' (*var.* 11.13.4) is a little puzzling.

[25] 'Rex Gothorum': Marcellinus, *chron. sa* 483, 487, 488 etc. (the same title is applied to Odovacer, *s.a.* 476, 489); Cassiodorus *Ordo generis Cassiodororum* (but never in his *chron.* or *var.*, and even in the *ordo* the forms 'rex Theodericus' and 'Theodericus rex' occur); Jordanes, *rom.* 24. The term applied by Procopius at *BV* 1.8.11 seems to me a description rather than a title.

as Cassiodorus, writing on his behalf, was able to employ terms such as 'king of the Franks' and 'king of the Burgundians' to describe other sovereigns.[26] Other authors called him 'king of Italy', and the Lombard king Agilulf was later referred to in an inscription as 'king of all Italy', but while the term 'kingdom of Italy' occurs in one of Cassiodorus' letters there is no sign that Theoderic sought to adopt this title, which in any case would not have covered all the territory subject to his rule.[27] Given that barbarian kingship was generally exercised over peoples rather than territories such a title would have been conceptually difficult; but Theoderic's styling himself 'king of the Goths', while accurate as far as it went, would have ignored his position relative to the Romans he governed, while any assertion that he was king of Romans as well as Goths would have been repugnant to Roman ears.[28] Jordanes carefully distinguished between the kingship over his people and the principate over the Roman people which Theoderic held (*rom.* 349, but cf. *rom.* 345), and when he described a position which Theoderic held with respect to both Goths and Romans he fell back on the vague term 'regnator' (*get.* 295). Use of the unadorned title 'rex' involved Theoderic in the employment of a barbarian term in a way Cassiodorus denied other barbarian kings and left open the question of his authority over the Romans he governed. We may well ask, then, to what degree Theoderic was perceived as having been dissimilar to other rulers of the states which succeeded the empire in the West, and to what extent it was felt possible to assimilate his authority to that exercised by emperors.

John of Nikiu wrote in the late seventh century, and his account of Theoderic is both obscure and riddled with errors. Nevertheless it is clear that he felt able to draw a distinction between Odovacer 'who bore the title of *rex*' and Theoderic who, John believed, resided in Rome (*sic*) for 47 years (*sic*) 'as its emperor' and refused to appoint a colleague, by which we are presumably to understand that he failed to exercise the imperial prerogative of appointing a

[26] *Var.* 2.40, 3.4; 1.46, 3.2; cf. 3.1, 3.3, 5.43f.

[27] 'Italiae rex': *chron. caesaraug. s.a.* 510, 513.2; Gregory of Tours, *hist. franc.* 3.31 (cf. 'rex Italicus', ibid. 3.5); Isidore of Seville, *hist. goth.* 36. Agilulf: Wickham, *Early Medieval Italy*, 34. 'Regnum Italiae': Cassiodorus, *var.* 2.41.3. Ennodius moved far enough in this direction to describe him as 'Italiae dominus' (*V. Epiph.* 163) and 'Italiae rector' (*pan.* 92).

[28] See the suggestive words of John the Lydian, *Powers*, 1.3. Despite Gregory of Tours, *hist. franc.* 2.27, Syagrius can scarcely have had the formal title 'Romanorum rex'.

caesar.[29] Procopius wrote that Theoderic had not adopted the garb or name of emperor but had been called king, like other barbarian leaders; nevertheless he commented that in spite of this 'he invested himself with all the qualities which appropriately belong to one who is by birth an emperor' (*BG* 1.1.26, trans. Dewing), and indeed we hear that because of the games he held in the circus and the amphitheatre the Romans called him Trajan and Valentinian;[30] elsewhere Procopius commented that although Theoderic was in name a usurper (τύραννος) in reality he was an emperor (*BG* 1.1.29). It came naturally to Procopius to see Theoderic in an imperial light; when he wrote of him as having become an object of terror to all his enemies (*BG* 1.1.31) he reproduced almost exactly a phrase he had elsewhere applied to the emperor Majorian (*BV* 1.7.14). Oddly enough, despite his having been compared with Trajan, there is no certain evidence that Theoderic led an army into battle subsequent to his defeat of Odovacer, behaviour closer to that of an emperor of his day than a barbarian king. Adoration of the purple was almost certainly practised at his court, and it is highly significant that when Cassiodorus arranged his official correspondence for publication as the *Variae* he placed a letter to the *vir spectabilis* Theon concerning the preparation of dye from murex for the robes of Theoderic immediately after the first letter in the collection, which was addressed to the emperor Anastasius and contained the assertion that Theoderic's kingdom was an imitation of the emperor's.[31] Theoderic seems to have received acclamations at a Roman synod in 499, and the surviving arms of figures in the mosaic representing Theoderic's palace at Ravenna, preserved in the church of Sant' Apollinare Nuovo at Ravenna, have been interpreted as performing the gesture of acclamation,[32] although unlike Odovacer he had not lived in the old imperial palace at Ravenna, preferring to build a

[29] John of Nikiu, *chron.* 50f.

[30] Anon. Vales. 60, cf. 71; note the evaluation of Trajan in Jordanes, *rom.* 267. The following of the example of Trajan was to figure in the propaganda of the Ostrogothic government immediately following the death of Theoderic in 526 (Cassiodorus, *var.* 8.3.5, 8.13.4f.).

[31] Adoration: Cassiodorus, *var.* 11.20, 11.31; cf. 9.24.8 and Ennodius, *pan.* 89. Letter to Theon: *var.* 1.2; cf. Procopius, *BG* 2.30.17, 26 for wearing of the purple. The defeated Vandal king Gelimer was wearing purple when brought into the presence of Justinian: Procopius, *BV* 2.9.12.

[32] Synod of 499: *MGH AA* 12.405.7; cf. Procopius, *BG* 1.6.4. Mosaic in Sant' Apollinare Nuovo: Deichmann, *Ravenna*, 2.1 (1974), 140ff. with 3, illustrations 107–10.

new one which nevertheless incorporated parts of the old.[33] Unfortunately Theoderic's palace, not to be confused with the 'Palace of the exarchs', of Byzantine or Lombard construction, which still stands in Ravenna, is no longer extant, but recent work has suggested, on the basis of what can be deduced from its general layout, placement of components, and specific features, that it was in part modelled on the great Palace of Constantinople.[34] His one visit to Rome is described by our most detailed source in frankly imperial terms, and it was probably some years later that Theoderic issued a medallion, in accordance with imperial precedent.[35] Needless to say such evidence does not prove that Theoderic was regarded as an emperor; it may simply have been the case that the ways in which power could be described and exercised in sub-Roman Italy were inevitably imperial, which would not have prevented a ruler described in such terms being considered as something less than an emperor. That statues of Theoderic were erected may not have implied anything significant;[36] the power Theoderic enjoyed of nominating consuls, but apparently not appointing them,[37] was one shared with Odovacer and was presumably based on control of Rome.

One sphere in which Theoderic may have appeared to have been operating in an imperial way was in his vast programme of public building. Under his reign, we are told, many cities were restored, well-fortified strongholds were built, wondrous palaces arose, and the wonders of the past were overcome by his great works.[38] Work was undertaken on palaces at Ravenna, Verona, Pavia, Monza, and

[33] Anon. Vales. 71 with Deichmann, *Ravenna*, 1.42f. The arguments of E. Dyggve, *Ravennatum palatium sacrum* (Copenhagen, 1941), go further than I would care to; cf. G. de Francovich, *Il palatium di Teoderico e la cosidetta 'architettura di potenza'* (Rome, 1970).

[34] M. J. Johnson, 'Toward a History of Theoderic's Building Program', *Dumbarton Oaks Papers*, 42 (1988), 73–96, at 80–92.

[35] Visit to Rome: Anon. Vales. 65–9. Medallion: see below, Ch. 6.

[36] Procopius, *BG* 3.20.29; cf. 1.24.22–7 for an image of Theoderic at Naples; Isidore of Seville, *hist. goth.* 39 for a statue covered with gold by the senate; see too Agnellus, *codex*, 94 (ed. Testi-Rasponi, 228f.); Procopius, *BG* 1.6.5 for statues of Theodahad; and Jordanes, *get.* 290 for an earlier statue erected by Zeno in Constantinople.

[37] E. K. Chrysos, 'Die Amaler-Herrschaft in Italien und das Imperium Romanum: Der Vertragsentwurf des Jahres 535', *Byzantion*, 51 (1981), 430–74 at 454ff.

[38] Cassiodorus, *chron. sa* 500; cf. Ennodius, *pan.* 56. See in general Lusuardi Siena, 'Sulle tracce della presenza gota in Italia: Il contributo delle fonti archeologiche', in *Magistra barbaritas* (Milan, 1984), 513–48, with excellent illustrations, and most recently Johnson, 'Theoderic's Building Program'.

Galeata,[39] an amphitheatre at Pavia,[40] baths at Pavia, Verona, Spoleto, and Abano,[41] and aqueducts in Ravenna and Verona.[42] His government was concerned with the state of *moenia*, perhaps to be taken as public buildings rather than walls, as well as defensive walls,[43] while marshes were drained and gardens planted at Ravenna.[44] It is an impressive range of activity, which also encompassed the building of Arian churches, possibly connected with rivalry between Arian and Catholic Christians;[45] one can only be struck by the similarities in the mosaics in the domes of the Catholic and Arian baptisteries at Ravenna, which make it almost certain that the latter was a copy of the former. Doubtless a good deal of Theoderic's activities would have fallen under the category of refurbishment rather than that of construction *de novo*, but he showed himself a not unworthy heir of emperors. Similarly Theoderic's founding of Theodoricopolis, a city which cannot now be identified, was in accordance with imperial activity, although it should be noted that the precedent was also followed by the Vandal king Huneric, a figure not likely to have been seen as an emperor in other contexts.[46]

[39] Anon. Vales. 71 and Fredegarius, *chron.* 2.57 (*MGH* SRM 2.82) mention palaces at Ravenna, Verona, and Pavia; see the discussion, with further references, of B. Ward-Perkins, *From Classical Antiquity to the Middle Ages: Urban Public Building in Northern and Central Italy* (Oxford, 1984), 158–66. Monza: Paul the Deacon, *hist. lang.* 4.21. Galeata. *V. Hiluri, AASS* Mai III, p. 473, with P. Leveque, 'Le Palais de Théoderic-le-Grand, à Galeata', *Revue archéologique*, 6th ser. 28 (1947), 58–61.

[40] Anon. Vales. 71.

[41] Pavia and Verona: anon. Vales. 71. Spoleto: Cassiodorus, *var.* 4.24. Abano: ibid. 2.39.

[42] Ravenna: anon. Vales. 71, Cassiodorus, *var.* 5.38 and *chron. s.a.* 502. The volume of evidence bearing on this may be explained by the notoriously bad water of the city, for which Sidonius Apollinaris, *ep.* 1.5.6 provides graphic evidence; see too *carm.* 9.298. Verona: Anon. Vales. 71. See too on Rome Cassiodorus, *var.* 7.6.

[43] G. Della Vale, 'Moenia', *Rendiconti della Accademia di archeologia lettere e belli arti*, 33 (1958), 167–76, with Ward-Perkins, *Urban Public Building*, 46 n. 39. For classical usage on 'walls', see Virgil, *aen.* 2.234: 'dividimus muros et moenia pandimus urbis', with which cf. 6.549. Defensive walls were built around Verona and Pavia: Anon. Vales. 71.

[44] Fiebiger/Schmidt, *Inschriftensammlung*, no. 179 (*CIL* 11.10), with perhaps Ennodius, *carm.* 2.111, De horto regis.

[45] Note in particular Fiebiger/Schmidt, *Inschriftensammlung*, no. 181 (*CIL* 11.280). Rivalry is suggested by Ward-Perkins, *Urban Public Building*, 72.

[46] Imperial precedents: A. H. M. Jones, *Later Roman Empire*, 719f. Theodoricopolis and its location: H. Ditten, 'Zu Prokops Nachrichten über die deutschen Stämme', *Byzantinoslavica*, 36 (1971), 1–24, 184–91 at 22f. O. R. Clavadetscher, 'Churätien im Ubergang von der Spätantike zum Mittelalter nach den Schriftquellen', *Vorträge und Forschungen*, 25 (1979), 159–78, esp. 164; Wolfram, *History*, 289. Unuricupolis: *Nomina episcoporum provinciae Byzacenae*, no. 107 (printed after Victor of Vita, *MGH AA* 3.1); note too a Reccopolis in Spain (K. F. Stroheker, *Germanentum und Spätantike* (Zurich, 1965), 150).

Despite this, John of Nikiu and Procopius saw Theoderic as one greater than a barbarian king. The latter comes close to basing this judgement on moral grounds, but some of Theoderic's subjects went beyond this. So Cassiodorus, in a famous passage in a letter he wrote in Theoderic's name to the emperor Anastasius:

Our royalty is an imitation of yours, modelled on your good purpose, a copy of the only empire, and in so far as we follow you do we excel all other nations . . . We think that you will not suffer that any discord should remain between two republics (*inter utrasque res publicas*), which are declared ever to have formed one body under their ancient princes.[47]

The position of this letter at the beginning of the *Variae* may encourage us to see in its lapidary formulations statements of general principles which would remain valid throughout the period of Ostrogothic government, but such a view would be erroneous. For example, in this letter Cassiodorus proceeds to comment 'Let there always be one will, one purpose in the Roman kingdom (*Romani regni*)' (1.1.5), implying that Italy and the empire both existed within the one *regnum*, yet in a later letter he expresses the wish 'that there should be no discord between the Roman realms (*inter Romana regna*)' (10.21.2), with the clear implication that they were separate *regna*.[48] The letter to Anastasius stands at the head of the *Variae* because it was written by Cassiodorus to the emperor on Theoderic's behalf while serving him as quaestor (507–11); Cassiodorus did not see fit to include any other letters directed to the emperor in this period in his collection. It was called into being by contingent circumstances. After Anastasius' granting of an honorary consulate to Clovis following his triumph over the Visigoths in 507 and more importantly the attack on the Italian coast by a Byzantine fleet in 507 or 508 (below, Ch. 6) there was a need to restore good relations, and the sentiments expressed in Cassiodorus' letter were designed to play on imperial vanity rather than provide a realistic description of a desire of Ravenna to model

[47] *Var.* 1.1.3f. I follow Hodgkin's translation, which seems to me to catch the meaning well, the construction placed on 'regnum nostrum . . . exemplar imperii' by Reydellet, *Royauté*, 208–10 being unacceptable (see N. Staubach, 'Germanisches Königtum und lateinischen Literatur vom fünften bis zum siebten Jahrhundert', *Frühmittelalterliche Studien*, 17 (1983), 1–54 at 42f.). See further W. Suerbaum, *Vom Antiken zum frühmittelalterlichen Staatsbegriff*, 3rd edn. (Munster, 1977), 248–52, and Claude, 'Universale und particulare Züge', 42–4.

[48] See Claude, ibid, 43f., and Maximianus, *eleg.* 5.3: 'Dum studeo gemini componere foedera regni' (ed. Baehrens, *Poetae Latini minores*, 5).

itself on Constantinople.[49] Elsewhere in his correspondence Cassiodorus does not consider the relationship of the Ostrogothic state to the empire in terms of imitation, model, and copy, while nevertheless expressing a conviction that they were like each other and dissimilar to other states: the Ostrogothic monarchs were not barbarian kings but lawful lords (*var.* 9.21.4; the word *dominus* is applied to emperors at *var.* 1.4.11, 3.51.4[50]), and Cassiodorus felt able to apply the terms *imperium* and *res publica* to the empire and the Ostrogothic state, but to no other Germanic kingdom.[51] It goes without saying that Cassiodorus sharply distinguished Theoderic from Odovacer. In his Chronicle he asserts that Odovacer had merely taken the name of king (*s.a.* 476) and persistently refers to him by his personal name without a title, whereas Theoderic is styled *dominus rex* (e.g. *s.a.* 489, 490) or indeed *domnus noster* (*s.a.* 508). In general, as with respect to the titles of Theoderic and other sovereigns in the West, so too the terms applied to states suggest that the Ostrogothic state was distinctive, and that this state was somehow on a par with the empire.

This suggestion of rough equality is even to be found in a letter of the emperor Anastasius seeking the help of the senate in 516, in which he spoke of the 'limbs of either republic', but there is no reason why a suppliant emperor in 516 would have been more likely to express true feelings than would the writer of a letter for a suppliant king in about 508; what is perhaps more interesting is that the senate used the same phrase in its reply.[52] It is persistent in Cassiodorus. For him, perhaps going a little further than Jordanes in this respect, Theoderic was a 'romanus princeps' (*var.* 3.16.3), and the extraordinarily frequent references to him as *princeps* in the *Variae* very often occur in contexts suggestive of imperial activity.[53] The *Variae* contain many references to emperors described as former *principes*,[54] and contemporary emperors are some-

[49] This would have been particularly the case if the letter was a protest against Byzantine recognition of Clovis: Wolfram, *Intitulatio*, 54 no. 103.

[50] But note on one occasion 'Burgundionum ... dominus': *var.* 1.45.2.

[51] Suerbaum, *Staatsbegriff*, 258–66. At *var.* 4.1.1 Cassiodorus writes to Herminifrid the Thoringian, on the occasion of his marrying a niece of Theoderic, of his shining 'claritate Amali *sanguinis*' (edd., yet most MSS read 'imperialis'; see Mommsen's edn. p. clxii).

[52] *Coll. avel.* 113.4, 114.7.

[53] The wearing of purple (*var.* 1.2.3, 6), care for the *res publica* (1.6.1, 1.20.1), choice of consuls (2.2.1), etc.

[54] e.g. 'antiqui principes', *var.* 1.1.4, 1.25.3; 'veteres principes', 4.20.1; 'priores principes', 4.33.1.

times so described,[55] thereby implying an equivalence between Theoderic and such people; indeed, if one chose to press references to both Anastasius and Justinian as 'Orientis princeps' (*var.* 11.1.10, 12.20.1) one could argue that there must have been a corresponding *princeps* of the West, just as the reference to the 'antiqui principes' of the two *res publicae* could be held to imply that there were now two *principes*, Anastasius and Theoderic (*var.* 1.1.4), although we read in Boethius of 'the emperor of the East, who is now called Anastasius' (*PL* 64.264A). If Cassiodorus could represent the senate as addressing Justinian as 'triumphator egregius' (*var.* 11.13.5), in one of his orations he could describe Theoderic as 'infatigabilis triumphator';[56] if emperors could be referred to as 'serenissimus' (*var.* 10.33.1, 10.34f.; cf. 1.1.4, 10.20.3, 10.23.2), 'serenitas' could be seen as a characteristic of Ostrogothic kings (e.g. *var.* 1.4.2, 1.33.1, 2.22.1, with 6.9.1). Cassiodorus went a considerable distance towards assimilating Theoderic into the model provided by emperors.

Ennodius seems to have gone further, but along a slightly different road. He was able to instruct Rome to acknowledge the clemency of its *dominus*, and describe Theoderic bluntly as 'our emperor',[57] but in various ways persistently asserted his superiority over emperors. 'Our king' could be compared favourably to 'the old *principes*' (*ep.* 9.30.6), and for Ennodius there was nothing to be ashamed of in the title *rex*, for on one occasion he referred to 'charity among kings' when the kings were the Visigoth Euric and the emperor Nepos (*V. Epiph.* 81). Theoderic's *pietas*, Ennodius felt, was greater than that of 'all the preceding emperors',[58] and a reference to him in the panegyric as 'one who gave and defended the diadem' and so compelled the love of the emperor Zeno, obliquely described as 'the ruler of those parts', implies his seniority to Zeno.[59] Ennodius felt able to observe that what diadems did for some people had been wrought in Theoderic by nature through the operation of God (in which context it may not be amiss to recall that

[55] e.g. Zeno (*var.* 1.43.2), Justinian (10.25.1).

[56] *MGH AA* 12.466.14.

[57] 'agnosce clementiam domini tui' (*pan.* 48); 'imperator noster' (*lib. pro syn.* 74); cf. 'imperialis auctoritas' (ibid. 36), 'imperiala scripta' (ibid. 73), and *V. Epiph.* 187: 'boni imperatoris est possessoris opulentia'.

[58] 'omnes retro imperatores': *V. Epiph.* 143.

[59] *Pan.* 14, on which consult S. G. MacCormack, *Art and Ceremony in Late Antiquity* (Berkeley, Calif., 1981), 230. For the phrase 'illarum rector partium', cf. *V. Epiph.* 88, and *pan.* 92.

the Vandal king Huneric felt able to refer to 'the provinces conceded to us by God'[60]) and to assert that his king was called Alamannicus by right, even though 'another', almost certainly the emperor Anastasius, bore the title.[61] Theoderic was perceived to enjoy moral superiority over emperors, but any comparisons were of like with like; Ennodius was content to let the term *res publica* stand for the empire in his *Life of Epiphanius* and, more or less, Theoderic's kingdom in the panegyric.[62]

Blurring of the distinctions between the emperor and a barbarian king becomes even stronger in two inscriptions. One Valerius Florianus, who may be identical with the Florianus to whom Cassiodorus wrote a letter on behalf of Theoderic within the period 507–11, erected an inscription which mentioned 'our lords Anastasius augustus for ever, and the most glorious and triumph-winning man Theoderic'.[63] The titles 'perpetuus' and 'augustus' are reserved for Anastasius, who properly comes first, but Theoderic is a 'gloriosissimus ac triumphalis vir' and they are together 'domini'. More startling is an inscription put up apparently more than once on the Via Appia by Caecina Mavortius Basilius Decius, which refers to 'Our lord the most glorious and renowned king Theoderic, victor and conqueror, ever augustus, born for the good of the state, guardian of liberty and defender of the Roman name, tamer of foreign peoples'. Theoderic is called 'rex', but apart from this one detail it is

[60] *Pan.* 91; note as well *pan.* 18. Theoderic did not wear a diadem, according to M. McCormick, *Eternal Victory: Triumphal Rulership in Late Antiquity, Byzantium and the Early Medieval West* (Cambridge, 1986), 270 n. 48, but see in addition J. Deér, 'Byzanz und die Herrschaftszeichen des Abendlandes', *Byzantinische Zeitschrift*, 50 (1957), 405–36 at 432. The Senigalla medallion shows him with no headdress, but coins of Theodahad indicate that this king wore a crown (Hahn, *Moneta*, 1, plate 41 no. 81). Ennodius frequently refers to the heavenly support Theoderic received: *pan.* 23, 52; *V. Epiph.* 125, 128. Huneric: Victor of Vita, *hist. pers.* 2.39 (cf. 3.14), with which cf. Jordanes on Geiseric: 'a divinitate, ut fertur, accepta auctoritate' (*get.* 169).

[61] 'rex meus sit iure Alamannicus, dicatur alienus': *pan.* 81. The title Alamannicus was claimed by Anastasius, the emperor at that time (*coll. avel.* 113), but its application to emperors was in any case conventional: G. Rösch, *Onoma basileus* (Vienna, 1978), 55f. I accept that the *alienus* was Anastasius (so, already, Hartmann, *Geschichte*, 171 n. 11); the arguments of Reydellet, *Royauté*, 172–5 are not compelling (cf. Staubach, 'Germanisches Königtum', 40f.). Cassiodorus, writing at about the same time, could apply the term 'aliena malignitas' to Byzantine activity: *var.* 3.4.4.

[62] S. Teillet, *Des Goths à la nation gothique* (Paris, 1984), 276.

[63] Best published by A. Bartoli, 'Lavori nella sede del senato romano al tempo di Teoderico', *Bullettino della Commissione archeologica communale di Roma*, 73 (1949–50), 77–90, failing which Fiebiger/Schmidt, *Inschriftensammlung*, no. 187 (*CIL* 6.1794). Cassiodorus' letter: *var.* 1.5.

hard to imagine a more imperial intitulature: the term 'victor ac triumphator' was to be applied to Justinian by Jordanes in the very context of his defeat of the Ostrogoths (*get.* 315); 'augustus' was a term not even Ennodius applied to his king, and the expression 'semper augustus' may be compared with the 'perpetuus augustus' applied to Anastasius in Florianus' inscription. He is described as 'custos libertatis', something which Cicero believed was proper to the Romans, and as we shall see it was an open question whether Italy enjoyed freedom under the Ostrogoths (below, Ch. 7); this inscription leaves no doubt as to the view of Basilius Decius. Pope Gregory the Great, writing to the emperor Phocas, was able to distinguish between emperors, who were lords of freemen, and kings of the nations (*reges gentium*), who were the lords of slaves; again, it is clear to which of these categories Theoderic would have been seen as belonging by Basilius Decius, particularly as the inscription describes him as 'domitor gentium'.[64]

Despite the degrees to which some of his subjects were prepared to assimilate Theoderic into the category of emperor, for official purposes he remained cautious.[65] He refrained from promulgating laws and contented himself with issuing edicts, a distinction which may appear trivial but was important, for while any magistrate with the *ius edicendi* could issue *edicta, leges* could issue from the emperor alone,[66] so that when Cassiodorus urged upon Romans 'reverence for the laws' (*var.* 4.10.1, cf. 1.27.1) he had in mind the enactments of emperors, not of Theoderic; while the Gothic king wished the laws of the Roman *principes* to be observed (*MGH AA* 12.391, no. vi), his own enactments were merely edicts (*var.* 9.18.12). A detailed study of Cassiodorus' vocabulary has shown that he described Theoderic's enactments in words which, in imperial laws, apply to the activities of magistrates rather than

[64] Fiebiger/Schmidt, *Inschriftensammlung*, no. 193 (*CIL* 10.6850–2), presumably erected in the context of the works mentioned in Cassiodorus, *var.* 2.32f. I agree with Wes, *Das Ende*, 163, that we are to see here merely an expression of the feeling of one aristocrat; see further the sensible comments of McCormick, *Eternal Victory*, 279f. Gregory the Great: *reg.* 13.4, cf. 11.4.

[65] A document purportedly issued in the name of 'Nos Theotericus Rex omnium Gothorum et Imperator Romanorum' is a recent forgery: R. Heuberger, 'Ein angebliches Edikt Theoderichs des Grossen vom Jahre 505 aus dem Castrum Maiense über dem Laureinberg', in *Festchrift Karl Piveč* (Innsbruck, 1966), 201–3.

[66] Mommsen, *Gesammelte Schriften*, 6.410f.; Claude, 'Universale und particulare Züge', 50. Ennodius uses the word *lex* loosely at *V. Epiph.* 122; the Vandal king Huneric on one occasion issued an enactment he characterized as a *lex* (Victor of *Vita, hist. pers.* 3.14).

emperors.[67] The coins struck by Theoderic were generally issued in the name of the reigning emperor and the imperial monopoly on the minting of gold coins accepted;[68] with the exception of the Senigalla medallion the portrait of no Ostrogothic sovereign appears on coins till the time of Theodahad, when relations with Constantinople were very bad.[69] And Theoderic may not have worn a diadem (above, n. 60). There seems, then, to have been a feeling of ambiguity about the precise status of Theoderic *vis-à-vis* the emperor, and it is a legitimate question whether this reflected any constitutional relationship which may have existed between them after the settlement arranged by Festus, if not from the time of Zeno. The emperor Anastasius wrote to the Roman senate in 516 referring to the power and care of ruling which had been entrusted to Theoderic, but his language is scarcely precise (*coll. avel.* 113.4); a letter written by Cassiodorus in the name of Athalaric to Justin mentioned the compacts and agreements concluded between Theoderic and the emperor's predecessors, presumably Zeno and Anastasius (*var.* 8.1.5), but any such *pacta* and *condiciones* have not survived; and while king Theodahad came to an agreement at the beginning of the Gothic war with Justinian's envoy Peter (Procopius, *BG* 1.6.15) it is doubtful whether one could infer from this the content or even the existence of similar agreements between Theoderic and Constantinople.[70] The lack of evidence has encouraged the multiplication of scholarly hypotheses: Theoderic has been seen as ruling by virtue of an office bestowed by the emperor, whence 'Italy remained part of the empire and was regarded as such officially both at Rome and Constantinople';[71] but also as the ruler of a state constitutionally no different from any other barbarian kingdom,[72] while others have suggested that the apparent confusion in

[67] Å. Fridh, *Terminologie et formules dans les Variae de Cassiodore* (Stockholm, 1956), esp. 109f.; see too G. Viden, *The Roman Chancery Tradition: Studies in the Language of Codex Theodosianus and Cassiodorus' 'Variae'* (Göteborg, 1984), 84, 88, 142 with n. 193.

[68] Hahn, *Moneta*, 79; Hendy, *Studies*, 484; for the Senigalla medallion see below, p. 187 f.

[69] Hahn, *Moneta*, 90.

[70] Chrysos, 'Amaler-Herrschaft', goes a little further than I would care to.

[71] J. B. Bury, *History of the Later Roman Empire*, 1 (London, 1923), 454, following Mommsen, *Gesammelte Schriften*, 6.362f. (Theoderic retained the office of *magister militum* conferred by Zeno in 483); see too Stein, *Histoire*, 40 (Theoderic given a new office, *magister militum per Italiam*).

[72] A. H. M. Jones, 'The Constitutional Position of Odoacer and Theoderic', *Journal of Roman Studies*, 52 (1962), 126–300; cf. Wes, *Das Ende*. Várady, *Epochenwechsel*, 43–61, suggests that Theoderic was regarded as having ruled over a client kingdom.

the sources reflects contemporary reality, and that Theoderic's position was never precisely defined.[73]

It is certainly true that contemporaries came to radically different conclusions as to the origin and nature of Theoderic's power. Procopius believed that the emperor Zeno had sent Theoderic to Italy to overcome the tyrant Odovacer and become the ruler of the Romans and Italians (*BG* 1.1.10f.). But this was susceptible of more than one interpretation, as Procopius brought out in a dialogue he represented as having occurred between Belisarius and three Gothic envoys during the Gothic war: the Goths claimed that following the defeat of the tyrant Odovacer they had assumed the government of Italy, whereas Belisarius claimed Theoderic had been sent not to rule Italy, which would merely have involved the replacement of one tyrant by another, but to restore it to the empire (*BG* 2.6.14–25, cf. 1.5.8). Procopius' own feeling was that Theoderic was only a tyrant in name; as he expressed it, in a formulation which owes something to Thucydides' description of Pericles, Theoderic was 'in name a tyrant but in reality a true emperor'.[74] Just as the data provided by Procopius could yield more than one interpretation, so could the narratives provided by Jordanes in his two works point in quite different directions. In the *Getica*, we find Zeno proposing to Theoderic that if he defeated Odovacer, he would hold the kingdom Zeno would give him as a gift and present (*get.* 291), but according to the *Romana* Zeno ordered him to go there as his *cliens* (*rom.* 348). Ennodius, for his part, was inclined to see God alone as responsible for Theoderic's power, and his observation that he came to Italy 'as the heavenly empire disposed' (*V. Epiph.* 109) seems almost designed to exclude the participation of the earthly empire. Indeed, it provided Ennodius with another way of assimilating Theoderic into the model of an emperor, for he felt that

[73] J. J. O'Donnell, *Cassiodorus* (Berkeley, Calif., 1979), 100f.; E. A. Thompson, *Romans and Barbarians* (Madison, Wis., 1982), 72–6. Cf. Hodgkin, *Italy*, 393: 'the whole matter seems to have been purposely left vague'.

[74] *λόγῳ μὲν τύραννος, ἔργῳ δὲ βασιλεύς*: *BG* 1.1.29; cf. Thuc. 2.65.9. See on this B. Rubin, *Das Zeitalter Iustinians I* (Berlin, 1960), 436 n. 489, 495 n. 830. Ennodius (*pan.* 24) and Jordanes (*get.* 291) affirm that Odovacer was a *tyrannus*; Cassiodorus found the word difficult and rarely used it: Teillet, *Des Goths*, 296. For the meaning of tyrant as 'usurper', cf. A. Cameron's commentary to her edition of Corippus, *laud.* praef. 11 (p. 120); but consult as well E. K. Chrysos, 'The Title *ΒΑΣΙΛΕΥΣ* in Early Byzantine International Relations', *Dumbarton Oaks Papers*, 32 (1978), 29–75 at 432, and note Anon. Vales. 94 (*iubente non rege sed tyranno*) as well as, more generally, Gregory of Rome, *mor. in Job* 12.38: 'tyrannus qui in communi republica non iure principatur' (CCSL 143A.654).

some years earlier divine ordination had entrusted Italy to the emperor Nepos (*V. Epiph.* 88). But in reality Nepos had come to Italy under imperial auspices, just as Theoderic did.[75]

Needless to say, the variety of interpretations current in the sixth century does not disprove there having been a formal understanding between Theoderic and Constantinople on the status of the former, whether concluded before his departure for Italy in 488 or in 497. But it does suggest that we would be hard pressed to get at whatever truth lies behind our sources. We may take it that Theoderic and Zeno may have had some understanding in 488, that if they did the death of Zeno in 491 and the action of the Goths in confirming Theoderic as king in 493 may well have complicated matters, but that good relations were restored by Festus in 497. I doubt whether we can go beyond this. All that the data of the sources entitle us to conclude is that radically different versions of Theoderic's constitutional position were circulating, and that his government was coy about giving it precise expression.

RELATIONS WITH THE BARBARIAN STATES

But Theoderic was not only concerned with his relations with Constantinople. He also had to deal with the barbarian states of the West, and one of his first accomplishments was the establishment of a network of marriage alliances. It could be that he was at least partly motivated by a search for protection against possible Byzantine attack,[76] but the bringing of concord between the barbarian kings seems at least as plausible a reason, and it is not impossible that in some of the marriages the initiative did not lie with him, it being certainly the case that Theoderic would have been a desirable ally after the defeat of Odovacer.[77] Already the father of two daughters by a woman variously described as his wife or concubine,[78] he had not been long in Italy when he took the hand of Audefleda, sister of Clovis, king of the Franks. Indeed, the mar-

[75] For Nepos and the emperor Leo, see e.g. Anon. Vales. 36; for Theoderic and God in Ennodius see further above, n. 60.

[76] Hartmann, *Geschichte*, 134; B. Behr, *Das alemannische Herzogtum bis 750* (Berne, 1975), 43.

[77] Cassiodorus, *var.* 5.43.1: 'Quamvis a diversis regibus expetiti pro solidanda concordia . . . '.

[78] Anon. Vales. 63, 'uxor'; Jordanes, *get.* 297, 'concubina'.

riage could have taken place even before the murder of Odovacer; if it did not, it certainly occurred shortly thereafter.[79] Clovis' power was rising, following his defeats of Syagrius at Soissons in 486–7 and of the Thoringians in 491–2; that his horizons extended further was shown by his marriage to Chlotilde, daughter of the Burgundian king Gundobad, probably at about the same time as the marriage of his sister to Theoderic,[80] a circumstance which may suggest that the initiative in the marriage of Theoderic came from north of the Alps. Following his own marriage to Audefleda one of Theoderic's daughters, Theodegotha, was married to the Visigothic king Alaric[81] and his other daughter, Ostrogotho, mysteriously also known as Areagni, married Sigismund, another child of the Burgundian Gundobad.[82]

It must have been shortly after this marriage that Theoderic sent bishops Epiphanius of Pavia and Victor of Turin with gold to ransom the captives Gundobad had taken in Liguria during his war with Odovacer. The bishops came before the king at Lyons and Epiphanius begged him to have mercy on the prisoners. After consulting with his adviser Laconius, Gundobad ordered the release of many prisoners without payment; others, however, were to be ransomed. The mission, in which Ennodius participated, was

[79] Anon. Vales. 63; Jordanes, *get.* 295 (where Audefleda is incorrectly said to have been Clovis' daughter); Gregory of Tours, *hist. franc.* 3.31. On the date, V. Bierbrauer, 'Zur ostgotische Geschichte in Italien', *Studi medievali*, 3rd ser. 14 (1973), 1–37 at 27 = *Die ostgotische Grab- und Schatzfunde in Italien* (Spoleto, 1975), 43.

[80] The traditional dating of Clovis' marriage, argued for by C. Binding, *Das Burgundisch-Romanische Königreich (von 443 bis 532 n. Chr.)* 1 (Leipzig, 1868), 113, and E. Zöllner, *Geschichte der Franken bis zur Mitte das sechsten Jahrhunderts* (Munich, 1970), 55 n. 2, does not seem to me to have been overthrown by Weiss' suggestion that it occurred in 500–2: *Clodwigs Tauf: Reims 508* (Berne, 1971).

[81] Anon. Vales. 63 (where she is misnamed); Jordanes, *get.* 297f.; Procopius, *BG* 1.12.22. On the date, Claude, 'Universale und particular Züge', 24.

[82] Anon. Vales. 63 (where misnamed); Gregory of Tours, *hist. franc.* 3.5; Jordanes, *get.* 297. Despite the suggestion of F. Wrede, *Über die Sprache der Ostgoten in Italien* (Strasburg, 1891), 65f. that her real name was Ariagne but that she was called Ostrogotho to differentiate her from Ariadne, the wife of Zeno, Ostrogotho is to be preferred, especially as the queen Suavegotta described by Flodoard as making a donation to the Church of Reims is presumably her daughter (*hist. eccl. rem.* 2.1); see Gregory of Tours, *hist. franc.* 3.5 (fin.) on the marriage of an unnamed daughter of Sigismund (by Ostrogotho?) to king Theuderic the Frank. Some confusion between Ostrogotho and Ariadne in Anon. Vales. 63 would be in accordance with this author's tendency to confuse material relating to East and West.

responsible for the liberation of more than 6,000 captives.[83] Epiphanius' diplomatic skill and eloquence were called on again two years later, when the taxpayers of Liguria considered themselves oppressed. The bishop made his way to Ravenna where he pleaded on their behalf before Theoderic. Years later, the people of Dalmatia were to send the poet Arator on an embassy to Theoderic, and the ambassador discharged his function 'not with commonplace words but with a rushing river of eloquence', and one wonders whether it was the case made by Epiphanius or the manner of its presentation which led the king to remit two-thirds of the tax due that year.[84] With the subsequent marriage of his sister to the Vandal king Thrasamund Theoderic had built up a remarkable system whereby all the major barbarian kings were his relations by marriage.

The practical advantages of this are difficult to establish. Bishop Epiphanius is represented asaving concluded his speech to Gundobad on behalf of his Ligurian captives by pointing out the *adfinitas* the lord of Italy now had with him: the freeing of prisoners could be the wedding-gift of a husband to his wife (*V. Epiph.* 163). But this seems to represent an appeal to the recently celebrated marriage between Ostrogotho and Sigismund rather than any lasting relationship it brought about between their parents. It was all very well at a later date for Theoderic representing himself as the father of one of the antagonists and the friend of the other ('iure patris . . . et amantis', *var.* 3.4.4) to forbid Clovis to make war on Alaric, and to remind Gundobad that the young kings were related

[85] Ennodius, *V. Epiph.* 136–77. I accept that Ennodius accompanied the bishops (so L. Navarra, 'Contributo storico di Ennodio', *Augustinianum*, 14 (1974), 315–42 at 331 f., contra P. Riché, *Education and Culture in the Barbarian West* (Columbia, SC, 1976), 54 n. 13); note that Laconius was subsequently a correspondent of Ennodius: *ep.* 2.5, 3.16, 5.24. The embassy is known to have set off one March (Ennodius, *V. Epiph.* 147), but the year is not supplied: Binding, *Burgundische-Romanische Königreich*, 109 n. 402 proposes 494, but better grounded is the suggestion of 496 made by L. C. Ruggini, *Economia e società nell 'Italia annonaria'* (Milan, 1961), 279 f. I have been unable to do more than glance at the richly documented study by M. Cesa, *Ennodio: Vita del beatissimo Epifanio vescovo della chiesa Pavese* (Como, 1988).

[86] Ennodius, *V. Epiph.* 182–9; see on Arator Cassiodorus *var.* 8.12.3. For the state's generosity, cf. the memorable sentiments expressed in a letter of Cassiodorus: 'we would rather that our Treasury lost a suit than it gained one wrongfully, rather that we lost money than that the tax-payer was driven to suicide' (*var.* 3.23, trans. Hodgkin). One cannot help noticing that each of Epiphanius' three last embassies was not entirely successful: some of Odovacer's supporters were to leave their lands, some of Gundobad's captives had to be ransomed despite the bishop's requesting their release *gratis*, and the tax on the Ligurians was not fully remitted.

to them both (*var*. 3.2.2); Clovis made war on Alaric regardless and before long enjoyed the support of Gundobad. Nor did the family relationship which Theoderic enjoyed with Thrasamund prevent the Vandal king from supporting Theoderic's enemy, the Visigoth Gesalic, in 511, much to the Ostrogoth's chagrin (*var*. 5.43.3). Hence, in a little more than a decade after the marriages into the Frankish, Burgundian, and Vandal royal houses, the claims of relationship had in each case been denied, and the conduct of Theoderic's fellow sovereigns is more than enough to refute the claim of Jordanes that 'for as long as he lived there was no people in the West which did not serve him in friendship or submission' (*get*. 303). Jordanes uses strong language, for even *amicitia* could be a relationship between unequal parties, and their friendship with Constantinople was one of the boasts of Theoderic's family,[85] but there is no sign that the major states of the West let themselves be swayed by friendship with Theoderic when issues of state were involved; still less that any regarded themselves as being in submission to him. Nevertheless, the various marriages can have done Theoderic's position in the West no harm.

THE PAPACY

Theoderic entered Italy during the pontificate of Pope Felix III (483–92), a *vir clarissimus* and ancestor of Pope Gregory the Great, who owed his election to the intervention of the praetorian prefect Caecina Decius Maximus Basilius, acting on behalf of Odovacer.[86] The Ostrogoths were Christians of the Arian persuasion, a fact which will require our attention when we consider relations between Goths and Romans, but for the time being we may note that the position of Felix was awkward for reasons unconnected with the advent of Theoderic. These reasons are complicated, and

[85] Claude, 'Universale und particulare Züge', 39–41, surveys the evidence for *amicitia* between Theoderic and barbarian states, but note as well this theme in his dealings with Constantinople: he had been acknowledged as Zeno's friend (Malchus, *frag*. 17 (Blockley, 18.4)), but *amicitiae* between Ostrogoths and Byzantines are asserted for an earlier period (Jordanes, *get*. 270f.) and the tradition of friendship was later recalled following the death of Theoderic: Cassiodorus, *var*. 8.1.5, 10.2.3.

[86] *MGH AA* 12.445–7; cf. Picotti, 'Relazioni'. For the relationship between Felix and Gregory, C. Pietri, 'Aristocratie et société cléricale dans l'Italie chrétienne au temps d'Odoacre et de Théoderic', *Mélanges d'archéologie et d'histoire*, 93 (1981), 417–67 at 434f.

require brief discussion. In 451 the Council of Chalcedon, in accordance with the teaching of Pope Leo as expressed in his Tome, a dogmatic letter to the patriarch Flavian of Constantinople, had confessed that there subsisted in Christ two natures in one person or hypostasis.[87] This formula has generally seemed to Christians an adequate expression of their faith, but it was badly received in Egypt, where there was strong support for the position later called Monophysite, according to which Christ possessed only one nature. Occupancy of the see of Alexandria fluctuated for some time between opponents of Chalcedon (Timothy 'Aelurus', i.e. the weasel; Peter Mongus) and its supporters (Timothy 'Salofaciolus', i.e. the white hat; John Talaia). In 482 Peter Mongus gained control of the see a second time and was able to enter into communion with Acacius, bishop of Constantinople; the dethroned bishop, John Talaia, made his way to Rome and subsequently became bishop of Nola. The emperor Zeno attempted to resolve the disputes by issuing his Henotikon, or edict of union, in 482.[88] At first glance this might seem an eirenic document, but by condemning any unorthodox professions of belief made at Chalcedon or elsewhere—surely a piece of disingenuous phraseology—and ignoring the question as to whether Christ had one nature or two, it was interpreted by the adherents of Chalcedon as fatally compromising their position.[89] Further, as it referred with favour to the pronouncements of Cyril of Alexandria but ignored the contribution of Leo of Rome, and was issued without consultation with Rome, it was felt to be of anti-papal tendency. At the beginning of his pontificate Felix sent two legates, bishops Misenus of Capua and Vitalis of Castrum Truentium, to Constantinople, but they were tricked into sharing communion with Acacius; on their return to Rome in 484 they were excommunicated, as was Acacius.[90] So began the Acacian schism, which was to outlast the death of Acacius in 489 and divide the Churches of Rome and Constantinople for thirty-five years.

By the time Felix was succeeded by Gelasius (492–6) relations

[87] For this and what follows, A. Grillmeier and H. Bacht, eds., *Das Konzil von Chalkedon*, I (Würzburg, 1951).

[88] Text in Evagrius, *hist. eccl.* 3.14. There is a convenient English translation in W. H. C. Frend, *The Rise of the Monophysite Movement* (Cambridge, 1972), 360–2; see further p. 177 n. 4 for other versions.

[89] For comparison, we may note that when Justinian condemned the teaching of Eutyches in 533 he contrived to do so without using the word *φύσις*: *cod. iust.* 1.1.8.

[90] *Lib. pont.* 252.7–12; Felix, *ep.* 11 (ed. Thiel).

between the papacy and Constantinople had, if anything, deteriorated. Zeno died in 491 and his successor, the former silentiary Anastasius (491–518) seems to have been sympathetic to Monophysite belief.[91] Gelasius, already a force in the Roman Church prior to the beginning of his pontificate, neglected the customary courtesy of writing to the emperor informing him of his accession, and the terms of the famous letter he dispatched to Anastasius in 494 were not such as to encourage the emperor: 'There are two powers, emperor augustus, by which this world is chiefly governed, the hallowed authority of bishops and the royal power. Of these the burden of priests is heavier, since at the divine judgement they will have to render account even for the kings of men themselves.'[92] In 495 a council of the Roman Church was held. It was attended by forty-five bishops and fifty-eight priests, and absolved bishop Misenus.[93] One could be excused for thinking that such events could hardly have been connected with the establishment of Theoderic's government, then under way at Ravenna. But such evidence as we have suggests that Theoderic was reluctant to interfere in the affairs of the Church and that Gelasius felt it worth his while to seek the intercession of the king's Catholic mother;[94] it may well be that the pope found the Arian king's reluctance to involve himself in ecclesiastical matters an attractive characteristic when compared to the conduct of some emperors of late antiquity.[95]

The accession of Gelasius' successor, Anastasius, on 24 November 496 cannot have been surprising,[96] but the new pope moved quickly to change the direction of papal diplomacy towards Constantinople. He wrote to the emperor asking that the name of Acacius be struck from the diptychs, and so no longer be recited in church, because of the scandal and offence he had caused the Church, and asked him also to prevail upon the Alexandrians to return to a sincere and Catholic faith. The letter, which addresses

[91] C. Capizzi, *L'imperatore Anastasio I (491–518)* (Rome, 1969). Contemporaries were prepared to believe the worst, his mother being described as a Manichee and his father's brother as an Arian: Theodore Lector, *epit.* 448.

[92] Gelasius' resonant language has attracted much analysis; see recently Ullmann, *Gelasius*, 198–212.

[93] *Coll. avel.* 103.

[94] Above, Ch. 1 n. 23.

[95] Usener, 'Das Verhältnis des Römischen Senats', 759–67, esp. 767; Ullmann, *Gelasius*, 218–22.

[96] He may have been the deacon of this name who read documents at the Roman council of 495, *coll. avel.* 103 (*CSEL* 35.475.12, 477.11), although there was still an Anastasius among the Roman deacons in 499; *MGH AA* 12.402 no. 3, 415 no. 2.

Anastasius as his 'most glorious and most clement son' ('gloriosissimus et clementissimus filius'),[97] breathes a mildness foreign to Gelasius' missive, and was sent to Constantinople with two bishops, Cresconius of Todi and Germanus of Pesaro, who travelled in search of an end to the Acacian schism.[98] We do not know whether the pope had in mind a weakening of the position adopted by his predecessors towards Constantinople, but at the same time Theoderic sent a legate to the emperor. Rufius Postumius Festus had already served as Theoderic's legate to Zeno in 490, and was alleged in Constantinople to have been confident that he could prevail upon the pope to accept Zeno's Henotikon when he returned to Rome;[99] suspicions that Theoderic may have been involved in the ecclesiastical content of the mission to Constantinople can only be strengthened by the circumstance that Germanus was one of two bishops sent by Theoderic with a message to a synod meeting in Rome in 502 (*MGH AA* 12.419.2). While in Constantinople Anastasius' legates came into contact with the priest Dioscorus and the lector Chaeremon, representatives of the Church of Alexandria, who expressed their position in a letter addressed to Festus and the bishops. They claimed that the Greek translation of Pope Leo had been falsified by followers of the Nestorian heresy, but that Photinus, a deacon of Thessalonica, had told them that Pope Anastasius had given satisfaction concerning things they found annoying in the translation. A confession of their faith followed. They accepted the councils of Nicaea, Constantinople, and Ephesus, which from a Roman point of view would have been unsatisfactory, as the papacy had yet to accept Constantinople but fervently accepted Chalcedon which, by implication, the Alexandrians rejected. They proposed various formulations of Christology which avoided mentioning Christ's natures, or nature, and defended the Alexandrian patriarch Dioscorus and his anti-Chalcedonian successors Timothy Aelurus and Peter Mongus. In short, it was a fairly uncompromising document.[100] We have no reason to believe

[97] Anastasius, *ep*. 1 (ed. Thiel).

[98] *Lib. pont*. 44. On the origins of the *fragmentum*, consult Duchesne's edn. of the *lib. pont*., pp. xxx f. All following citations of this source are to the text printed by Duchesne.

[99] Theodore Lector, *epit*. 461 (ed. Hansen, 130.13–15), qualified by *ὡς λόγος*. Theodore wrote as a member of the staff of the Great Church in the time of Justinian, but he was not in Constantinople at the time of Festus' visit.

[100] *Coll. avel*. 102. Frend, *Monophysite Movement*, 198, goes too far in finding here 'a surprising degree of conciliation'.

that Anastasius would have conceded so much ground, but elements within the Roman Church were fearful and withdrew from communion with him, claiming that he had shared communion with Photinus, a deacon of Thessalonica who had been in communion with Acacius and who, presumably, had subsequently been in touch with the Alexandrians in Constantinople, and that he secretly wished to 'call back' Acacius, a false charge if it meant that the pope wished to reinstate Acacius in the diptychs, as his letter to the emperor indicates.[101] There is a possibility that the division in the Roman Church reflected the continuation of uneasiness which already existed during the pontificate of Gelasius, for the synod he convened in 495 to rehabilitate Misenus was attended by only fifty-eight priests, whereas a synod convened by Pope Symmachus in 499 was attended by seventy-four priests, sixty-seven subscribing to its proceedings.[102] But before further developments took place Pope Anastasius died, 'struck down by the will of God', his enemies felt, on 17 November 498.[103] It was later believed that Theoderic's ambassador Festus, having assured the emperor that he could bring about papal adherence to the Henotikon, arrived back in Rome on the day Anastasius died.[104] Whether or not this was so, Festus was certainly involved in the incidents which followed. Five days after the death of Anastasius two men were ordained bishop of Rome on the one day, the deacon Symmachus in the Lateran basilica and the archpriest Laurentius in the nearby basilica of Sta Maria Maggiore.[105] Later we shall have to enquire in some detail as to the issues that were at stake in the now-apparent schism; for the time being we may merely note that there had been a schism during the pontificate of Anastasius which seems to have lasted until his death, and there is no need to regard the parties which became manifest in the subsequent Laurentian schism, so-called from the name of the defeated party, as anything other than a continuation of those which already existed. It is particularly significant that in the collection of accounts of popes which forms the *Liber Pontificalis* a hostile

[101] *Lib. pont.* 258, with Duchesne's n. 3 on Photinus.
[102] Priests in 495: *coll. avel.* 103 (*CSEL* 35.474f.). Priests in 499: *MGH AA* 12.401f., 410–15.
[103] *Lib. pont.* 258.
[104] Theodore Lector, *epit.* 461 (ed. Hansen, 130.15). But see G. Pfeilschifter, *Der Ostgotenkönig Theoderich der Grosse und die katholische Kirche* (Munster, 1896), 40.
[105] *Lib. pont.* 260.2f. The emperor felt that Symmachus had been consecrated *non ordine*.

account of Anastasius is followed by an extremely favourable exposition of the pontificate of Symmachus, while the *Fragmentum Laurentianum*, as it survives, provides the end of what was apparently a favourable discussion of Anastasius followed by a negative account of Symmachus,[106] and that Ennodius felt moved to condemn both the followers and forerunners of Laurentius (*lib. pro synodo*, 125). An interpretation of the schism along these lines can only be strengthened by the convergent data which indicate that the supporters of Laurentius felt that the conciliatory attempts of Anastasius which had led some members of the Roman Church to withdraw from his communion had been in fact worthwhile (*lib. pont.* 44), that Festus was said to have boasted in Constantinople that he would bring about acceptance of the Henotikon by the pope (i.e. Anastasius), and that the same Festus was the most prominent supporter of Laurentius.[107] The dispute came to the attention of Theoderic[108] who, on grounds which are obscure but which may have been rendered more compelling by the disbursement of bribes at his court, decided for Symmachus.[109] On 1 March 499 Symmachus presided over a well-attended synod at St Peter's basilica, at which were present sixty-six bishops of the suburbicarian dioceses, together with seventy-four priests and the seven deacons of the Roman Church; Laurentius attended in his capacity as the priest of the *titulus Praxidae*.

The synod passed legislation concerning the manner in which future papal elections were to be conducted, and after it was com-

[106] Cf. below, Ch. 4 n. 91.

[107] Theodore Lector, *epit.* 461 has him bringing about Laurentius' election by bribery; see further *lib. pont.* 260.10, 13, 19 for his backing of Laurentius. When Laurentius acknowledged defeat he retired to the estates of Festus: ibid. 46.

[108] I see no way of deciding between *lib. pont.* 260.5 (both parties agreed to accept his adjudication) and *frag. laur.*, ibid. 44 (the two sides were forced to submit to his decision). If the former be accepted we may agree with Bury that the referring of the dispute to the Arian king was 'a remarkable episode in the history of the Church' (*History*, 464f.), but Duchesne's judicious verdict is preferable: 'le gouvernement royal se vit obligé d'intervenir' (L. Duchesne, *L'Église au VIe siècle* (Paris, 1925), 114).

[109] Again, it is difficult to decide between *lib. pont.* 260.6f. (Theoderic decided in favour of the one ordained first and accepted by the greater part) and *frag. laur.*, ibid. 44 (Symmachus obtained the decision by much bribery). Passages in Ennodius' correspondence make it clear that Symmachus was not quick to settle expenses incurred by his supporters at Ravenna on his behalf (*ep.* 3.10, 4.11, 5.13, 6.16, 6.33), although they need not refer to payments made on this occasion. For bribery to obtain episcopal office cf. Cassiodorus, *var.* 9.15; Gelasius, *ep.* 22, ed. Löwenfeld (*PLS* 3.758f.); Gregory of Tours, *hist. franc.* 6.38.

pleted Laurentius was awarded the see of Nocera in Campania, an appointment which may answer to a wish of Symmachus to get his rival out of Rome and hence point to his fear of future trouble.[110] But it must have seemed that an unpleasant interlude in papal history had been completed. As was often the practice, participants of the synod of 499 gave voice to frequent acclamations. Of these, the most frequent was 'Hearken, O Christ! Life to Theoderic!' (*MGH AA* 12.405.6). Earlier Roman synods held during the reigns of Odovacer and Theoderic had refrained from acclaiming the sovereign, but this acclamation, uttered thirty times, was testimony to a large measure of goodwill towards Theoderic. Festus had recently brought about peace with the emperor Anastasius concerning his taking of the kingdom, and now that peace had been restored to the Church there would never be a better time for Theoderic to visit Rome. This he proceeded to do.[111]

THEODERIC IN ROME

Theoderic paid his only known visit to Rome in 500, in celebration of his *tricennalia*, following in the footsteps of emperors such as Constantine, who had made trips to Rome in 315 and 326, to celebrate his *decennalia* and *vicennalia* respectively, and Constantius, whose one recorded trip, in 357, occurred twenty years after the death of Constantine.[112] The event from which the thirty years was counted cannot be identified with certainty, but his defeat over Babai shortly after his return from being a hostage at Constantinople seems more likely than anything else.[113] The fullest

[110] Synod of 499: *MGH AA* 12.399–415; Laurentius' participation is known from p. 410 with 401, which invalidates the assertion of *lib. pont.* 260.8f. that Symmachus appointed him bishop 'eodem tempore'. The *Fragmentum Laurentianum* implies that Laurentius' appointment to Nocera arose from Symmachus' bribery of Theoderic's court (*lib. pont.* 44), surely erroneously.

[111] Note the context for Theoderic's visit provided by anon. Vales.: 'facta pace cum Anastasio imperatore' (64), and 'post facta pace in urbe ecclesiae' (65).

[112] The year is clear from Cassiodorus, *chron. s.a.* 500. See for Constantine T. D. Barnes, *Constantine and Eusebius* (Cambridge, Mass., 1983), 66, 221. The latter trip to Rome was a year late, but the emperor had already celebrated the *vicennalia* with his bishops following the Council of Nicaea. For Constantius, see below, n. 126. More distantly, cf. Pliny's account of Trajan's entry into Rome in 99: *pan.* 22f.

[113] So e.g. Claude, 'Königserhebung', 1f., against e.g. Stein, *Histoire*, 133 (30 years from accession to Ostrogothic monarchy) and Reydellet, *Royauté*, 236 and n. 163 (30 years from death of father). The anonymous's first editor, Henricus Valesius,

account occurs in Anonymous Valesianus, according to whom proceedings began with Theoderic visiting St Peter's 'most devoutly and as if he were a Catholic'.[114] Outside the city he was met by Pope Symmachus and the entire senate and Roman people 'with all joy'.[115] Then, having crossed the Tiber and entered the city, he proceeded to the senate and spoke to the people at the Palma, promising that, God being his helper, he would maintain unbroken all things which the previous Roman emperors (*retro principes Romani*) had ordained.[116] He entered in triumph the palace on the Palatine and entertained the Romans with circus games. This in itself may seem of little importance, but when Totila was in Rome in 549 he is known to have held horse races, and it may be that this activity was felt by Gothic kings in some way to legitimize their presence in the city. He gave the Roman people and the poor *annonae* of 120,000 *modii* of grain annually, a figure which is not generous but may represent an increase of this amount from an existing sum, and ordered that 200 pounds were to be set aside annually from the tax on wine for the repair of the palace and the restoring of the public buildings (*moenia*) of the city.[117] During his stay in Rome Theoderic made Liberius, whom he had appointed

sought to overcome the problem by reading 'decennalem' for 'tricennalem' at anon. Vales. 67, and has been followed by e.g. J. Richards, *The Popes and the Papacy in the Early Middle Ages 476–752* (London, 1979), 70, and Wickham, *Early Medieval Italy*, 15, but the solution is drastic.

[114] Anon. Vales. 65. Unlike P. Scardigli, *Die Goten: Sprache und Kultur* (Munich, 1973), who sees 'beato Petro' as referring to the pope, I take it to refer to the basilica (cf. Anon. Vales. 91, where Pope John is met '*ac si* beato Petro'). Other visitors to Rome who proceeded by way of the basilica were Sidonius Apollinaris in 467 (*ep.* 1.5.9) and the emperor Constans II in 663 (*lib. pont.* 343.9; his visit provides interesting parallels to that of Theoderic).

[115] Anon. Vales. 65, cf. Ferrandus, *V. Fulg.* 13 (*PL* 65.130C) for the *gaudium* produced by Theoderic's visit.

[116] Anon. Vales. 66. There is no evidence for Boethius' having given an oration of welcome in the senate (so Sundwall, *Abhandlungen*, 101 f., followed by O. Bertolini, *Roma di fronte a Bisanzio e ai Longobardi* (Bologna, 1941), 51 f.), still less for Cassiodorus, *var.* 3.43 having been a speech delivered at an enthronement of Theoderic in 500 (so L. Musset, *The Germanic Invasions* (London, 1975), 48 f.). The African monk Fulgentius also saw Theoderic 'in the place called Palma Aurea' (*V. Fulg.* 13, PL 65.130D); on the identity of this place, P. de Francisci, 'Per la storia del Senato Romano e della curia nei secoli V e VI', *Atti della Pontifica accademia Romana di archeologia rendiconti*, 22 (1946–7), 275–317 at 306 ff.

[117] Anon. Vales. 67; for Totila, Procopius, *BG* 3.37.4 and for the *annonae* consult J. Rougé, 'Quelques aspects de la navigation en Méditerranée au Ve siècle et dans la première moitié du VIe siècle', *Cahiers d'histoire*, 6 (1961), 129–54 at 143, with Jones, *Later Roman Empire*, 1295 n. 56.

praetorian prefect at the beginning of his reign, patrician, appointed Theodorus the son of Basilius to succeed Liberius in the praetorian prefecture, and, apparently, detected count Odoin plotting against him. The count was beheaded in the Sessorian palace.[118] Anonymous Valesianus concludes his account of Theoderic's achievements in Rome by stating that at the request of the people he ordered the words of the promise he had made before them, presumably at the Palma, to be inscribed on a bronze tablet which was to be displayed in public. After six months he returned to Ravenna.[119]

An African monk, Fulgentius, later to become bishop of Ruspe, was in Rome for at least some of Theoderic's visit, perhaps making the acquaintance of the aristocrats, among them the prefect Theodorus, to whom he was later to address letters of spiritual counsel. He found the splendour of the city on this occasion conducive to pious thoughts: if the earthly Rome was so resplendent, how beautiful must the heavenly Jerusalem be![120] But this was probably not the kind of feeling Theoderic was trying to create. In the brief account of his visit preserved in Cassiodorus' *Chronicle*, which obviously stands in some relation to Anonymous Valesianus at this point although it is impossible to demonstrate how they are linked, we are told that Theoderic treated his senate with wonderful courtesy.[121] It is clear that Theoderic was trying to befriend the senate, but by the use of the possessive 'his', Cassiodorus contrives to suggest that Theoderic's relation to it was the same as that claimed by the emperor Anastasius, who referred to it in 516 as 'his senate',[122] and it cannot be accidental that on the only occasion when Cassiodorus used the word 'courtesy' (*affabilitas*) in his *Variae* he applied it to an empress.[123] Such data may, of course, simply provide more evidence for the care Cassiodorus took to assimilate

[118] Anon. Vales. 68. See further on Liberius J. J. O'Donnell, 'Liberius the Patrician', *Traditio*, 37 (1981), 31–72, who interprets his subject as being influenced by religion to a degree I cannot accept, and on Odoin *auct. haun. s.a.* 504, which supplies the date 4 May, although the year is contrary to the 500 implied by Anonymous Valesianus.

[119] Anon. Vales. 69f.

[120] *PL* 65.130f.

[121] 'senatum suum mira affabilitate tractans': Cassiodorus, *chron. s.a.* 500. Cessi noted the connection between Cassiodorus and the Anonymous (see the introduction to his edn. of the latter, p. cvii), although I would put more distance between them than he did.

[122] 'senatus suus', *coll. avel.* 113 (*CSEL* 113.506.24f.).

[123] Theodora, in her dealings with Queen Gudeliva: *var.* 10.24.1.

Theoderic into the model of an emperor, but this can hardly be the case with the factual narrative of Anonymous Valesianus: the visit to the tomb of Peter, for which there were imperial precedents, celebration of *tricennalia*, entry into the Palace, holding of circus games, and giving of *annonae* (*panem et circenses!*) must inescapably have suggested the activity of an emperor.[124] Theoderic's devotions at the tomb of Peter are particularly interesting in the light of a comment made towards the middle of the sixth century by Bishop Nicetius of Trier, who observed that while the Goths venerated the twelve disciples, they did not enter the basilicas where their bodies were venerated.[125] If there is any truth in this remark, it suggests that the king's devotions were of political rather than religious inspiration. We may compare Theoderic's one known visit to Rome in 500 with that of Constantius, who similarly is not known to have visited the city more than once, as described by Ammianus Marcellinus: not only is there a general suggestion of great pomp, but Constantius' speaking to the *nobilitas* in the curia and the people from the tribunal, being received into the Palace, and giving games is remarkably similar to the conduct of Theoderic.[126] One has the feeling that the Gothic king was behaving in a very imperial way.

Theoderic's stay in Rome was marked by one more event. Thrasamund, king of the Vandals, had recently lost his wife. As it happened, Theoderic's sister Amalafrida, the mother of two children, Theodahad and Amalaberga, had been widowed, and a match was made.[127] Here, it may be suggested, was the culmination of the series of marriage alliances embarked on by Theoderic early in his reign, for the territory occupied by the Vandals was not only wealthy, but located in a sensitive area; further, the Vandals had hitherto been able to deploy a powerful fleet and were notorious for their power and ferocity. Writing almost a century later, Jordanes

[124] Imperial visits to Peter's tomb: Augustine, *enarr. in ps*. 65.4, 86.8; *ep*. 232.3. Note, however, that despite Anon. Vales. 67 'per tricennalem triumphans', he does not seem to have celebrated a triumph in Rome: McCormick, *Eternal Victory*, 272.

[125] *MGH Ep*. 3.121.14–18.

[126] Ammianus Marcellinus, 16.10.13, in particular.

[127] Procopius, *BV* 1.8.11. Amalafrida's charms may have been heightened by her proven fertility (Jordanes, *get*. 299), but if this was so Thrasamund was destined to be disappointed, as far as we know. Despite the testimony of Procopius that the initiative was Thrasamund's (*BV* 1.8.11), it is not certain that this was so: H.-J. Diesner, 'Die Auswirkungen der Religionspolitik Thrasamunds und Hilderichs auf Ostgoten und Byzantiner', *Sitzungsberichte der Säschischen Akademie der Wissenschaften zu Leipzig, phil.-hist. Klasse*, 113.3 (Berlin, 1967), 4.

could still describe Geiseric as 'very well known' for his sack of Rome in 455,[128] and in 468 the emperor Leo went to extraordinary expense in mounting what turned out to be a disastrous campaign against the Vandals.[129] Among the islands conquered by the Vandals was Sicily, but part of this was conceded to Odovacer in return for the payment of tribute shortly after he came to power in Italy.[130] In 491 the Vandals came to terms with Theoderic and we are told that their depredations of Sicily ceased;[131] presumably they had taken advantage of the confused situation in Italy to cause trouble. With the benefit of hindsight modern historians of the period are prone to emphasize the danger posed to Ostrogothic Italy by the rising power of the Franks, but in 500 it must have seemed to Theoderic, and still more the Romans, that the Vandals posed a greater threat. When the marriage between Thrasamund and Amalafrida took place, almost certainly during Theoderic's six months in Rome, and the bride left for Africa with a bodyguard of 1,000 noble Goths and 5,000 warriors as attendants, having been given as a dowry Lilybaeum, the westernmost promontory of Sicily, the outlook for relations with Africa was sunny.[132] In his panegyric on Theoderic, probably written in 507, Ennodius could refer to the ties of friendship and marriage which existed with the Vandals: 'your friendship suffices instead of an annual payment', and the Vandals, wise enough not to roam about beyond their power, 'deserve to be related to you by marriage, since they do not refuse obedience'.[133] The time when the Vandals had been 'a savage and swift enemy', as Cassiodorus put it in a letter possibly written in the same year as Ennodius' panegyric (*var.* 1.4.14), had passed; Thras-

[128] *Get.* 168; Sidonius Apollinaris was aware of the Vandals as a 'ferus hostis': *carm.* 5.385.

[129] Procopius, *BV* 1.6.2, 'they say' £130,000 were spent; see further C. Courtois, *Les Vandales et l'Afrique* (Paris, 1955), 201 n. 7.

[130] Victor of Vita, *hist. pers.* 1.14, with Courcelle, *Vandales*, 192.

[131] Cassiodorus, *chron. s.a.* 491.

[132] Time of marriage: Anon. Vales. 68. Gothic followers and dowry: Procopius, *BG* 1.8.12f.; Fiebiger/Schmidt, *Inschriftensammlung*, no. 41 (*CIL* 10.7232).

[133] *Pan.* 70. It has been suggested that Thrasamund may have been something like a *cliens* of Theoderic: S. J. B. Barnish, 'Pigs, Plebians and *Potentes*: Rome's Economic Hinterland, *c.*530–600 A.D.', *Papers of the British School at Rome*, 42 (1987), 157–85 at 181. I am unable to account for the fact that a sentence from the 'Formula praefecturae vigilum urbis Romae' of Cassiodorus (*var.* 7.7.2, 'Eris igitur . . . decipere gloria') also occurs, with the word 'igitur' omitted, in an inscription from Vandal Africa (Fiebiger/Schmidt, *Inschriftensammlung*, supplement (1939), no. 12 (*CIL* 8.2297).

amund was married to one of Theoderic's family and he could rely on her advice (cf. *var.* 5.43). The passing of time would reveal whether this would always be followed. But at the time of Amalafrida's marriage Theoderic had placed his relations with all the major barbarian powers of the West on a sound footing, and had come to terms with Constantinople. Further, he had secured the support of at least large sections of the aristocracy, some of whom obviously regarded him as all but an emperor, and the Roman Church. In 500 the future of his kingdom must have seemed bright.

3

Goths and Romans

It is difficult to establish what kind of people the Ostrogoths were. They have been described by modern scholars as 'huge, fair-skinned, beer-drinking, boasting thanes'[1] on the one hand, and as 'an insecure barbarian tribe trying very hard to prove that they were worthy to lord it over the ancient heart of Empire' on the other.[2] The question involves problems of evidence, for our literary evidence for the Goths comes from Romans, and frequently is more revelatory of the perspectives of its authors than of any objective reality, while non-literary sources, in particular archaeological data, tend to be difficult to evaluate. Furthermore, lurking behind the two quotations which stand at the head of this paragraph is the unstated assumption that it is possible to sum up a whole people with a blanket description. Yet one of the themes emerging in our investigation into Theoderic is that Romans chose to take different stances towards Gothic rule; surely it is no less likely that various Goths could have taken different attitudes towards the Romans, some being boastful and others insecure.

GOTHIC POPULATION AND SETTLEMENT

Theoderic presided over a state composed of two peoples: barbarians, who may for convenience be described as Ostrogoths, although as we have seen his invading host included other peoples, and there were doubtless already Rugians and other barbarians who lived in Italy prior to his coming, and Romans. He clearly hoped that relations between them would be good, enjoying the thought that the two nations would live in common and concur in the same desires (Cassiodorus, *var.* 2.16.5, amended trans. Hodgkin), without going as far as the Visigoths of the early fifth century, who are reported to have expressed the wish so to live in Italy with the

[1] C. S. Lewis, *The Discarded Image* (Cambridge, 1964), 79.
[2] Ward-Perkins, *Urban Public Building*, 105.

Romans that they could be believed to be of one race.[3] There was no disguising the fact that power lay with the Goths, a situation which our sources attest by the order of words in the expression 'Goths and Romans' generally used to describe the inhabitants of Italy.[4] Yet it is clear that the Goths constituted a tiny minority of the population. The question of the number of Goths who accompanied Theoderic to Italy is a difficult one and has elicited a disconcerting variety of responses.[5] We may quickly dismiss references to such things as 'uncountable bands of Goths' as nothing more than literary clichés,[6] and in principle there are two classes of evidence from which it might be thought possible to deduce how many of them there were. References to Ostrogothic manpower before the migration to Italy have been taken to indicate that some 35,000 to 40,000 Goths accompanied Theoderic on his trek to Italy.[7] Yet it is difficult to work from these figures, relating as they do to parts of the manpower of parts of the Ostrogothic people, and the grand total of 35,000 to 40,000 seems low in the light of later evidence. We are told that when Amalafrida married Thrasamund in 500 she was accompanied to Africa by 1,000 noble Goths and 5,000 warriors

[3] Jordanes, *get.* 152.

[4] Cassiodorus, *var.* 1.17 superscript, 2.16.5, 2.19 superscript, 3.48 superscript, 4.47.2, 5.5.2, etc., and *chron. s.u.* 519; Jordanes, *get.* 137, 142, 165, 213, 295 (but note 166, 271) and *rom.* 370; Procopius, *BV* 1.14.5 and *BG* 1.1.25, 1.4.6, 2.30.26 ('Goths and Italians', but note *BG* 1.4.28). Anon. Vales. prefers 'Romans and Goths' (60 *bis*, 61); on 'Romans and barbarians' in the *Edictum Theodorici*, G. Vismara, *Edictum Theoderici* (*Ius romanum medii aevi*, pts. 1, 2b.aa.α; Milan, 1967), 92 ff.

[5] Among recent authors: T. S. Burns, 'Calculating Ostrogothic Population', *Acta Antiqua Academiae Scientiarum Hungaricae* 26 (1978), 457–63 estimates 35,000 to 40,000; Wolfram, *History*, 279 estimates about 100,000; V. Bierbrauer, 'Aspetti archeologici di Goti, Alamanni e Longobardi', in *Magistra barbaritas*, 445 ff. at 446 proposes 100,000 to 125,000; K. Bosl, *Gesellschaftsgeschichte Italiens im Mittelalter* (Stuttgart, 1983), 3 estimates about 300,000. The estimates of earlier scholars are summarized by K. Hannestad, 'Les Forces militaires d'après la Guerre gothique de Procope', *Classica et mediaevalia*, 21 (1960), 136–83 at 155–67. For comparison, the size of the population of the Visigoths is discussed by A. M. Jimenez Garnica, *Origenes y desarrollo del reino Visigodo de Tolosa (a. 418–507)* (Valladolid, 1983), 193–6; for the size of Odovacer's following, Stein, *Histoire*, 41 and n. 4.

[6] 'innumerae Gothorum catervae', Ennodius, *ep.* 9.23.5, but cf. *V. Caes. Arel.* 1.27, Sedulius, *carm. pasch.* 1.148. For 'in populo harenae aut sideribus conparando' (Ennodius, *pan.* 29), cf. Gen. 22: 17. Is it coincidental that both Cassiodorus and Jordanes apply the term 'innumerabilis multitudo' to the Visigoths (*var.* 3.1.1, *get.* 190)? Cassiodorus also regarded the Alamanni as an 'innumerabilis natio' (*var.* 2.41.2). Ennodius describes the 'inmensa multitudo' Theoderic led to Italy (*V. Epiph.* 109); but there is little to be gained by multiplying such references.

[7] Burns, 'Calculating Ostrogothic population', on the basis of Malchus, *frag.* 17f. (Blockley 18.4, 20); John of Antioch, *frag.* 211 (5).

(Procopius, *BV* 1.8.12), which, however we compute the proportion of the population comprised of males of an age for military service, would have left Theoderic's kingdom in a militarily precarious position. Perhaps some returned home before long, but some still seem to have been with Amalafrida when Hilderic became king in 523 (Procopius, *BV* 1.9.4). The figure of 35,000 to 40,000 also seems low in the light of the second body of numerical evidence for the population of the Goths, the data provided by Procopius in his detailed narrative of the Gothic war, which has been plausibly interpreted to suggest a figure of a little over 30,000 for the Gothic army at the beginning of the war. It is not impossible that the population was growing so quickly that the male population of an age to bear arms had come to be almost as large as the entire people less than fifty years earlier, but this seems unlikely. Perhaps it would be safer to see the population of the Goths at the time of their migration to Italy as having been, in very general terms, in the order of 100,000. But what is more important is the undoubted fact that the Romans in Italy, who may conceivably have numbered something like four million, were far more numerous than the Goths.[8]

One pointer to the role the Ostrogoths played in Italian society is provided by the pattern of their settlement.[9] There are various classes of evidence on which one could draw to establish this, and they converge remarkably: the distribution of grave goods, which reveals concentrations of Ostrogoths in central Italy (the modern coastal provinces of Ascoli, Piceno, and Ancona) and northern Italy (the provinces of Pavia and Milan) and a relative frequency along the Adriatic coast of modern Pesaro and Ravenna and northern Italy south of the Po (the modern Reggio Emilia and Parma), on the one hand, and a total absence of such goods south of a line from Rome to Pescara, in Sicily, in the Western hinterland from Rome to

[8] Hannestad, 'Forces militaires', 157–62, 180. On the total population of Italy, see the discussion of J. C. Russell, *Late Ancient and Medieval Population* (Philadelphia, 1958), 71–3, which involves a good deal of extrapolation.

[9] By far the best treatment is that of Bierbrauer, 'Zur ostgotische Geschichte', 12–26 (*Ostgotische Grab- und Schatzfunde*, 27–41), see too *Magistra barbaritas*, 445–68, with maps on pp. 491 ff., but note already Hartmann, *Geschichte*, 96, with Z. V. Udal'tsova, *Italiia i Vitzantiia v VI Veke* (Moscow, 1959), 24f. There is no evidence that Ostrogoths settled in Raetia (Clavadetscher, 'Churrätien', 164f.); note too the lack of Ostrogothic grave goods in Viennensis, Narbonensis, and the Alpes maritimae (Bierbrauer, *Ostgotische Grab- und Schatzfunde*, 211 and map 20). Cassiodorus, *var.* 3.24 is addressed 'Universis barbaris et romanis per Pannoniam constitutis'; I take the barbarians in question not to have included Ostrogoths.

Genoa, and north of Trent on the other hand;[10] the copious references to Goths in literary sources such as Cassiodorus' *Variae*, which includes letters addressed to Goths residing at Tortona, at Verruca (near Trent), in Picenum and the Tuscanies and in Picenum and Samnium (although the appointment of counts of Syracuse, Naples, and the islands of Curitana and Celsina is suggestive of the presence there of military contingents),[11] and the *Wars* of Procopius, which contains the blank assertion that at the beginning of the Gothic war no Goths were to be found in Calabria and Apulia;[12] the overwhelming concentration of modern Italian place-names containing Gothic components north of the Po, particularly dense clusters being around Cremona, Brescia, and Como;[13] the occurrence of Ostrogothic personal names on grave inscriptions;[14] the establishment by Theoderic of palaces at Ravenna, Verona, Monza, and Galeata and the distribution of Arian Gothic churches, six of which were in Ravenna and had to be 'reconciled' by bishop Agnellus following the Byzantine conquest of the town,[15] combine to suggest that the Goths

[10] Bierbrauer, *Ostgotische Grab- und Schatzfunde*, 209–15; on Trent as the northern point of Gothic settlement, V. Bierbrauer, 'Frühgeschichtliche Akkulturationprozesse in den Germanischen Staaten am Mittelmeer (Westgoten, Ostgoten, Langobarden) aus der Sicht des Archäologen', in *Atti dei 6° Congresso internazionale di studi sull'alto medioevo*, 1978 (Spoleto, 1980), 89–105.

[11] Respectively *var.* 1.17, 3.48, 4.14 (where the Tuscanies are respectively Tuscia annonaria and Tuscia suburbicaria: F. Schneider, *Die Reichsverwaltung in Toscana von der Gründung des Langobardenreich bis zum Ausgang der Staufer (568–1268)* (Rome, 1914), 9 f.), 5.26, 6.22, 6.23, 7.16. See further on Naples S. Palmieri, 'Reminiscenze gotiche nelle fonti napoletane d'età ducale', *Koinonia*, 6 (1982), 61–72.

[12] *BG* 1.15.3, cf. 1.11.16. In Sicily, the only resistance Belisarius encountered from the Goths was from a garrison at Palermo: *BG* 1.5.12. But the use in this context of references in Procopius to the stationing of Gothic troops during the Gothic war (Bierbrauer, 'Zur ostgotische Geschichte', 13 (*Ostgotische Grab- und Schatzfunde*, 29))) seems to me of doubtful validity, for it reflects a wartime situation, not settlement or even the disposition of the army in time of peace.

[13] E. Gamillscheg, *Romania Germanica*, 2 (Berlin, 1935), 5–16 is the classic discussion; see too W. von Wartburg, *Die Entstehung der römanischen Völker* (Halle, 1939), 98–100; G. B. Pellegrini, in *Verona in età gotica e longoborda* (Verona, 1982), 18 ff., 22, 41; Bierbrauer, 'Zur ostgotische Geschichte', 23 (*Ostgotische Grab- und Schatzfunde*, 40). The limitations of Ostrogothic evidence compared to that provided by Lombard place-names are brought out by M. G. Arcamone, in *Magistra barbaritas*, 404.

[14] Bierbrauer, 'Zur ostgotische Geschichte', 24 (*Ostgotische Grab- und Schatzfunde*, 41.

[15] The evidence collected by Ullmann, *Gelasius*, 219 f., belies his assertion that the churches were spread out 'über ganz Italien'. Bishop Unscila (Cassiodorus, *var.* 1.26) presumably resided in Rome; that the Goths had a bishop in Rome is indicated by Tjäder, *Papyri*, 2.198 n. 20, with 301 n. 17. That another bishop, Gudila, was involved in a dispute with people from the Umbrian town of Sarsina (Cassiodorus, *var.* 2.18) suggests but does not prove that he lived in that area. On the activities of Agnellus, see Agnellus, *Codex*, 216.

were overwhelmingly concentrated in the north of Italy, and there more to the east than the west. A proposal later made by the Byzantines that the Gothic king Witigis was to rule over territory north of the Po (Procopius, *BG* 2.29.2) constituted simple recognition of this fact.[16]

Broadly speaking, there would have been two reasons for Goths living in particular areas: they could have made homes there, or been quartered in military camps. It has been well observed concerning the Visigoths in Gaul, although in a slightly different connection, that 'military occupation was one thing, and settlement another', and the *Edictum Theodorici*, certainly Italian even if it be denied to Theoderic, distinguishes between barbarians living at home and those in army camps.[17] But surely we are entitled to see the concentration of the Goths in the north of Italy as answering to military priorities, for the series of invasions of Italy in the fifth century begun by Alaric and completed by Theoderic himself had all proceeded from the north. The *notitia dignitatum* recorded the existence during the later empire of military workshops in Italy under the control of the *magister officiorum* at Concordia, Verona, Mantua, Cremona, Pavia, and Lucca,[18] and the Romans developed a *tractus Italiae circa Alpes*, to which Theoderic's northern frontier stands in some relation.[19] Hence the great importance of Verona for Theoderic: commanding, as it does, the Adige valley and the road from the Brenner Pass, but also a natural target for any force moving into Italy eastward by way of the Lombard plain, as that of Theoderic himself did in 489, it was an ideal place for the king to establish himself when he feared trouble from other nations, as he is known to have done in about 519, and it seems likely that the city saw a lot of its king.[20] Goths

[16] But see too 2.29.35, 3.2.15.

[17] J. M. Wallace-Hadrill, *The Long-Haired Kings and Other Studies in Frankish History* (London, 1962), 32; cf. *Ed. Theod.* 32: 'barbari . . . sive domi sive in castris fuerint constituti'.

[18] *Not. dig., occ.* 9.23–9, ed. Seeck, 145.

[19] See in particular H. Zeiss, 'Die Nordgrenze des Ostgotenreiches', *Germania*, 12 (1928), 25–34, and most recently Lusuardi Siena, in *Magistra barbaritas*, 513ff., with V. Bierbrauer, 'Frühmittelalterliche Castra im Östlichen und Mittleren Alpengebiet: Germanische Wehranlagen oder Romanische Siedlungen?', *Archäologisches Korrespondenzblatt*, 15 (1985), 497–513. The suggestion of Picotti, 'Relazioni', 379f., that Odovacer's troops were all over Italy reads too much into Ennodius, *pan.* 51. On the continuing importance of the Alpine passes after the fall of the Ostrogoths, G. Fingerlin, J. Garbsch, and J. Werner, 'Die Ausgrabungen in langobardischen Kastell Ibligo-Invillino (Friaul)', *Germania*, 46 (1968), 73–110, esp. 73f.

[20] Anon. Vales. 81, 'Verona consistente propter metum gentium'; cf. Ennodius, *pan.* 39, 'Verona tua'; evidence for the palace, baths, and aqueducts built by Theoderic at Verona has been presented above (Ch. 2 nn. 39, 41, 42). See in general

stationed at Tortona, Verruca, and Aosta may be presumed to have been there for military purposes,[21] and it may well be that some garrisons were located with an eye to road junctions.[22] In general, the areas where the Goths are known to have resided imply that the purpose of their residence was military, and here we have a very clear indication of their role in Italian society.

DIVISION OF LABOUR

'While the army of the Goths makes war, the Roman may live in peace', Cassiodorus observed on one occasion (*var.* 12.5.4, oddly enough written during the opening stages of the Gothic war), and the theme of the Goths fighting while the Romans enjoyed the benefits of peace is one frequently encountered in his letters (*var.* 6.1.5, 7.3.3, 7.4.3, 8.3.4, 9.14.8). One easily forms a picture of Goths and Romans as having been engaged in different pursuits, and as Theoderic is reported to have maintained the civil service for the Romans as it had been under the emperors, while Gothic ambassadors were reported to have told Belisarius in 537 that the Romans had held all the offices of state, with no Goth having a share in them, modern scholars have frequently envisaged a situation in which the Goths fought and the Romans manned the civil service.[23] A cleavage between civilian and military careers had already existed in the later empire,[24] and so a strict division of employment between Goths and Romans could be held to have constituted no

G. P. Bognetti, 'Teodorico di Verona e Verona longobardica, capitale de regno', in *Studi giuridici in onore de Mario Cavalieri* (Padua, 1960), 1–39, where *inter alia* are stimulating comments on Theoderic's medieval title 'Dietrich von Bern', and most recently, Lusuardi Siena, in *Magistra barbaritas*.

[21] Cassiodorus, *var.* 1.17, 3.48; cf. 2.5 for soldiers at the *Augustanae clausurae*, i.e. Aosta (see *var.* 1.9).

[22] See e.g. on Dertona, at the junction of the via Aemilia and the via Postumia, G. Schmiedt, 'Città scomparse e di nuovo formazione in Italia', *Topografia urbana e vita cittadina nell'alto medioevo in occidente* (settimane di studio, 21; Spoleto, 1974), 503–607 at 569f.

[23] Civil service for Romans: anon. Vales. 60 ('militiam' cannot mean 'military service', as translated by J. C. Rolfe). Gothic ambassadors: Procopius, *BG* 2.6.19. The classic statement is that of Mommsen, *Gesammelte Schriften*, 6.436, followed with minor reservations by Stein, *Histoire*, 120, but queried by R. Soraci, *Ricerche sui conubia tra Romani e Germani nei secoli IV–VI* (Catania, 1974), 85–9. For comparison, Victor of Vita sometimes refers to the Vandal *exercitus* where one would expect *gens* (*hist. pers.* 1.13, 3.60).

[24] Jones, *Later Roman Empire*, 101.

more than the imposition of race on a dichotomy which had already existed, and indeed, as the army towards the end of the imperial period in the West had become increasingly barbarized, to have been a thoroughly natural development. But care is called for. To judge by his name, the Count Colosseus who was in charge of Pannonia Sirmiensis and apparently had troops under him was a Roman, although his commission was both military and civilian;[25] so too Servatus, *dux Raetiarum*, who again seems to have had a jurisdiction which was both military and civilian.[26] The absence of Ostrogothic grave goods in Raetia, however, suggests that Servatus may not have commanded Gothic troops.[27] Liberius, when praetorian prefect of Gaul, could be described as an army hero whose commanding appearance was made the more attractive by his wounds, and another Roman, Cyprian, is unambiguously stated to have served Theoderic in both military and civilian capacities.[28] Romans are definitely known to have held arms under the Visigoths during the span of the kingdom of Toulouse,[29] Count Pierius was killed fighting the Ostrogoths on behalf of Odovacer in 490,[30] and Mummolus, one of the most famous military commanders in Gaul towards the end of the sixth century, was a Roman,[31] so we would do well to accept that the bearing of arms was open to them under Theoderic. On the other hand, the civilian office of *comes patrimonii*

[25] I agree with Soraci, *Ricerche*, 87 n. 12, against Ensslin, *Theoderich*, 373 n. 25, Wolfram, *History*, 321, and *PLRE* 2.305. The last argues for his being a Goth from *var*. 3.23.3, 'ut . . . Gothorum possis demonstrare iustitiam', but the *iustitia* is presumably that of the Gothic government. That he had troops under him is indicated by *var*. 4.13. His commission was both military and civil: 'armis protege, iure compone', *var*. 3.23.2. Note as well the Roman *dux* Ursus who lived in Teurnia: Wolfram, *History*, 316f.

[26] *Var*. 1.11. See Clavadetscher, 'Churrätien', 162, against Ensslin, *Theoderich*, 373 n. 25.

[27] Bierbrauer, *Ostgotische Grab- und Schatzfunde*, 211 with map 20; cf. Wolfram, *History*, 316.

[28] Liberius: Cassiodorus, *var*. 11.1.16; see too *V. Caes. Arel*. 2.10–12 (discussed by O'Donnell, 'Liberius', at 48–50) and, for his unusual military career under Justinian, *PLRE* 2.680. Cyprian: 'in utraque parte', Cassiodorus, *var*. 8.21.3 (on which *PLRE* 2.332); a summary of his achievements follows. Cassiodorus, *var*. 9.23.3 mentions some young Romans trained in arms.

[29] Best known is the participation of 'maximus . . . Arvernorum populus' led by Apollinaris against the Franks in 507 (Gregory of Tours, *hist. franc*. 2.37), but see as well D. Claude, *Adel, Kirche und Königtum im Westgotenreich* (Sigmaringen, 1971), 46, Jimenez Garnica, *Origenes y desarrollo*, 232f., and Wolfram, *History*, 466 (467) n. 379.

[30] *Auct. haun. sa* 491, where he is described as 'qui bellicis rebus praeerat'.

[31] Gregory of Tours, *hist. franc*. 4.42, 44f., 5.13, 7.34–9.

was held by Wilia, clearly a Goth, at the very end of Theoderic's reign, despite having been held by at least two Romans earlier during his rule,[32] and Gothic *cubiculares* are known, as we shall see.

And yet, despite their continuing to hold an effective monopoly of high civil offices of state, the Romans must have been aware that there were non-military areas of power from which they were excluded. Theoderic, while he could be described in an inscription as 'glorious in both war and peace', remained a Goth and a military man; hence the expressions 'our Goths' and 'our army' are often found in the letters written on his behalf by Cassiodorus.[33] Needless to say all high offices were in the gift of the king,[34] and in a letter written on behalf of Tuluin this noble Goth was made to claim that his intercession with Theoderic had often been responsible for the appointment of consuls, patricians, and prefects (Cassiodorus, *var.* 8.11.3); if there is any truth in this the king, even in matters most dear to the hearts of Roman aristocrats, must have paid heed to the advice of his military cronies.

Goths were able to exercise power over Romans in various ways. The Goth Triwila, who towards the end of Theoderic's reign held the office of *praepositus sacri cubiculi*, the only person known to have held it during the Ostrogothic period,[35] was on one occasion instructed, together with Ferrocinctus, to see that no lesser person than Faustus *niger*, a member of the *gens Aniciana*, the praetorian prefect and a former consul and ambassador of Theoderic to Constantinople, return any property he could be proven to have taken from one Castorius together with property of equal value; any further attempt to harm the interests of Castorius would result in Faustus' having to pay 50 pounds of gold.[36] Triwila was a powerful figure whose services Ennodius found it worthwhile to seek when trying to

[32] Cassiodorus, *var.* 5.18–20, dated to 526 by Ruggini, *Economia e società*, 555f. and S. Krautschick, *Cassiodor und die Politik seiner Zeit* (Bonn, 1983), 70; he is perhaps to be identified with the Wilia of *var.* 1.18. The office had earlier been held by Julianus (Cassiodorus, *var.* 1.16; Ennodius, *ep.* 4.7) and Senarius, who, while his name may perhaps be Germanic (Schönfeld, *Wörterbuch*, 202), was a Roman (Ennodius, *ep.* 1.23.3) and a Catholic (Avitus of Vienne, *ep.* 39).

[33] Inscription: Fiebiger/Schmidt, *Inschriftensammlung*, no. 179 (*CIL* 10). 'Gothi nostri': *var.* 3.24.4, 3.42.1, 5.27.1. 'Exercitus noster/noster exercitus': *var.* 2.8, 2.15.3, 3.38.2, 3.43.2, 4.36.2, 5.29.1. Note, however, that legislation of Justinian refers to 'noster exercitus' (*novella* 130). So too 'saio noster': *var.* 1.24.2, 3.48.1, 4.28, etc.

[34] 'Exeunt a nobis dignitates relucentes quasi a sole radii': *var.* 6.23.2.

[35] Anon. Vales. 82; cf. Boethius, *phil. cons.* 1.4.10.

[36] Cassiodorus, *var.* 3.20.

obtain a house at Milan, but was not invariably benevolent towards Romans: he is reported to have helped the Jews of Ravenna make charges against the Catholics of the city before Theoderic, and Boethius described him as one who began and perpetrated wrongs, presumably against Romans.[37] Again, members of the *comitiva Gothorum* were making their presence felt.[38] While Count Arigern could be described to the senate as almost one of its members, together with his fellow-Goths Gudila and Bedeulf he was named in 502 by bishops meeting in Rome as a witness of the wounds borne by Pope Symmachus, which had almost certainly been inflicted by the agents of the pope's senatorial enemies.[39] On another occasion Arigern was ordered to obtain from the powerful patrician Venantius, a member of the Decii family, a promise that he would submit to judgement with respect to a charge one Firminus was bringing against him, and when the senators Basilius and Praetextatus were accused of practising magic it fell upon Arigern to bring them to trial before the *iudicium quinquevirale*.[40] In no case is there any indication that these officers of the king overrode the provisions of the law. But the servants of the count of the Goths in Savia were alleged to have taken property from the provincials, perhaps as a form of blackmail, and after the death of Theoderic, Gildilas, the count of Syracuse, was accused of extorting money from the provincials for the repair of walls but failing to proceed with any building, of claiming the properties of the dead for the fisc, of demanding heavy payments from those who sought justice in the courts, of hearing cases involving two Romans before his court, and of corrupt dealings with the cargoes of ships.[41] Even without such men such as Gildilas, the fact that Goths were put in positions of authority over even the most powerful Romans would have been conducive to feelings of ill-ease. Further, Goths were certainly becoming in-

[37] Ennodius: *ep.* 9.21. Jews of Ravenna: anon. Vales. 82. Boethius: *phil. cons.* 1.4.10.

[38] Wolfram, *History*, 291 ff. See too W. G. Sinnigen, 'Administrative Shifts of Competence under Theoderic', *Traditio*, 21 (1965), 456–67.

[39] Almost one of its members: Cassiodorus, *var.* 4.16.1. Witness of wounds: *MGH AA* 12.429.3–6; on Symmachus' wounds see below, Ch. 4 n. 16.

[40] Cassiodorus, *var.* 3.36 where the Venantius is probably to be seen as the consul of 508 (Sundwall, *Abhandlungen*, 167), 4.23; it appears from the latter that the 'disciplina Romanae civitatis' had been entrusted to Arigern (4.23.1). See further Stein, *Histoire*, 124.

[41] Cassiodorus, *var.* 9.14; cf. the less explicit *var.* 9.11f., and the list of abuses in *var.* 5.14. See too the general comments of Stein, *Histoire*, 120f.

fluential at court. We know of one Gothic eunuch who held the office of *cubicularis*,[42] and Triwila was not the only Goth to have played a part in the network of contacts Ennodius built up at Ravenna: Count Tancila informed him what had come to pass concerning the estates of his nephew Lupicinus, whereupon Ennodius wrote to Faustus with the news; he was in touch with one Alico, whom he addressed as *dominus*, about matters concerning the Church; on another occasion he felt that Trasimund (or Thrasimund), a relative of Theoderic, could render him assistance; and he thought it worth his while writing to Gudilevus with congratulations on the success and power God had given him.[43] The consulship of Theoderic's son-in-law Eutharic in 519 and the advent of Goths to membership of the senate[44] cannot have given pleasure to Romans, for whom such dignities were the highlights of a life well lived; whatever the distinction between the roles of Goths and Romans in society which could be drawn in broad terms, the Romans may have felt that their position was being eroded. And this quite apart from the general cultural considerations which prompted Sidonius Apollinaris to comment that he avoided barbarians even when they were considered good.[45]

LAW, AREAS OF TENSION

Just as Goths and Romans were nominally kept apart with respect to their basic functions in society, military and civilian respectively, so they were divided before the law. Legal matters are complicated by the existence of a code legislating for 'barbarians and Romans' substantially based on Roman law which, since the time of the

[42] Seda, described as 'ignucus et cubicularis', Fiebiger/Schmidt, *Inschriftensammlung*, no. 183 (*CIL* 11.310). One other Gothic eunuch is known, Wiliarit: Fiebiger/Schmidt, *Inschriftensammlung*, no. 205 (*CIL* 6.9379); no office is specified. Tuluin is reported to have proceeded at an early age 'ad sacri cubiculi secreta': Cassiodorus, *var*. 8.10.3.

[43] Ennodius, *ep*. 2.23 (Tancila), 4.2 (Alico), 4.10 (Trasimund, including the interesting phrase 'vultis quasi aequales tractare famulos'), 6.28 (Gudilevus, presumably a Goth: Schönfeld, *Wörterbuch*, 115).

[44] Theoderic had himself been consul in 484, but note Tuluin's becoming patrician and senator shortly after the accession of Athalaric (*var*. 8.9–11). Arigern's membership of the senate is indicated by *var*. 4.16.1f.; other Gothic *illustres* included Marabad (4.12, 4.46), Osuin (1.40, etc.), Suna (2.7, 3.15), and Wiliarius (5.23).

[45] *Ep*. 7.14.10.

edition published in 1579 by Pierre Pithou, has been referred to as the *Edictum Theodorici* and ascribed to Theoderic the Ostrogoth. Recently voices have been raised querying its paternity and attributing it to Odovacer or Theoderic the Visigoth, among others.[46] Where experts in legal history differ, others will be reluctant to commit themselves. But it must be accepted that there is some support in our sources for Theoderic's having issued a code in addition to individual edicts;[47] that there would have been nothing surprising in his having issued such a code, given the existence at this time of codes among the Franks, Visigoths, and Burgundians; and that references in the code to 'the venerable city' ('urbs venerabilis', c. 10) and the burying of corpses in the city of Rome (*cap.* 111) point inescapably to an Italian origin. However, the greater part of the *Edictum* is simply a restatement or reworking of existing Roman legislation, and the society which it reveals is that of late antiquity in general rather than Ostrogothic Italy in particular, which makes it difficult to use as a source for the reign of Theoderic, except in the cases where earlier legislation is modified.

That the provisions of the *Edictum* were to apply to barbarians and Romans suggests that law in Ostrogothic Italy was territorial rather than personal, with the exception of the Jews, for the Edict envisages disputes occurring between those of their number who lived in accordance with their own laws; needless to say, in such cases they were to have their own judges (*cap.* 143). So it was that

[46] P. Rasi, 'Sulla paternità del c.d. Edictum Theodorici Regis', *Archivio giuridico*, 145 (1953), 105–62 (Odovacer); G. Vismara, 'Romani e Goti di fronte al dirotto nel regno Ostrogoto', in *I Goti in occidente problemi* (Settimane di studio, 3; Spoleto, 1956), 409–63; id., *Edictum Theoderici*; and A. Cavanna, in *Magistra barbaritas*, 356 (Theoderic the Visigoth). It is not entirely excluded that the document was forged by Pithou. The traditional attribution is upheld by B. Paradisi, 'Critica e mito dell'editto teodericano', *Bollettino dell'Istituto di dirotto romano*, 68 (1965), 1–47 and H. Nehlsen, *Sklavenrecht zwischen Antike und Mittelalter* (Göttingen, 1972), 120–3; agnostic is C. l. Schott, 'Der Stand der Leges-Forschung', *Frühmittelalterliche Studien*, 13 (1979), 29–55. The provisions of the edict are fully discussed by Ensslin, *Theoderich*, 220–36. See further A. R. Korsunskij, 'K Diskusii ob "Edikte Theodericha"', in *Europa v sredie veka: Ekonomika, politika, kul'tura* (*Fest. C. D. Skazkin*) (Moscow, 1972), 16–31. One would welcome a detailed study of barbarian laws; S. Weber, 'Die Leges Barbarorum aus germanischer, römischer und byzantinischer Sicht: Ein Beitrag zu ihrer historischen Analyse', in V. Vavrinek, ed., *From Late Antiquity to Early Byzantium* (Prague, 1985), 167–71.

[47] Anon. Vales. 60, 'edictum suum, quo ius constituit'; cf. *chron. pasc.*, ed. Dindorf, 605, *διάταξις πὲρι ἑκάστου νόμου*. It is true that Procopius makes some Goths assert that neither Theoderic nor his successors issued a *νόμος*, written or unwritten (*BG* 2.6.17), but their testimony is of no weight (see above, Ch. 1).

Cassiodorus could write of the legal system (*ius*) which the Goths and Romans had in common (*var*. 8.3.4). Nevertheless, it is clear that Goths and Romans were to be judged in different ways. This is made clear by the formula for the appointment in various cities of the *comes Gothorum*, an official whose power extended beyond the merely military to some degree. In the case of disputes between Goths this official would put an end to them in accordance with Theoderic's edicts; when a Goth and a Roman fell out he would consult with a Roman learned in the law and come to a fair decision; when a case arose between two Romans it would be heard by Roman judges sent by the government into the provinces: in this way each person would keep the laws of his own people, and despite the diversity of judges there would be one justice for all.[48] So much in theory; as the case of Gildilas shows, a Gothic count was quite capable of summoning two contending Romans before his court.

Indeed, areas of tension between Romans and Goths abounded. On one occasion the extraordinarily large sum of 1,500 solidi was awarded as compensation for the damage done as the army moved through a district,[49] and Cassiodorus' letters contain numerous references to the devastation wrought by the Gothic army in Italy in time of peace;[50] small wonder that the government was afraid that the contingent dispatched to Avignon following Theoderic's occupation of some of the portion of Gaul formerly held by the Visigoths would oppress the people and act with violence, whereas the Gothic army was expected to live *civiliter* with the Romans (*var*. 3.38). Given that Ulfilas is alleged to have failed to translate the Old Testament books of Kings into Gothic because he felt that the

[48] Cassiodorus, *var*. 7.3.1, but with 'ut unicuique sua iura serventur' cf. 8.3.4 ('Gothis Romanisque apud nos ius esse commune'), 3.13.2 (not to be 'discretum ius' for Goths and Romans). On the powers of the counts, Wolfram, *History*, 290ff. O. J. Zimmermann, *The Late Latin Vocabulary of the 'Variae' of Cassiodorus* (Washington, DC, 1944), 204f. K. F. Drew, 'Law, German: Early Germanic Codes', in J. R. Strayer, ed.. *Dictionary of the Middle Ages*, 7 (1986), 468–75 provides a recent discussion of this complex topic.

[49] Cassiodorus, *var*. 2.8, dated by Mommsen to 508. The see of the addressee, Bishop Severus, cannot be identified.

[50] *Var*. 4.36 (devastation of fields in the Alpes Cottiae), 5.13 (army to be given proper *annonae* so provincials would not be plundered), 5.26 (Goths of Picenum and Samnium not to lay waste fields and meadows on their way to Ravenna), 12.5.3 (destruction and rapine caused by the army in Lucania and Brutium); cf. 5.10f. for fears occasioned by a force of Gepids crossing Venetia and Liguria on their way to Gaul. *Var*. 8.27.2, however, seems to envisage pillaging undertaken by both Goths and Romans.

Gothic people was already øιλοπόλεμος,[51] and that for decades tnıprior to their arrival in Italy the Goths had largely been living on what they could extort or plunder from the Romans, it is scarcely surprising that violent behaviour continued, despite the allotment of thirds and the donatives which Theoderic paid his troops.[52] Theoderic's army often treated the civilians badly, and the terror among the people which followed the dispatch of Gothic troops to Rome in 535 may have been justified (cf. *var.* 10.14).

Against this background we may locate the violence which sometimes broke out when Gothic *saiones* were detailed to provide personal protection (*tuitio*) for individual Romans. In his formula for the bestowing of *tuitio*, Cassiodorus wrote that it was granted to deal with both 'the hot-headed onslaughts [of the Goths] and the ruinous chicanery [of the Romans]',[53] but there can be no doubt that the former constituted the greater problem, as elsewhere it alone is mentioned as a reason for *tuitio* being extended.[54] The system had obvious drawbacks. People to whom *saiones* were assigned could employ them in violent attacks on personal enemies (*var.* 7.42), and *saiones* could sometimes turn against the people whom they had been appointed to protect. The latter danger is illustrated by the case of the *saio* Amara who, having drawn his sword and wounded in the hand the *vir spectabilis* Peter, the man he was assigned to protect, then demanded money from him (*var.* 4.27f.). Peter, known to have been a wealthy landowner,[55] was doubtless a tempting target, but the story may stand, and the government's injunction to the Romans that they should have a keen love for the Goths (*var.* 7.3.3) may indicate a fear that the reverse could be expected.

The frequently expressed desire that Goths would live *civiliter* with Romans and the steps that needed to be taken against their hot-headed onslaughts (*inciviles impetus*) raises the question of the

[51] Philostorgius, *hist. eccl.* 2.5 (ed. F. Winkelmann (Berlin, 1981), 18).

[52] The *donativum* is known from *var.* 4.14.2, 5.26f., 5.36, 8.26.4. See too, perhaps, the evidence of the *Edictum Theodorici*, summarized by Burns, *Ostrogoths*, 127.

[53] *Var.* 7.39.2, 'contra inciviles impetus et conventionalia detrimenta'. Hodgkin's expanded translation, quoted here, is certainly valid; cf. further in this passage, 'adversus Gothos illa, adversus Romanos illa', and the comments of R. Morosi, 'I *saiones*, speciali agenti di polizia presso i Goti', *Athenaeum*, 59 (1981), 150–65 at 161 n. 102. Morosi's study, and Mommsen, *Gesammelte Schriften*, 6.410–12, treat the subject well.

[54] *Var.* 1.37.5, 'contra incivilium impetus'; *var.* 4.27.5, 'contra inciviles impetus'.

[55] His estates at Polentia are known from Gelasius, *ep.* 7 (ed. Löwenfeld).

precise meaning to be attached to *civilitas*, one of the key words in the vocabulary of Cassiodorus' *Variae*. In the 'index rerum et verborum' he compiled to accompany the edition of Mommsen, Traube proposed for it the meaning 'the proper (or just) condition of the state', and such a definition excellently suits the vague uses to which Ennodius often puts the word.[56] But Cassiodorus often uses it in contexts suggestive of a more precise meaning. Writing to Speciosus in 509 he makes Theoderic observe: 'If the practices of foreign peoples are held in check by the law (*sub lege*), if whatever is associated with Italy serves Roman law (*ius Romanum*), how much more should the dwelling place of *civilitas* herself have reverence for the laws (*leges*)?' (*var.* 1.27.1). Writing to the Jews of Genoa he asserts that 'keeping the laws is the sign of *civilitas*' (*var.* 4.33.1); writing to Theoderic's nephew Theodahad he noted a complaint that some of Theodahad's men had seized the possessions of another in neglect of the laws, whereas they should have been taken over *civiliter*, if indeed the demand for them had been legal (*var.* 4.39.3). Following Theoderic's death an edict was published in the name of his successor, Athalaric, in response to a situation in which 'certain people, despising *civilitas*, affect to live with the savagery of beasts; having returned to their boorish origin in the manner of wild animals they consider human law hateful' (*var.* 9.18 proem, *MGH AA* 12.282.26–8). In letters bestowing *tuitio* the terms 'unlawful presumptions' and 'the assaults of the *inciviles*' seem synonymous (*var.* 3.27.2, 4.27.5, 7.39.2) and, as Traube himself pointed out, the terms 'saving the laws' and 'saving *civilitas*' are synonymous (*var.* 2.29.2, 4.27.5, 1.15.2). Seen in this light, *civilitas* and its cognates take on a more precise meaning. They indicate the quality of abiding by the laws.[57] Hence, in his letter to Gildilas, Cassiodorus could observe that 'the praise of the Goths is *civilitas* defended' (*var.* 9.14.8). This is not, as a hasty reading of the text might lead one to believe, merely a general utterance to the effect that it was the job of the Goths to guard the *civilitas* of the Romans, but

[56] 'status rei publicae iustus' (*MGH AA* 12.521); cf. Ennodius, *pan.* 11, 15, 56, 87, and Sidonius Apollinaris, *ep.* 1.2.1, 3.8.2. See the general discussions of Ensslin, *Theoderich*, 217–20; Wes, *Das Ende*, 45f.; Reydellet, *Royauté*, 222–4.

[57] Note too the occasions on which *civilitas* is mentioned in connection with *leges* (*var.* 4.33.2, 5.4.2, 5.12.3, 5.37.1, 7.1.4, 7.3.2), *praesumptiones illicitae* (4.17.2f., 8.33.1), and *illicitae seditiones* (1.44.2; cf. *MGH AA* 12.445.5f. for *civilitas* being called into question by *seditio*). The *novellae* issued by Justinian do not use the word *civilitas*, but *civiliter* occurs once (*app.* 7.23), where it has the meaning 'per civiles iudices'. Cf. *ed. Theod.* 152: 'civiliter actionem proponere'.

rather that with them protecting the state the Romans could carry on with their lawsuits, for Cassiodorus immediately continues: 'All fame will come your way if the litigant sees you only occasionally. Defend the laws with arms, and allow the Romans to litigate in the peace of the laws' (*var.* 9.14.8). But as with the injunction to the Romans to love the Goths, the statement of such a principle may imply that something different was expected.

GOTHS AS BARBARIANS

Behind these concerns stands the more general issue of what kind of people the Goths were, and specifically the extent to which they were open to the influences of the numerically superior Romans amongst whom they lived. Doubtless they retained some of the trappings of a barbarian people. Writing in the mid-sixth century, Jordanes was aware of the old songs of the Goths,[58] and early in the seventh century Isidore of Seville wrote of Visigothic boys singing 'the songs of their ancestors, by which the hearers are provoked and stimulated to glory'; he had no trouble in accepting that the word 'Goth' meant 'fortitude'.[59] A word suggestive of characteristics of the Goths we may see as barbarian is *virtus*. It is derived from the Latin *vir*, which means 'man' in the sense of an adult male, sometimes a hero, rather than 'man' in the sense of human being. The word has strong connotations of power, so that, for example, in hagiographical texts the plural *virtutes* when applied to saints often means not 'virtues' but 'miraculous acts of power' by which a holy

[58] Jordanes, *get.* 28 (*prisca carmina*), 43 (*cantus maiorum*), 72 (*cantiones*); cf. 79 (*fabulae*). Such songs may stand in some relation to the 'maiorum notitia cana' of Cassiodorus, *var.* 9.25.4. Note, however, that Jordanes professed to prefer written evidence to the 'fables of old women' (*get.* 38), and that Cassiodorus' description of Gersimundus as 'ille toto orbe cantabilis' (*var.* 8.9.8) is no evidence for his being celebrated in song: E. Curtius, *Europäische Literatur und Lateinisches Mittelalter*, 3rd edn. (Berne, 1961), 170. Ulysses is similarly described as *cantabilis*: *var.* 2.40.11. The 'metamorphosis of Ostrogothic society' is discussed, with particular reference to archaeological evidence, by Burns, *Ostrogoths*, 108 ff.

[59] Visigothic boys: Isidore, *Institutionum disciplinae*, ed. P. Paschal, *Traditio*, 13 (1957), 426; cf. Julian of Toledo, *Historia Wambae*, 1 (CCSL 115.218). Such evidence has been connected, implausibly perhaps, with later Spanish epic: R. Menendez Pidal, 'Los Godos y el origen de la epopeya española', in *I Goti in occidente problemi* (Settimane di studio, 3; Spoleto, 1956), 285–322. Fortitudo: Isidore, *hist. goth.* 2; 'et re vera', he comments.

man effects his will in various situations.[60] In a famous passage the proceedings of the third council of Toledo, held in 589 at the time of the conversion of the Visigoths to Catholicism, refer to the true *virtus* of the race of the Goths,[61] and our narrative sources for the history of the Ostrogoths persistently associate the word, with its overtones of force and masculinity, with both the people in general and Theoderic in particular.[62]

This theme is strongest in the letters of Cassiodorus. In 508 he wrote: 'To the Goths a hint of war rather than persuasion to the strife is needed, since a warlike race such as ours delights to prove its courage. In truth, he shuns no labour who hungers for the renown of valour (*virtutis gloriam*).' The theme is frequent in his letters.[63] Writing to Count Colosseus in Theoderic's name, Cassiodorus advised him: 'Show forth the justice of the Goths, a nation happily suited for praise, since it is theirs to unite the prudence of the Romans and the *virtus* of the nations (*gentes*)' (*var*. 3.23.3, amended trans. Hodgkin). Exactly the same concepts occur in a description of Theodahad as 'dear to the Romans for his prudence, revered by the nations for his *virtus*' (*var*. 11.13.4). There is here the hint of a tension in the thought of Cassiodorus: presumably the Goths stood in the same relation to the Romans and the nations, combining the merits of both. No one would suppose that the Goths were Romans; were they therefore not to be seen as a *gens*? On one occasion Cassiodorus affirmed that by living with the Romans in accordance with the laws the Goths had something which the other *gentes* lacked (*var*. 7.25.1). Here the Goths are only mentioned as a *gens* with reference to a characteristic which allowed them to be

[64] On the word in hagiographical writing, H. Beumann, 'Die Historiographie des Mittelalters als Quelle für die Ideengeschichte des Königtums', *Historische Zeitschrift*, 80 (1955), 449–88 at 472–4. Note the titles of two works by Gregory of Tours: *Liber de passione et virtutibus sancti Iuliani martyris*, and *Liber de virtutibus sancti Martini episcopi*.

[65] 'Adest enim omnis gens Gothorum inclyta et fere omnium gentium genuina virtute opinata': *Concilios*, ed. Vives, 110. For *virtus* in Isidore of Seville, Teillet, *Goths*, 478f.; an interesting reference in Paul the Deacon, *hist. lang*. 1.27 (*ad fin*.).

[62] Cassiodorus, *chron. s.a.* 451, 504; Ennodius, *pan*. 32, 88 (and cf. 39 on Odovacer: 'maxima in luctaminis promissione virtus'); Jordanes, *get*. 272. Marcellinus, on the other hand, felt that some of Theoderic's victories were won 'ingenio magis quam virtute' (*chron. s.a.* 479.2). Note as well Jordanes' summary of the contents of his *Getica*: 'Getarum origo ac Amalorum nobilitas et virorum fortium facta' (*get*. 315).

[63] *Var*. 1.24.1, trans. Hodgkin; cf. 1.38.2, 5.27.2, 9.1.2, 10.18.2, 10.31.2 for the *virtus* of individual Goths or the Gothic people. *Virtus* could also be predicated of Clovis the Frank: *var*. 2.41.1, 3.4.2.

distinguished from the rest of the nations, and it must be said that Cassiodorus' application of the word is confused and uneasy.[64] Theoderic, described on the Senigalla medallion as 'victor gentium', could hardly be described unambiguously as the member of a national *gens* himself.

Cassiodorus, then, was prepared to see the Goths as possessors of *virtus*, but was uncertain as to whether they were a *gens*. He was, however, absolutely certain that they were not barbarians, a word which, in the opinion of an African writer who lived under the Vandals, connoted ferocity, cruelty, and terror.[65] Opinions on this matter differed, however, and the interrogation of our sixth-century sources as to whether the Goths were barbarians reveals a fascinating spectrum of views. At one end stands Cassiodorus. In his chronicle and *Variae* he had occasion to refer to the Goths on a multitude of occasions. Not once are they described as barbarians, which is the more remarkable because on one occasion he implied that the Franks were barbarians.[66] Indeed, Cassiodorus sometimes intimated clearly that the Goths were not barbarians,[67] and on one memorable occasion altered his source, which described the emperor Claudius driving off Goths, to represent him as having driven off barbarians, presumably the last word any reader of Cassiodorus would associate with Goths.[68] It is not surprising that the *Getica* of Jordanes, which professes to be an abbreviation of Cassiodorus' lost work in 12 books on the origin and deeds of the Goths, with a certain amount of additional material, implicitly denies that the Goths were barbarians. The point is neatly made when the emperor

[64] See Traube's *Index verborum, s.v.* (*MGH AA* 12.543f.); *gens Gothica* is frequent, yet more often *gentes* are opposed to Goths and Romans; sometimes the *gentes* include Goths and Romans, and on one occasion the Romans are the *gens Romulea* (*var.* 8.10.11). Jordanes describes Theoderic as taking over the 'regnum gentis sui et Romani populi' (*rom.* 349), although note the term 'gens Romanorum' in the full title of the *Romana*. It is certainly interesting that Cassiodorus once lets slip the expression 'Geticus populus' (*var.* 10.31.2); cf. references to Goths and Romans as 'utraeque nationes' (7.3.2), 'uterque populus' (7.3.3), and 'populi' (2.16.5; cf. *de anima* 28 (CCSL 96.574) for an apparent characterization of Goths and Romans as 'magni populi'). See in general Teillet, *Goths*, 281–303.

[65] Victor of Vita, *hist. pers.* 3.62.

[66] *Var.* 3.17.1; cf. Jordanes, *get.* 176 and, perhaps, Teillet, *Goths*, 285 n. 36. I do not take Goths to have been among the *antiqui barbari* of Pannonia who took Roman wives (*var.* 5.14.6); in the light of Cassiodorus' usage elsewhere the 'praesumptor barbarus' of *var.* 1.18.2 cannot have been a Goth.

[67] *Var.* 2.5.2 (Gothic army has to deal with *barbari*), 9.21.4 (*barbari reges* contrasted with *legales domini*).

[68] *Chron. s.a.* 271; see O'Donnell, *Cassiodorus*, 38.

Maximinus, the son of a Gothic father and an Alan mother, is described as *semibarbarus* (*get.* 83 f.),[69] and if the Goths are described at an early stage of their development in language which suggests, without necessarily asserting it, that the Goths were barbarians, before long they are found curbing barbarous customs.[70] For Ennodius Odovacer's followers were barbarous, yet he could call Ricimer a barbarian while acknowledging that he was a Goth, and when one Jovinianus grew a beard in the Gothic style Ennodius wrote an epigram referring to his 'barbarica facies'.[71] Pope Gelasius seems to have considered the Goths barbarous,[72] the *Edictum Theodorici* deals with a state in which the inhabitants were divided into Romans and barbarians,[73] and Procopius had no doubt that the Goths were barbarians.[74] As a final witness we may interrogate Boethius. Imprisoned for treason and looking back on his activities while in office he recalled acting against Conigast, Triwila, and 'the avarice of the barbarians'; there can be no doubt that the two Goths named were regarded as barbarians.[75] Needless to say the circumstances in which Boethius found himself while writing the *Consolation* were not such as to encourage a dispassionate appraisal of Gothic officials, and the sentiments expressed in it may not have tallied with Boethius' feelings in early years,[76] but the Boethius of the *Consolation* may stand as representative of an opinion far removed from that of Cassiodorus.

INTERMARRIAGE, LANGUAGE

Different Romans, then, looked on the Goths in different ways, at least when they were writing for publication. But beyond the ways in

[69] So too Jerome describes Stilicho, the son of a Roman mother and a barbarian father, as *semibarbarus* (*ep.* 123.16).

[70] *Get.* 40 (*pene omnibus barbaris Gothi sapientiores*), 69.

[71] Odovacer's followers: *V. Epiph.* 97, 100. Ricimer: *V. Epiph.* 64, although the term by which his race is denoted, *geta*, elsewhere refers to the Visigoths of Toulouse (ibid. 80, but note *pan.* 83), Ennodius preferring to term the Ostrogoths *Gothi*. Jovinianus: *carm.* 2.57, on which see I. Opelt and W. Speyer, 'Barbar', *Jahrbuch für Antike und Christentum*, 10 (1967), 251–90 at 281. This is by far the best study of the topic.

[72] Gelasius, *ep.* 6.1 (ed. Theil; *barbarici incursi*), *frag.* 9.35. See too *MGH AA* 12.391 *ep.* 8.

[73] ed. Bluhme, *MGH* Leges, 5.145–79.

[74] *BG* 1.1.26, 1.2.8, 17, 18, etc.

[75] *Phil. cons.* 1.4.10, cf. 'barbara cupiditas' in Eugippius, *V. Sev.* 44.1.

[76] H. Chadwick, *Boethius: The Consolations of Music, Logic, Theology and Philosophy* (Oxford, 1981), 92, presses a passage of the *de musica* more closely than I would care to when he sees there a pointed reference to 'tough Gothic ears'.

which people regarded each other there were objective social realities which conditioned and expressed relations between the races. A law of Valentinian and Valens had forbidden marriage between provincials and barbarians on pain of capital punishment, and the *interpretatio* provided for this title decreed capital punishment for marriage between Romans in general and barbarians.[77] This would seem to have ruled out marriage between Romans and Ostrogoths, at least those among them who were not members of the officer class, who may be presumed to have obtained Roman citizenship and the *ius conubii*, and indeed among the Visigoths a ban on miscegenation was not abrogated until the reign of Liuvigild, a long way into the sixth century.[78] There was a strong tradition of barbarian rulers marrying Roman women, even though compulsion was sometimes involved, although the husbands were of different types; on the one hand one thinks of the marriage of Serena to Stilicho, a highly Romanized barbarian with rights of citizenship, and on the other of the marriages of Galla Placidia to Athaulf and of Eudocia to Huneric, who were little more than tribal leaders. In 467 the barbarian *magister militum* Ricimer took to wife Alypia, the daughter of the emperor Anthemius, and in 478 the emperor Zeno, finding himself in difficulties, offered Theoderic the hand of Anicia Juliana, the daughter of the emperor Olybrius. Nothing came of this suggestion, but Juliana went on to marry another barbarian, the highly Romanized Areobindus Dagalaifus Areobindus, by whom she had a son, significantly named Olybrius. Areobindus was later to have the embarrassment of being acclaimed emperor by a crowd in Constantinople in 512, and Juliana lived in Constantinople where she followed the progress of negotiations with the emissaries of Pope Hormisdas, became a famous builder and restorer of churches on a scale such as to suggest to one recent scholar that Justinian's building of Hagia Sophia was a response to her works at Hagios Polyeuctos, and sponsored the production of a copy of the *Materia medica* of Dioscorides which still survives.[79]

[77] *Cod. theod.* 3.14. See in general Soraci, *Ricerche*, and R. C. Blockley, 'Roman–Barbarian Marriages in the Late Empire', *Florilegium*, 4 (1982), 63–79.

[78] P. D. King, *Law and Society in the Visigothic Kingdom* (Cambridge, 1972), 13f.

[79] Ricimer's wife: Sidonius Apollinaris, *carm.* 2.484–6, *ep.* 1.5.10. Theoderic and Juliana Anicia: Malchus, *frag.* 16 (Blockley, 18.3). Juliana's marriage: John Malalas, *chron.* 398. Areobindus Dagalaiphus' names were both Germanic: Schönfeld, *Wörterbuch*, 27, 68f. Juliana's later career: *PLRE* 2.636. Hagia Sophia as response: Averil Cameron, *Procopius and the Sixth Century* (London, 1985), 104 n. 151. Juliana's wealth was known in the West: Gregory of Tours, *glor. mart.* 102.

Juliana outlived Theoderic by one or perhaps two years, and her extraordinary career furnishes a reminder of a Byzantine ambience which was Theoderic's early in his life and from the influence of which he may never have freed himself. Later, female members of the Ostrogothic royal family were to marry Romans: in 535 one of them married Flavius Maximus, himself a descendant of the emperor Petronius Maximus, and Theoderic's granddaughter Matasuentha, following the death of her husband Witigis, was to marry Germanus, the nephew of the emperor Justin,[80] whose own wife, if Procopius' *Anecdota* are to be trusted, was of barbarian origin.[81] In short, there is copious evidence for intermarriage between Romans and barbarians at the highest levels of political life.

Inevitably it is more difficult to detect marriages across racial lines between those of humbler station, and it seems clear that 'old barbarians' known to have taken Roman wives in Savia were not Goths.[82] But Theudis, sent by Theoderic to command the Gothic army in Spain, married a wealthy Hispano-Roman woman from whose estates he raised a following of about 2,000 soldiers;[83] a sad case of domestic discord brought to Theoderic's attention involved two married couples whose names, Brandila and Procula, and Patza and Regina, imply mixed marriages;[84] and several inscriptions seem to indicate marriages between barbarians and Romans,[85] while a lost papyrus is known to have recorded a gift made to the Catholic bishop of Ravenna by Hildevara, the wife of John, in 523.[86] The evidence is not as strong as one would like, but it suggests that, whether or not Goths held Roman citizenship, in a modest way Goths and Romans were marrying each other in Theoderic's Italy, and that the male partner tended to be a Goth, a circumstance which could be taken to imply that some Goths were seeking heiresses. Such behaviour would be in no way surprising, for the soldiers of Belisarius' army who took Vandal women as

[80] Flavius Maximus: Cassiodorus, *var.* 10.11.3, 10.12.3. Matasuentha and Germanus: Procopius, *BG* 3.39.14; Jordanes, *get.* 81, 251, 314, *rom.* 383.

[81] *Anec.* 6.17, 9.48.

[82] Cassiodorus, *var.* 5.14.6, correctly interpreted by Hartmann, *Geschichte*, 127, no. 4. See too Šašel, 'Antiqui barbari', 135, Soraci, *Ricerche*, 155–8, and Várady, *Epochenwechsel*, 99–104. On 'Savia', Ditten, 'Prokops Nachrichten', at 8f.

[83] Procopius, *BG* 1.12.50–4.

[84] Cassiodorus, *var.* 5.32f.

[85] Tzitta and Honorata, Fiebiger/Schmidt, *Inschriftensammlung*, no. 220, dated to 568 (*CIL* 5.7793); Amora and Antonina, ibid. no. 224 (*CIL* 5.1583); while Theodosus and Dumilda shared a tomb, ibid. no. 225.

[86] Marini, *Papiri*, 132, 283f., with Tjäder, *Papyri*, 1.53.

their wives following the defeat of the Vandals acted in precisely this way.[87]

If some Goths were taking Roman wives, others were adopting Roman names. Few would have gone as far as the Goth known from a papyrus of 539 to have been called Latinus, but two clergy, Peter and Paul, mentioned in a papyrus of 551 may have been Goths.[88] Sometimes people were known by both Gothic and Roman names, such as the Ademunt qui et Andreas mentioned in a papyrus of 553;[89] sometimes we hear of children with Roman names whose parents bore Gothic names, such as Agata, the daughter of Count Gattila, who died at the age of about 40 in 512 and was buried in Milan.[90] Not all the names our sources record for Goths are easy to understand. Theoderic's son-in-law Eutharic Cilliga bore names which were respectively Germanic and not Germanic,[91] and as we have seen, one of his daughters, Ostrogotho Areagni, seems to have borne a simple barbarian name to which a source appended a quasi-Roman addition by mistake. However, while it is clear that Roman names were being taken by some Goths, no Roman is known to have taken a Gothic name.[92]

As with personal names, so with language. Evidence for Romans in Ostrogothic Italy learning Gothic is confined to the members of one family, that of Cyprian, and as Cyprian is also one of the few Romans known to have served Theoderic in the army we may conclude that we are dealing with a family of strong pro-Gothic

[87] Marriage to Vandal women: Procopius, *BV* 2.14.8–10.

[88] Latinus: Tjäder, Papyri, pap. 30.102 (vol. 2, p. 62) with n. 31. Peter and Paul: Tjäder, *Papyri*, pap. 34.83f., 98ff., 108ff. (vol. 2, pp. 100, 102), with the most interesting discussion of G. Kampers, 'Anmerkungen zum Lateinisch-gotischen Ravennater Papyrus von 551', *Historisches Jahrbuch*, 101 (1981), 141–51. Cf., too, the Arian cleric 'Felix, natione barbarus' (Ferrandus, *Vita Fulgentii*, 17 (PL 65.125)).

[89] Tjäder, *Papyri*, pap. 13.21f. (vol. 1, p. 304); so too 'Gundeberga qui (*sic*) et Nonnica' (Fiebiger/Schmidt, *Inschriftensammlung*, no. 45 (*CIL* 11.941), an inscription of 570 at Modena).

[90] Fiebiger/Schmidt, *Inschriftensammlung*, 223 (*CIL* 5.6176); so too Basilius the son of Guntelda (Fiebiger/Schmidt, *Inschriftensammlung*, no. 232 (*CIL* 5.5415); Basilius' own son was named Guntio); John the son of Ustarric (Fiebiger/Schmidt, *Inschriftensammlung*, no. 234 (*CIL* 10.7116)). To judge by an inscription found at Serdica this process was already occurring while the Goths were still pagan: Fiebiger/Schmidt, *Inschriftensammlung*, no. 171 (*CIL* 3.12396).

[91] Schönfeld, *Wörterbuch*, 82f., 62f.

[92] A different situation obtained in Gaul, where one of the uncles of Gregory of Tours' mother bore the name Gundulf: Gregory of Tours, *hist. franc.* 6.11, with 5.5.

[93] Cassiodorus, *var.* 5.40.5 (Cyprian spoke three languages, presumably Latin, Greek, and Gothic), 8.21.7 (his sons spoke Gothic; cf. 8.22.5). Sidonius Apollinaris' friend Syagrius was skilled at Gothic: *ep.* 5.5.1.

leanings.[94] The impact of barbarian vocabulary on the Latin of the day, as far as it can be deduced from written evidence, was minimal, and while some words may have passed into colloquial speech but escaped being written they cannot have been many.[95] On the other hand, to judge from the evidence of Ulfilas' translation of the Bible, Gothic was wide open to borrowings from Latin,[96] and it is clear that the Goths were coming to employ their own language less. Priscus had been made aware by his dealings with the Huns in the mid-fifth century of barbarians who knew Latin as well as Hunnic and Gothic.[97] Cassiodorus' *Variae* include many letters addressed to Goths, and while these tend to be short and free of the scholarly digressions and etymologies often encountered in his letters to Romans there is no indication that he expected his Gothic readers to have trouble with Latin. While Gothic remained the language of the army, of eleven Gothic clergy of the church of Sta Anastasia in Ravenna who signed a document in 551, seven did so in Latin and four in Gothic.[98] Some members of Theoderic's family were becoming extremely learned: as we shall see later, Theoderic himself, despite the testimony of a source which seems to indicate the contrary, may well have been able to read and write in both classical languages and must have had better Greek than almost all his Roman subjects; his daughter Amalasuintha, who may just possibly have owed something of her learning to a noble Roman matron Barbara, known to have been offered a position at Theoderic's court in the middle of 510, was credited with knowing Greek, Latin, and Gothic, and did her best to see that her son was given an education along Roman lines, although the opposition of some of the Goths forced her to abandon this project;[99] his nephew

[94] Note in Cassiodorus *carpa* (*var.* 12.4.1) and *saio* (*passim*); in Ennodius *flasco* (*carm.* 2.147.1), concerning which Gregory the Great observes 'lignea vascula, quae vulgo flascones vocantur' (*dial.* 2.18). But only the second of these cases is indubitably a Germanic word. A list of Italian words of Gothic origin is provided by G. Bonfante, *Latini e Germani in Italia*, 3rd edn. (Genoa, 1965), 31 f. The words include *albergo* ('place of lodging'), which has been derived from the Gothic **haribergôn*. As there are similar words in modern French and Spanish (*auberge, albergue*) one may be tempted to see here a sign of a method of military settlement imposed during the Gothic hegemony. But consultation of dictionaries suggests that the borrowing from Gothic occurred during the Roman period.

[95] Wolfram, *History*, 113.

[96] Ed. Blockley, 266, 288.

[97] Army: Procopius, *BG* 1.10.10. Document of 551: Tjäder, *Papyri*, pap. 34.88–139 (vol. 2, pp. 100–4).

[98] Cassiodorus, *var.* 11.1.6f.; Procopius, *BG* 1.2.6ff. On Barbara, Ennodius, *ep.* 8.16 with *PLRE* 2.209f.

Theodahad was famed for his learning, which seems to have extended to ecclesiastical writings, Latin literature, and the Platonists, and wrote Latin verse;[99] his niece Amalaberga was 'learned in letters';[100] and Amalafrida may have been the sister we know to have been at the Byzantine court in 487.

It must be admitted that such evidence is a little difficult to square with what we know of the progress of Latin among other barbarian peoples. King Euric the Visigoth seems to have been able to understand the words of Bishop Epiphanius when he came to plead on behalf of Nepos, and if the saint's words bore any resemblance to the convoluted periods with which Ennodius credits him one would have thought the king likely to have been an accomplished Latinist. But when it came time to reply he spoke in 'some foreign tongue' and relied on the services of an *interpres*, which we may take to mean 'interpreter'. Nevertheless, Euric is known to have been a king of anti-Roman disposition, and he may be contrasted with his predecessor, Theoderic, who was in his youth a student of Virgil; Theoderic's predecessor, Thorismod, had a silver plate depicting the story of Aeneas.[101] On one occasion the Vandal bishop Cyrila found it expedient to deny that he knew Latin, and while his Catholic antagonists were not deceived the bishop must have felt that his claim was at least plausible.[102] But Dracontius reports that in Vandal Africa Romans were joined to barbarians in the auditorium,[103] and while we have no evidence for this happening on a wide scale in Ostrogothic Italy it seems clear that Latin was becoming current among the Goths in Italy. Needless to say, Latin could be expected to have progressed most rapidly among Goths who were literate or members of the royal family, and Gothic presumably remained the language of the army (cf. Procopius, *BG* 1.10.10). But it is clear that Latin was

[99] Cassiodorus, *var*. 10.3.4f.; Procopius, *BG* 1.3.1, 1.6.10, 16. On Theodahad as a poet, A. Fo, 'L'Appendix Maximiani', *Romanobarbarica*, 8 (1984–5), 151–230.

[100] Cassiodorus, *var*. 4.1.2. Note as well that a sister of Theoderic is described as having been the 'companion' of the empress Ariadne in 487: John of Antioch, *frag*. 214.8.

[101] Euric: Ennodius, *V. Epiph*. 89f. On the meaning of *interpres*, I accept K. F. Stroheker, *Euric König der Westgoten* (Stuttgart, 1937), 6, no. 10; cf. Sidonius Apollinaris, *ep*. 9.2.2, 'Hieronymus interpres'. But note that Sidonius was the author of Latin verses which were to be inscribed on a basin to be presented to Euric's wife Ragnahilda: *ep*. 4.8. For Thorismod, see Riché, *Education and Culture*, 208 n. 203. Theoderic: Sidonius Apollinaris, *carm*. 7.495–9.

[102] Victor of Vita, *hist. pers*. 2.55.

[103] Dracontius, *rom*. 1.14.

advancing. One naturally wonders whether the same was true of Catholic Christianity.

ARIANS AND CATHOLICS

The question of religion is both complicated and comparatively well documented, and will require extended discussion. At the time of their arrival in Italy the Ostrogoths were Christians of a belief we may conveniently term Arian, although the relationship between their belief and the doctrines promulgated early in the fourth century by the Alexandrian cleric Arius is not as clear as one would wish.[104] Jordanes reports that the Visigoths were converted shortly after their entry on to Roman soil in 376, during the reign of the heretical emperor Valens, and that they in turn evangelized the Ostrogoths and Gepids.[105] This account has been widely discussed and there seems to be no consensus as to how much, if any, credence should be placed in it.[106] Nevertheless, when the Goth Radagaisus invaded Italy in 405 he was a pagan,[107] and our sources give us no reason to doubt that by the time of their entry into Italy the great bulk of the Ostrogoths were at least nominally Christians, although the continued depiction in Ostrogothic art of the Italian period of birds of prey, apparently a pagan symbol in earlier times, is suggestive of continuing pagan sympathies.[108] But at what time within the period 405–89 they were converted seems an open question. That Theoderic's Gothic mother Erelieva took a new name when baptized a Catholic suggests that she had hitherto been a pagan, but

[104] See the different assessments of K. Schäferdiek, 'Germanenmission', *Reallexikon für Antike und Christentum*, 10 (1978), 492–548 (moderate Arianism), and M. Simonetti, 'Arianesimo Latino', *Studi medievali*, 3rd ser. 8 (1967), 663–744 (radical Arianism of Ulfilas and his followers).

[105] Jordanes, *get.* 131–3.

[106] Wolfram, *History*, 75–85; Schäferdiek, 'Germanenmission'; E. A. Thompson, 'Christianity and the Northern Barbarians', in A. Momigliano, ed., *The Conflict between Paganism and Christianity in the Fourth Century* (London, 1963), 56–78.

[107] Orosius, *adv. paganos*, 7.37; Marcellinus, *chron. s.a.* 406.2; Augustine, *civ. dei*, 5.23.

[108] See on birds of prey, Bierbrauer, 'Aspetti archeologici', 450. Note the eagle brooch found in the Cesana treasure, deposited 30 km south of Ravenna, conveniently to be seen on p. 161 of D. A. Bullough, 'Germanic Italy: The Ostrogothic and Lombard Kingdoms', in D. Talbot Rice, ed., *The Dark Ages* (London, 1965), 157–74. Little can be built on Ennodius' assertion that Theoderic demanded a cup 'causa . . . auspicii' in 488 (*pan.* 33).

we have no information as to when she was baptized, and a description of Theoderic as one who had worshipped the highest God from the beginning of his life, while suggestive, does not necessarily mean that he had been baptized as an infant,[109] and agnosticism seems called for. We do not know how many years of Christian practice lay behind the Goths when they marched into Italy.[110]

In so far as they were adherents of different creeds, relations between Goths and Romans need not have been good. In Africa the Vandals had revealed themselves staunch persecutors of Catholics, and in 484 Huneric finally demanded that all his subjects become Arians; by 488 Pope Felix was aware of the problems caused by Catholics who had submitted to rebaptism as Arians.[111] On the other hand, Vandals were not immune to the attractions of Catholicism. Huneric was said to have placed special police armed with grappling hooks outside the cathedral in Carthage to attack Vandals or people in Vandal clothing who tried to enter. Some people lost their eyes and others were killed, but, despite such disincentives, epigraphic evidence suggests that some Vandals were indeed converted.[112] The encounter between the animosity of the Vandals and the combative traditions of African Catholicism was to produce some memorable disputes, raising issues which interested King Thrasamund, Theoderic's brother-in-law, and one has the feeling of a situation where theology was a matter for public controversy as it had been in the days of St Augustine.[113] While the Visigoths were more benign, there can be no doubt that Euric left some bishoprics vacant as a matter of policy, with a consequent decline in the number of priests—as ordinations could not be held to compensate for those who died—and physical decay in churches, although a pas-

[109] Ennodius, *pan*. 80.

[110] Scholars may be divided into those who date their conversion before the break-up of the empire of the Huns (J. Zeiller, *Les Origines chrétiennes dans les provinces danubiennes de l'empire romain* (Paris, 1918), 534–7; H.-E. Giesecke, *Die Ostgermanen und der Arianismus* (Leipzig, 1939), 117; Burns, *Ostrogoths*, 150) and those who favour a later date (Thompson, 'Christianity', 73; Schäferdiek, 'Germanenmission', 513; E. Demougeot, *La Formation de l'Europe et les invasions barbares*, 2 (Paris, 1979), 823). In neither case are the arguments conclusive. See recently B. Krüger, ed., *Die Germanen*, 2 (Berlin, 1983), 281 f.

[111] Huneric's decree: Victor of Vita, *hist. pers*. 3.12; cf. Procopius, *BV* 1.8.3f. Felix's distress: PL 58. 924–7. See too *col. avel*. 86.5 (Pope Agapetus, writing in 535).

[112] Huneric's police: Victor of Vita, *hist. pers*. 2.9. Epigraphic evidence: Courtois, *Vandales*, 224 n. 3.

[113] Note e.g., among the controversial works of Fulgentius, those written in connection with King Thrasamund: CCSL 91.65–94, 95–185.

sage in Gregory of Tours dealing with this does no more than provide evidence for that author's extreme hostility towards the Visigoths.[114] Avitus of Vienne was kept busy expounding Catholicism to Gundobad the Burgundian and was believed to have convinced him of the truth of his position, although Gundobad refused to confess Catholicism in public; his son Sigismund, however, was converted.[115] Works definitely by Bishop Caesarius of Arles, and others attributed to him, indicate animosity towards Arianism.[116]

In Italy, on the other hand, one has the impression that all was sweetness and light, and this despite an interesting piece of evasiveness in Cassiodorus' Chronicle which suggests that this author did not wish to draw attention to the division between Arians and Catholics and an anti-Arian letter by Bishop Agnellus of Ravenna.[117] Far from imposing his will on the Church Theoderic was content to let it see to its own affairs, to such an extent that in 502 he refused to become involved in the judgement of Pope Symmachus and insisted that the bishops who had assembled in Rome for this purpose come to a decision.[118] It is true that he did intervene in 499, but this may have been on request, and Theoderic's involvement was much more low-key than that of Charlemagne in a similar

[114] Sidonius Apollinaris, *ep*. 7.6.7–10, developed by Gregory of Tours, *hist. franc*. 2.25. But whereas Sidonius describes the doors of churches being torn away and the entrances to basilicas coming to be blocked by briars, Gregory, citing Sidonius, claims that Euric ordered the doors to be blocked by briars so that few people would enter the churches and the faith would be forgotten.

[115] See in particular the 'Dialogi cum Gundobado rege vel librorum contra Arrianos reliquiae', *MGH AA* 6.2.1–15. Sigismund had been converted by the time Pope Symmachus died in 514: Avitus of Vienne, *ep*. 29.

[116] The evidence of Caesarius' sermons is difficult to evaluate: some of the most hostile discussions of Arianism are borrowed from other sources (e.g. *serm*. 123), sometimes Arianism occurs in a list of heresies of which at least some could have posed no threat to the faith of his congregation (Donatists, Manichaeans, Arians, and Photinians are all condemned at *serm*. 96.5), Jews and 'heretics' are accused of a merely literal understanding of Scripture (*serm*. 83.1, 107.1). Note too among the works attributed to Caesarius the *De mysterio sanctae trinitatis*, with its reference to people of 'another religion' asking Catholics awkward questions (ed. Morin, 2.165), the *Breviarium adversus haereticos*, and an interpretation of the *potestas* of Apoc. 13: 4 as being that of heretics (2.245), and the suggestive circumstance that he is an early witness to something very similar to the *Quicunque vult*: G. Morin, 'L'Origine du symbole d'Athanase', *Revue bénédictine*, 44 (1932), 207–19.

[117] With Cassiodorus, *chron. sa* 380, 'Ambrosius episcopus de Christiana fide multa sublimiter scribit', cf. his source, Prosper Tiro, *chron. sa* 380: 'Ambrosius episcopus multa pro catholica fide sublimiter scribit'. Agnellus of Ravenna: PL 68.381–6.

[118] *MGH AA* 12.424–6, cf. Anon. Vales. 60, 'nihil contra religionem catholicam temptans'.

situation three centuries later, although he may have felt constrained by his adherence to Arianism. Just as Theoderic could be distinguished from emperors by his failure to build Catholic churches, so too he could be distinguished from them by his extreme reluctance, at this stage of his reign, to intervene in the affairs of the Church, and ecclesiastics of the generation which saw Gelasius make his famous distinction between priestly and royal spheres of activity must have regarded this as a desirable characteristic. That Goths and Romans were of different persuasions was acknowledged on both sides. In a form of address to Catholic bishops meeting in Rome in 502 Theoderic mentioned 'many bishops of your religion and ours', and Pope Gelasius had no doubt that Count Teias was of a different communion.[119] But Gelasius had no difficulty in appealing to Theoderic's 'Christian mind' and in referring to the king as his son, while Theoderic was happy to describe himself as the son of the Catholic bishop Caesarius of Arles,[120] and if there may be a hint of trouble in a carefully worded note from Gelasius to Theoderic in which the Pope observes that, just as the king had ordered that the laws of the emperors were to be maintained, he should maintain reverence for the apostle Peter much more, so that his felicity would be increased, it is not referred to elsewhere.[121] Anonymous Valesianus observed that Theoderic, despite being an Arian, did nothing against the Catholic religion (anon. Vales. 60), and described him as the son of a Catholic mother (58), going to St Peter's 'most devoutly and like a Catholic' on his one recorded trip to Rome (65). That Theoderic was a heretic only becomes an issue for this author towards the end of his work (91–5). In the same way the only occasions when the *Liber pontificalis* refers to Theoderic as a heretic occur in its account of Pope John, presumably written after John's death, which describes the period in the closing years of Theoderic's reign when his relations with Catholics took a sudden turn for the worse, although it must be admitted that the primitive *abrégés* of this document both term Theoderic a heretic as early as the pon-

[119] *MGH AA* 12.425.16f., 390.2.28.

[120] 'Christianae mentis vestrae pietate': *MGH AA* 12.389.1.1. 'Filius meus': ibid. 390.3.1; Gelasius, *frag*. 12f. (ed. Thiel). For an Arian king described as the son of a pope, see also Pope Hilary to Bishop Leontius of Arles, *Epistolae Arelatenses genuinae*, 15 (*MGH Ep*. 3.21.1). Theoderic as son of Caesarius: *V. Caes*. 1.26.

[121] *MGH AA* 12.391.6.

tificate of Symmachus.[122] Ennodius, who had a fine ear for the niceties of correspondence, did not find alluding to Scripture in a letter to a Goth at all out of place, and was content to describe Theoderic arguing against Bishop Epiphanius on scriptural grounds.[123] The king, he believed, had been a worshipper of the true God from the beginning of his life.[124] One is hardly surprised to find that Zachariah of Mytilene believed that Theoderic had converted from Arianism, but the general description of Theoderic encountered in this source stands in some relation to a passage in Anonymous Valesianus, and his assertion that Theoderic converted seems based on a misunderstanding of Anonymous Valesianus, or some lost source which the latter drew on or which drew on him.[125] Nor is one surprised by the story told by Theodore Lector of how Theoderic killed a Catholic deacon who apostatized with a view to gaining royal favour.[126]

The Ostrogoths, then, were Arians of a type with whom the Catholics of Italy could coexist peacefully. It cannot be accidental that Theudis, an Ostrogoth who came to rule Visigothic Spain, was remembered there for his benevolence towards the Catholic Church while Eutharic, who came from Spain to Italy, was regarded by Italians as an enemy of the Catholic faith.[127] As the outpourings generated by the Acacian schism showed, the Italian Church was quite capable of producing controversial literature, yet there is scarcely a sign of writing against Arianism. Pope Gelasius is known to have written two books against Arius, but they have not survived;[128] Boethius' theological works contain only two references to Arianism, both tangential;[129] and while Eugippius' *Life of Severin* is

[122] *Lib. pont.* 275.6.17; 276.3.10; see too 104.26f. But note on the other hand 96f. On the date and composition of this source, C. Vogel, 'Le *Liber pontificalis* dans l'édition de Louis Duchesne: État de la question', in *Monseigneur Duchesne et son temps* (Collection de l'École française de Rome, 23; Rome, 1975), 99–127.

[123] Ennodius, *ep.* 6.28; *V. Epiph.* 132.

[124] Ennodius, *pan.* 80; cf. *ep.* 9.30.7.

[125] Zachariah, *hist. eccl.* 7.12 (trans. F. J. Hamilton and E. W. Brooks, 84), but cf. Anon. Vales. 57–60; in particular, for an alleged conversion, 'dum ipse quidem Arrianae sectae esset, tamen nihil contra religionem Catholicam temptans' (60); note too that Zachariah's assertion that Theoderic rebelled against Anastasius would answer to a misreading of the data lying behind anon. Vales. 57.

[126] Theodore, ed. Hansen, 131.16ff.; see too Theophanes, *chron.* AM 5991.

[127] Theudis: Isidore of Seville, *hist. goth.* 41. Eutharic: Anon. Vales. 80.

[128] *Lib. pont.* 255.14.

[129] *Theological Tractates*, ed. Stewart and Rand, 1.1.10–13, 4.32–4.

hostile to the Arianism of the Rugians[130] we have no reason to believe that this author would have felt similarly about the Ostrogoths. It is very difficult to see any trace of religious controversy in the mosaics produced during Theoderic's life at Ravenna.[131] Optatus, a correspondent of Fulgentius of Ruspe, had to deal with the questions of heretics.[132] But by the standards of the Vandals or even the Burgundians, such discussions were as nothing.

It is difficult to establish what Arianism meant to the Ostrogoths. Some scholars have been condescending in their dismissal of Germanic Arianism.[133] But the community of faith which produced the *Codex argenteus*, an exquisite gospel book of which 188 folios remain, copied in gold and silver ink on purple parchment so fine it may have come from the hides of unborn calves, in a workshop which may be identifiable,[134] and which patronized the beautiful mosaics still to be seen in the Arian baptistery and the church of S Apollinare Nuovo at Ravenna[135] cannot be accused of lack of sophistication, and Arians in general seem to have given generously to their churches.[136] But it is arguable that the massive building programme undertaken by Catholics and Arians at Ravenna during and after the life of Theoderic was partly prompted by rivalry,[137] and we may suspect that for the Goths an essential feature of their Arianism was simply that it was not the faith of the Romans. In common with the peoples who were to be converted directly or indirectly from Constantinople in the Middle Ages their use of a vernacular liturgy and scriptures would have allowed them to regard their Christianity in a national light rare in the medieval

[130] Eugippius, *V. Sev*. 8.1, in particular.

[131] The arguments of R. Sörries, *Die Bilder der Orthodoxen im Kampf gegen den Arianismus* (Frankfurt, 1983), are not persuasive.

[132] Fulgentius, *ep*. 8.2 (CCSL 9.258).

[133] G. Pepe, *Le Moyen Âge barbare en Italie* (Paris, 1956), 29: 'un christianisme plus simplifié, plus adapté à leur esprit grossier'.

[134] J.-O. Tjäder, 'Der Codex argenteus in Uppsala und der Buchmeister Viliaric in Ravenna', in V. E. Hagberg, ed., *Studia Gotica* (Stockholm, 1972), 144 64, argues in fascinating detail that the Wiliarit mentioned in a Ravenna papyrus of 551 (Tjäder, *Papyri*, pap. 34.85, 136 (vol. 2, pp. 100, 104)) directed a workshop which provided both the *Codex* and a manuscript of Orosius' *Adversus paganos*. See, too, on an innovation in the layout of tables of canons, C. Nordenfalk, *Die spätantike Kanontafeln* (Göteburg, 1938), 283f.

[135] Best treated by Deichmann, *Ravenna*.

[136] Procopius, *anec*. 11.16–20 stresses the wealth of Arian churches in the East; that Visigothic churches were well endowed can be deduced from Gregory of Tours, *hist. franc*. 3.10.

[137] So Ward-Perkins, *Urban Public Building*, 72.

West; that the Christianity they practised was deemed heretical by the numerically dominant population among whom they settled can only have strengthened this feeling. At a time when barbarian Arians in Gaul, Spain, and Africa are known to have called Catholics 'Romans',[138] the Arian clergy of Ravenna could describe themselves as adherents of 'the law of the Goths'.[139] By the time of Theoderic, Arianism could afford to be tolerant because, having become a mark, perhaps something of a defiant one, of national identity, it had no aspirations to universality. The aspirations of Catholicism were quite different.

It is therefore not entirely surprising that such evidence as we have indicates that there was a steady flow from Arianism to Catholicism among the Ostrogoths in Italy. Procopius represents Gothic envoys talking with Belisarius, declaring that under their rule no Roman had changed belief, whereas Goths had converted and suffered no penalty, and with the exception of the last phrase there is no reason not to accept this assessment of the situation.[140] As early as 523 Hildevara, a Gothic woman married to one John and so presumably involved in a mixed marriage, made a gift to Ecclesius, the Catholic bishop of Ravenna, for his church, while in 553 the *sublimis femina* Ranilo made a generous donation, apparently to the church of Ravenna, although donations made to Catholic churches after the apparent defeat of the Ostrogoths in 540 reflect a political situation very different to that which existed in the time of Theoderic.[141] During the reign of Theoderic's son Athalaric the Gothic woman Ranilda was converted to Catholicism and appar-

[138] 'Romanos enim vocant nostrae homines religionis': Gregory of Tours, *glor. mart.* 24. See too ibid. 78 f.; John of Biclarum, *chron. s.a.* 580.2 (where the 'Romana religio' is Catholicism and the 'nostra Catholica fides' Arianism); the same usage in Victor of Vita, *hist. pers.* 3.1; cf. 1.44 and 3.62, and perhaps Gregory of Tours, *hist. franc.* 9.15, where Reccared speaks of the bishops 'qui se Catholicus [*sic*] dicunt'. In the light of this, Theoderic's words at Anon. Vales. 88 ('ut reconciliatos in Catholica restituat religione') are ambiguous.

[139] 'lex Gothorum': Tjäder, *Papyri*, pap. 31.1, 7, 8, 10 (vol. 2, pp. 84 f.), 34.108, 122 (vol. 2, p. 102). I accept Tjäder's interpretation of this phrase (vol. 2, p. 268 n. 3) against that of Scardigli, *Goten*, 295 f.; note that the 'lex Gotica' of *Passio Sigismundi*, 4 (*MGH SRM* 2.335) is certainly Arianism, and that Pope Vigilius probably had this form of belief in mind when he wrote to the bishop of Arles describing Totila as 'aliene legis' (*MGH Ep.* 3.68.24).

[140] Procopius, *BG* 2.6.18. The issue of conversion is sensibly discussed by Zeiller, *Origines*, 568–71.

[141] Hildevara: Marini, *Papiri*, 132, 283 f.; cf. Tjäder, *Papyri*, 1.53. Ranilo: Tjäder, *Papyri*, pap. 13 (vol. 1, pp. 304–8); see too pap. 20 (vol. 1, pp. 346–52) for a donation by Sisivera to the Church of Ravenna in about 600.

ently made to regret it.[142] Archaeology reveals the burial of Ostrogothic women in the Catholic cemeteries of St Valentine in Rome and SS Gervasius and Prostasius in Milan.[143] The example of Theoderic's mother Erelieva may stand as a reminder that some of this evidence may relate to a Catholic minority which existed among the Goths before their coming to Italy, just as there was certainly a tradition of Catholicism among the barbarians in Italy before the Ostrogoths arrived and heretical, presumably Arian, clergy living in Gaul are known to have been converted.[144] The treasure discovered at Galognano in 1963 contains a chalice and patten donated by Himnigilda and Sivegerna, who were probably Goths rather than Lombards,[145] while in 523 the praepositus of St Peter's basilica bore the name Transmund, and St Benedict was to welcome a Goth, presumably a convert, as a monk.[146] Apart from Theodore Lector's apocryphal story of the apostatizing deacon, there is nothing to suggest that a single Catholic was converted to Arianism in Ostrogothic Italy.

Perhaps, then, as time passed religion would have become less satisfactory as a focus for awareness of being different from Romans, as more Goths, with perhaps women leading the way, adopted the Roman religion, just as they were adopting the Latin language and Roman names. This phenomenon will require our attention again later, but for the time being we may note that, compared with both other barbarian kings and Roman emperors, Theoderic was non-interventionist in the affairs of the Church. This characteristic is also to be seen, but even more clearly, in another aspect of religious life in Ostrogothic Italy, the status of the Jewish community, and it will be worth our while interrupting our general

[142] Cassiodorus, *var.* 10.26.3.

[143] Bierbrauer, *Ostgotische Grab- und Schatzfunde*, 50f.; *id.*, 'Frühgeschichtliche Akkulturationprozesse', at 103f.

[144] Valila qui et Theodovius, *magister utriusque militiae* from 471, was presumably a Catholic (*PLRE* 2.1147), and provides another case of someone known by Germanic and Roman names; in 462 Count Herila was burned in Rome 'in pace fidei Catholic(a)e': Fiebiger/Schmidt, *Inschriftensammlung*, no. 301 (*CIL* 6.31,996). Clergy in Gaul: Council of Orleans (511), can. 10 (*Concilia Galliae*, 8f.), cf. Council of Epanone (517), can. 16 (*Concilia Galliae*, 28).

[145] O. von Hessen, W. Kurze, and C. A. Mastrelli, *Il tesoro di Galognano* (Florence, 1977).

[146] Fiebiger/Schmidt, *Inschriftensammlung*, no. 190; see nos. 227 (*CIL* 11.2885), 227a, and (1939) 48 for more clergy with apparently Gothic names, but they could have been Arians. Gregory of Rome, *dial.* 2.6 tells of a Gothic monk; a monk Bonosus, 'barbarus genere', is mentioned in Eugippius, *V. Sev.* 35.1.

discussion of Goths and Romans in the interest of pursuing this theme.

THE JEWS

Nowhere is Theoderic seen more attractively than in his policy towards the Jews. The Jews of Genoa were given permission to rebuild their synagogue in a letter from him which concluded with the famous observation, 'We cannot command adherence to a religion, since no-one is forced to believe unwillingly',[147] and were assured, in language reminiscent of the *Edictum Theodorici*, that the provisions of the old laws concerning them were to continue to apply, which may have been a comfort to Jews at the beginning of the sixth century.[148] Indeed, in 526 the Jewish *scholasticus* Symmachus issued orders on behalf of Theoderic, and it is possible that whatever function he held was being exercised contrary to a *novella* of Theodosius excluding Jews from civil administration.[149] Christians responsible for the burning of a synagogue at Rome were to be punished, although the cause of their anger towards the Jews was to be investigated, and the synagogue at Milan was to be protected from the attacks of clerics.[150] It must be said that the language used in these letters is not always particularly friendly, although it is more positive than the one reference to the religion of the Samaritans, 'Samarea superstitio'; even the Samaritans, however, were allowed to state their case in a dispute with the Roman Church over property.[151] The extent to which Theoderic was responsible for the

[147] 'religionem imperare non possumus, quia nemo cogitur ut credat invitus' (*var.* 2.27.2). Casaubon adorned the frontispiece of his text of the *Variae* with these words and another tag, 'cum divinitas patiatur diversas religiones esse, nos unam non audemus imponere' (*var.* 10.26.4), according to A. Momigliano, 'Un appunto di I. Casaubon dalle "Variae" di Cassiodoro', *Tra Latino e volgare: Per Carlo Dionisotti* (*Medioevo e Umanesimo*, 17f.; Padua, 1977), 615–17. Theoderic's policies are discussed at length by B. Saitta, '"Religionem imperare non possumus": Motivi e momenti della politica di Teoderico il Grande', *Quaderni Catanese*, 8 (1986), 63–88.

[148] *Var.* 4.33; cf. 'deposcitis privilegia vobis debere servari, quae Iudaicis institutis legum provida decrevit antiquitas' (4.33.2) with 'Circa Iudaeos privilegia legibus delata serventur' (*Ed. theod.* 143).

[149] Anon. Vales. 94, with *nov. theod.* 3.2 (in *Codex Theodosianus*).

[150] Rome: *var.* 4.43. Milan: *var.* 5.37.

[151] *Var.* 3.45, assuming that the 'Samarea superstitio' does not refer to Judaism, as suggested by e.g. Schmidt, *Ostgermanen*, 332. Gregory the Great mentions Samaraei at Catania and Syracuse (*reg.* 6.30, 8.21).

sentiments with which Cassiodorus credited him in the *Variae* is difficult to determine, and the question naturally arises as to how much the generally benign attitude taken towards the Jews in the *Variae* reflects Theoderic's views.[152] But the way the government responded to trouble which broke out in Ravenna, probably in 519 or 520, when Cassiodorus did not hold office, makes it clear that the line taken in the *Variae* was that of Theoderic himself.[153]

Our only source for the incidents in Ravenna is a difficult passage in Anonymous Valesianus (81 f.), according to which the Jews of the city, 'wishing ill to those who had been baptized, as they chanted psalms, often threw *oblata* into the water'.[154] It is not clear how we are to understand *oblata*, but when we recall that Gregory of Tours saw in a vision an altar with *oblationes* on it we may take it that some connection with the eucharist is likely.[155] The people, paying no attention to the king, his son-in-law Eutharic, or Bishop Peter, rose up and burned the synagogues; the same thing happened on another occasion. The text of the anonymous is awkward here, the second occasion being described in surviving manuscripts of the text in the opaque words 'quod et in cena eadem similiter contingit'. Editors have been puzzled by the reference to a supper (*cena*), and have generally replaced this word by 'Roma', the emendation being widely followed by scholars who have concluded that the second

[152] See e.g. F. Schneider, *Rom und Romgedanke im Mittelalter* (Munich, 1926), 43, 86f.; Ensslin, *Theoderich*, 93f.; B. Blumenkranz, *Juifs et Chrétiens dans le monde occidentale 430–1096* (Paris, 1960), 98; Reydellet, *Royauté*, 185; and Ullmann, *Gelasius*, 221, whose position is close to mine, despite his unexpected error on the religion of Cassiodorus. Recall that Cassiodorus considered the quaestor 'the voice of the king's tongue': *var.* 6.5.1. It should be noted that 'religionem imperare non possumus' (*var.* 2.27.2) is closely parallelled by 'unam [religionem] non audemus imponere' (*var.* 10.26.4).

[153] Assuming that the order of the paragraphs in the text of Anonymous Valesianus is that in which they were originally written (see below, Appendix 2), the troubles occurred after the consular celebrations of Eutharic in 519 (Anon. Vales. 80) and while Bishop Peter of Ravenna was still alive (on the date of his death, Stein, *Histoire*, 248 n. 2). Evidence for the Jewish community at Ravenna has been collected by L. Ruggini, 'Ebrei e Orientali nell'Italia settentrionale fra il IV e il VI secolo d. Cr.', *Studia et documenta historiae et iuris*, 25 (1959), 186–308 at 228 n. 103.

[154] Here and elsewhere my interpretation of passages of this text is indebted to the careful reading of C. Morton, 'Marius of Avenches, the *Excerpta Valesiana* and the Death of Boethius', *Traditio*, 38 (1982), 107–36, here 119.

[155] Gregory of Tours, *hist. franc.* 7.22 (*MGH SRM* 1.304.9f.); for another 6th-cent. example, the Council of Arles (554): 'Ut oblatae, quae in sancto offeruntur altario . . .' (can. 1, CCSL 148A.171). J. C. Rolfe's translation only obtains 'holy water' by adding 'aquam' to the text, producing the bizzare 'oblatam aquam in aquam fluminis iactaverunt'.

outbreak of trouble occurred at Rome.[156] Mommsen, the most learned editor of this text, defended his emendation by pointing to the expression 'omnis populus Romanus' which follows shortly (anon. Vales. 82), but the defence is inadequate, for the word 'Romanus' is immediately qualified 'Ravennatis', and surely means 'people of Roman race living in Ravenna'.[157] The meaning of 'cena' is not as clear as one would like; perhaps 'Lord's Supper', which would preserve an association with the eucharist, but 'Passover' would be fitting in the Jewish context.[158] After the second burning of synagogues the Jews made their way to Verona where, with the help of the heretical *praepositus cubiculi* Triwanis, almost certainly to be identified with Boethius' enemy Trigguilla and Ennodius' friend Triggua, they made a charge against the Christians before the king.[159] Theoderic conveyed an order by means of Bishop Peter and Eutharic that the entire Roman populace of Ravenna was to pay for the rebuilding of the synagogues; those who lacked the means to contribute were to be led about in public naked.[160]

The reign of Theoderic was obviously a turbulent period for the Jews of Italy, although the language used by Cassiodorus in connection with the outbreak in Rome suggests that this particular disturbance can be located more profitably in the context of the civil disorders which were endemic in large cities during the period than in that of specifically anti-Jewish activity, and this may hold true for the incidents which took place in Ravenna as well.[161] But Theoderic

[156] The text has been amended to 'quod et in Roma in re eadem similiter contingit' (Mommsen, Cessi, Rolfe) and 'quod et Romae . . .' (Moreau/Velkov). Trouble in Rome is accepted by Stein, *Histoire*, 249; Ensslin, *Theoderick*, 302; Chadwick, *Boethius*, 58.

[157] Note too Agnellus, *Codex* 39: Theoderic conceded peace 'non solum Ravenenses cives [*sic*] sed eciam omnibus Romanis' (ed. Testi-Rasponi, p. 109). Note the comments of Morton, 'Marius of Avenches', 119f. n. 58, and S. J. B. Barnish, 'The *Anonymous Valesianus II* as a Source for the Last Years of Theoderic', *Latomus*, 42 (1983), 572–96 at 586.

[158] Lord's Supper: T. Hodgkin, *Italy and her Invaders*, 2nd edn. (Oxford, 1896), 269; in the same sense Saitta, '"Religionem imperare"', 79. Passover: G. B. Picotti, 'Osservazioni su alcuni punti della politica religiosa di Teoderico', in *I Goti in occidente problemi* (Settimane di studio, 3; Spoleto, 1956), 173–226 at 203f.

[159] 'insinuans regi factum adversus Christianos'; Rolfe's translation 'cajoled the king into taking action against the Christians' misses the legal sense of *insinuans* (cf. Anon. Vales. 85).

[160] On this, Morton, 'Marius of Avenches', 119f. n. 58.

[161] Cassiodorus, *var.* 4.43.1, *levitates*, cf. 1.30.4; 4.43.1, *seditiones*, cf. 1.20.2, 1.31.1, 1.32.1, 4, 1.44.2, 4; 4.43.2, *inflammata contentio*, cf. 1.31.4; 4.43.3, *romana gravitas*, cf. 1.44.4. Cf. perhaps, B. S. Bachrach, *Early Medieval Jewish Policy in Western Europe* (Minneapolis, 1977), 31f.

seems in general to have stood by the Jews, and it is scarcely surprising that when the forces of Belisarius besieged Naples in 536 the Jews were among the supporters of the Goths in the city.[162] We should not exclude the possibility that reasons of state may have led Theoderic, whose Ostrogoths formed such a small minority of the population in Italy, to have accorded his support to another minority,[163] although one wonders whether prudence would not have rather dictated supporting the Catholics in their controversies with the Jews; Ambrose of Milan, who took such a firm line with Theodosius I following the burning of the synagogue at Callinicum, would not have approved of Theoderic's treatment of the Jews of Rome and Ravenna. Doubtless application of the concept of tolerance to him would be anachronistic. But it is pleasant to think that the Jews enjoyed from Theoderic protection which may have owed at least something to sheer good-will.

WEALTH AND ROMANIZATION AMONG THE GOTHS

As we have seen, the Goths, despite holding political power in Italy, were strongly outnumbered by the Romans. So it was that they found themselves culturally giving ground: they were taking Roman names, coming to speak Latin, and beginning to convert to Catholicism. As the experience of the Visigoths in Spain was to show, the Gothic language and Arianism were doomed. It may well be that mixed marriages aided this process of assimilation, for the child of a Goth and a Roman would be likely to adopt the culturally dominant forms more readily than would the child of a Gothic couple. But Romanization cannot have proceeded at an equal pace everywhere among the Goths, and it will be worth our while exploring this issue.

It is clear that the period of their settlement in Italy was one of increasing social differentiation among the Ostrogoths.[164] Some

[162] Procopius, *BG* 1.8.41, 1.10.24–6.

[163] Stein, *Histoire*, 248. We may doubt whether any affinity which may have existed between Arianism and Judaism, or any economic motives, were important, despite Ruggini, *Economia e società*, 353 n. 419.

[164] Hartmann, *Geschichte*, 93. I agree with Burns, *Ostrogoths*, 174 that there was a 'widening gulf between the lower class and the Gothic nobility', although he reads rather a lot into Tjäder's *Papyri*, pap. 8 (T. S. Burns, *The Ostrogoths: kingship and society* (Wiesbaden, 1980), 109–12, substantially reproduced in *Ostrogoths*, 132–5. D. Claude, *Adel, Kirche und Königtum im Westgotenreich* (Sigmaringen, 1971) studies an aristocracy which is, alas, better documented than that of the Ostrogoths.

Goths were being forced to perform servile obligations during Theoderic's reign,[165] and while this may not have been a novelty others, who are needless to say more abundantly documented, were doing better for themselves. Personal attendance on the sovereign was one path to advancement: Theudis, Theoderic's *armiger*, was sent to govern Spain as guardian of Theoderic's grandson Amalaric. He subsequently married a wealthy woman, making himself virtually independent of Theoderic. He refused to act on the suggestion that he visit the king in Ravenna, and following the death of Amalaric in 531 he become king of the Visigoths;[166] in 540, following the agreement their king, Witigis, made with Belisarius, the Ostrogoths declared Hildibad, the nephew of Theudis, their new king.[167] Witigis had followed a similar path: a man of undistinguished family, the Goths elevated him from being Theodahad's *armiger* to the kingship following the deposition of Theodahad in 536.[168] In his *Dialogues* Pope Gregory the Great told the story of how Riggo, an *armiger* of King Totila, pretended to be the king in an attempt to deceive St Benedict. Needless to say the holy man was not taken in, but Gregory's tale at least metaphorically suggests how close the *armiger* could be to being king.[169]

Goths of noble family were able to increase their power by military leadership. Such a noble was Tuluin, who was not only astute enough to marry into Theoderic's family, the Amals, but who also served Theoderic in three campaigns, against the Gepids and then the Byzantines in 504–5, against the Franks and Burgundians in 508 and subsequently, and in territory the Burgundians had held in 523. Following the last campaign Theoderic made him the lord (*dominus*) of the lands he had acquired during it, and shortly after Theoderic's death, in a move which may reflect the weakness the government found itself in, he was appointed *patri-*

[165] Cassiodorus, *var.* 5.29f.; cf. 5.39.15.

[166] Jordanes, *get.* 302; Procopius, *BG* 1.12.51–4, 1.13.13; Isidore of Seville, *hist. goth.* 41. For the implication of personal attendance, note Sidonius Apollinaris' observation on the court of Theoderic the Visigoth: 'circumstitit sellam comes armiger' (*ep.* 1.2.4). On the offices of *armiger* and *spatarius*, Mommsen, *Gesammelte Schriften*, 6.454. L. Schmidt, *Die Ostgermanen* (Munich, 1934), 361, contrasted a *Geburtsadel* and a *Dienstadel* among the Ostrogoths, with reference to the *comites* and *primates* of Jordanes, *get.* 304; I am not sure this is helpful.

[167] Procopius, *BG* 2.30.14–17.

[168] Jordanes, *get.* 309; Procopius, *BG* 1.11.5.

[169] Gregory, *dial.* 2.14.

cius praesentalis and received membership of the senate.[170] Tuluin's estates would have been in Provence, and one wonders whether, like Theudis, he was building up a zone where personal power could be exercised far from Theoderic. More difficult to assess is another noble whose very name is in doubt. Theoderic's forces in the campaign of 504–5 were led by a man referred to in one source as Count Pitzia, one of the most noble of the Goths, and in another in two separate places as Count Pitzamus and Petza.[171] In 514 Theoderic came to Milan and put to death a man described as Count Petia, but in 523–6 Theoderic mentioned a judgement made by Count Pitzia.[172] Clearly there were two, and just possibly three, counts with similar names, but it is not at all clear whether the general of 504–5 was put to death in 514, in which case he would have fallen into the same category as Odoin, who had been killed by Theoderic in Rome some years earlier,[173] or the man referred to something like a decade later, who is possibly to be identified with the Goth Pitzas who defected to the Byzantines in 537.[174] It is impossible to be certain whether Theoderic found it necessary to dispose of the noble count in 514, or whether the count deserted his fellow Goths early in the Gothic war, or whether neither of these is true, but the data concerning this man can only confirm the suspicions created by Theudis and Tuluin that Theoderic's control over Goths distinguished in war was limited.

Another way in which Goths could accumulate power was by ownership of land. Even if Goths were not marrying Roman heiresses,[175] the *tertiae* they gained following the distribution conducted by Liberius would have given them sufficient income to enter

[170] Cassiodorus, *var.* 8.10 summarizes Tuluin's career; his lordship is obscurely indicated, 'quem et ille arbiter rerum largitione redituum iudicavit esse prosequendum, ut ibi fieret dominus possessionum, ubi utilitati publicae procuravit augmentum' (8.10.8). For his advancement after Theoderic's death, *var.* 8.9–11; his marriage to an Amal, *var.* 8.9.7, 8.10.1.

[171] Pitzia: Ennodius, *pan.* 62–8; MS variants Pizia, Pitziah, Picia (see Hartel's apparatus, 278.6). Pitzamus: Jordanes, *get.* 300, with Petzamin among MS variants. Petza: Jordanes, *get.* 301, with Pizza and Pitza among the variants.

[172] Count Petia: *auct. haun. s.a.* 514. Count Pitzia: Cassiodorus, *var.* 5.29.2; Pithia is the only variant.

[173] Anon. Vales. 68f., where the context implies a date of 500; *auct. haun. s.a.* 504.

[174] *Πίτζας*: Procopius, *BG* 1.15.1f. No variants are recorded in the critical edition of Haury and Wirth. The various difficulties have been seen by Krautschick, *Cassiodor*, 69f. n. 3, and Wolfram, *History*, 502f. n. 233, but are not properly brought out by *PLRE* 2.886f.

[175] On the 'antiqui barbari' of *var.* 5.13.6, see above, n. 82; but as we have seen, Gothic men were marrying Roman women.

the land market. There is a good deal of evidence bearing on one member of the land-owning class, Theoderic's nephew Theodahad. His rapacity is described in colourful terms by Procopius: Theodahad 'had gained possession of most of the lands of Tuscany, and he was eager by violent means to wrest the remainder from their owners. For to have a neighbour seemed to (Theodahad) a kind of misfortune.'[176] The implication that his estates were in Tuscany is borne out by some remarkable pieces of converging evidence: he built baths near Viterbo, he took over the property of a man known to have had interests in the area around Spoleto, he was summoned by Amalasuintha from Tuscany so that he could be associated with her in the government of the kingdom, probably in 534, but subsequently exiled her to Lake Bolsena where she was killed, he offered to hand Tuscany over to Justinian, and is referred to in a later Frankish source as 'king of Tuscany'.[177] It is all too easy to see Theodahad as a typical Roman owner of *latifundia*, even down to his intellectual interests, which are known to have extended to the reading of secular and ecclesiastical authors. One cannot help thinking that the latter group of authors was comprised of Catholics.[178] Such a man could easily attract scorn from other Goths (cf. Cassiodorus, *var*. 10.31.2). The only citation of Scripture in a letter written to a Goth by Cassiodorus on Theoderic's behalf occurs in a letter to Theodahad, and the text to which his attention was drawn was indeed relevant (*var*. 4.39.1, citing 1 Tim: 6.10).

What is known of Theodahad suggests the truth of at least the second half of Theoderic's famous dictum, 'The poor Roman imitates the Goth, the well-to-do Goth the Roman.'[179] Scholars have

[176] Procopius, *BG* 1.3.2, trans. Dewing, testimony fully confirmed by Cassiodorus, *var*. 4.39, 5.12, 10.5.

[177] Baths: Fiebiger/Schmidt, *Inschriftensammlung*, 2, no. 41. Property of Domitius: Cassiodorus, *var*. 4.39; his interests near Spoleto: *var*. 2.21. Amalasuintha: Jordanes, *get*. 306. (Note, however, that contrary to what is sometimes believed Theodahad did not have a fortified residence at Bolsena: Fo, 'L'Appendix Maximiani', 218f.) Offer to Justinian: Procopius, *BG* 1.3.4, 29, 1.4.17. *Rex Tusciae*: Gregory of Tours, *hist. franc*. 3.31. On the extent of 'Tuscany': R. Thomsen, *The Italic Regions from Augustus to the Lombard Invasion* (Copenhagen, 1947), 230f.

[178] Cassiodorus, *var*. 10.3.5; Procopius, *BG* 1.3.1, 1.6.10, 16.

[179] 'Romanus miser imitatur Gothum et utilis Gothus imitatur Romanum', Anon. Vales. 61. Scardigli's reading 'vilis' for 'utilis' (*Goten*, 140) is fanciful. For 'utilis' used in a similar sense in 6th-cent. Gaul: M. Bonnet, *Le Latin de Grégoire de Tours* (Paris, 1890), 288; J.-P. Bodmer, *Der Krieger der Merowingerzeit und seine Welt* (Zurich, 1957), 50f.; H. Grahn-Hoek, *Die fränkische Obersicht im 6. Jahrhundert* (Sigmaringen, 1976), 189–91. The nature of our sources would make it difficult in principle to

sought to see behind this pithy saying biblical or Roman sources, perhaps implausibly;[180] but even if such sources did lie behind Theoderic's words there is no reason to doubt that they described a social reality. The rich Goth Tuluin was admitted to the senate;[181] another rich Goth, Theodahad, wrote Latin verse and devoted himself to the intellectual pursuits characteristic of wealthy Romans, as indeed did other members of Theoderic's family; rich Goths were increasingly coming to be buried with Roman grave-goods;[182] and the Goths who could afford to make generous donations to Catholic churches were presumably moderately wealthy. One suspects that the poorer Goths would have had neither the resources nor the religious allegiance implied by such gifts. In short, at a time when the Goths were being subject to pressures of Romanization, the wealthy seem to have led the way.

THEODERIC'S CHARACTER

The wealthiest of the Goths was presumably Theoderic, and we may conclude this chapter by enquiring in more detail into what he was like. The sources are not nearly as helpful as one could wish, and it is salutary to recall that we do not even know whether he was literate. One of our sources reports that he was 'inlitteratus' and tells how in ten years he had not learned to sign his edicts. A golden stencil with four letters, perhaps making up the word 'legi' ('I have read it') was prepared, and when it was placed on a document Theoderic moved a pen through it (anon. Vales. 79; cf. 61 'Although he was *inlitteratus*'). But Anonymous Valesianus' story occurs immediately after a description of how the emperor Justin came to power, and as it happens Procopius, after providing an account of how Justin succeeded Anastasius, relates how the new emperor's officials pre-

demonstrate the truth of the first four words of Theoderic's dictum. Nevertheless, given the barbarian nature of the army in Ostrogothic Italy, one wonders whether the large-scale presence of slaves and *coloni* in the army of Totila in the 540s does not point in this direction. See on this Hartmann, *Geschichte*, 305 f., a discussion somewhat forced at points.

[180] Such sources are urged respectively by N. Tamassia, 'Sulla seconda parte dell'anonymo Valesiano', *Archivio storico italiano*, 71 (1913), 3–22, and by B. Pferschy, 'Das Probleme der Getreidpreise unter Theoderich', in *Siedlung Macht und Wirtschaft: Festschrift Fritz Posch* (Graz, 1981), 481–6 at 482.

[181] Cassiodorus, *var.* 8.10f.

[182] Bierbrauer, 'Frühgeschichtliche Akkulturationprozesse'.

pared a stencil with the Latin word 'legi', 'I have read it' cut into it for his use, for the emperor was illiterate (*Anec.* 6.14–16 with 11). It seems highly unlikely that both Theoderic and Justin would have made use of such an implement, and it has been argued that a story originally told of Justin was incorrectly applied to Theoderic; further, given that letters sent by Theoderic to bishops meeting in Rome in 502, presumably prepared by the royal chancery, conclude with requests for prayers 'alia manu' ('in another hand') it has been suggested that Theoderic could write as well as read.[183] Some support for the story of Anonymous Valesianus may be derived from the words of some Goths which occur in Procopius, according to which Theoderic had 'not so much heard of letters', and some scholars have not hesitated to affirm that Theoderic was, indeed, illiterate.[184] But as we have seen, the sentiments Procopius attributes to Goths are suspect (above, Ch. 1), and this evidence simply shows that Procopius felt the claim would have been a plausible one for a faction of Goths to have made. The question of Theoderic's literacy is not one which can be resolved, although it seems likely that he could both read and write;[185] the learning possessed by his near relatives Amalasuintha, Theodahad, and Amalaberga may incline us to believe that such education as Theoderic received in Constantinople endowed him with the ability to read and write, possibly in Latin as well as Greek. When his grandson Athalaric was about 8 his mother Amalasuintha was sending him to a grammarian; Theoderic had been sent to Constantinople at the same age, and he may be presumed to have had the same experience.[186]

[183] *MGH AA* 12.420.12f., 422.13f.; cf. 424.25. The case for Theoderic's being able to read and write has been argued by W. Ensslin, 'Rex Theodericus inlitteratus?', *Historisches Jahrbuch*, 60 (1940), 391–6. But see already R. Cessi, 'Theodericus inlitteratus', in *Miscellanea di studi critici in onore di Vincenzo Crescini* (Cividale, 1927), 221–6, and cf. H. Steinacker, 'Die römische Kirche und die griechischen Sprachkenntnisse des Frühmittelalters', *Mitteilungen des Instituts für Österreichische Geschichtsforschung*, 62 (1954), 28–66 at 38 n. 47. Among the Visigoths, the proceedings of the third council of Toledo (589) were signed by Reccared 'mea dextra' and his queen Baddo 'mea manu' (*Concilios*, ed. Vives, 116, cf. 'manu sua' of Reccared, 120).

[184] Procopius, *BG* 1.2.16. See, arguing for illiteracy, Stein, *Histoire*, 719f., although he holds that Theoderic ultimately learned to write, and H. Grundmann, 'Litteratus-illitteratus', *Archiv für Kulturgeschichte*, 40 (1958), 1–65 at 24–30.

[185] See H. Helbling, *Goten und Wandalen* (Zurich, 1954), 31 with n. 221. But B. Baldwin, 'Illiterate Emperors', *Historia*, 38 (1989), 124–6, is sensible.

[186] Theoderic in Constantinople: Ennodius, *pan.* 11, 'Educavit te in gremio civilitatis Graecia praesaga venturi'; John Malalas, *chron.* 383. Athalaric: Procopius, *BG* 1.2.6.

But, literate or not, Theoderic was popular among the Romans. The comparisons they made between him and the emperors Trajan and Valentinian imply recognition of his military successes, and also a positive evaluation of these, but perhaps more than this.[187] One reason for his popularity must have been his genuine respect for Roman ways and intellectual curiosity concerning classical civilization; a man who could be described in a letter to the senate as seeming to be a philosopher wearing the purple and could be recalled in another letter to this body as having been outstanding for his *sapientia* must have been impressive.[188] We are told that Italians as well as Goths had great love of him.[189] Anonymous Valesianus comments that Theoderic was a man of such *sapientia* that some of his utterances were still proverbial at the time he wrote. Besides that already quoted regarding Goths and Romans, another, 'Someone with gold and a demon cannot hide the demon', may be seen as obscurely reflecting the same truth.[190] Stories were told to illustrate his wisdom. Anonymous Valesianus himself tells of a widow who, under pressure from her intended second husband, denied her son. When the son appealed to Theoderic the king ruled that the woman could take no man for her husband other than the one she refused to acknowledge as her son, whereupon she broke down and told the truth; 'there are many other things he did', concludes the narrator.[191] Another was described by Theodore Lector, when he told the story of the Catholic deacon who converted to Arianism in the hope of advancement. Theoderic commented that if he had not been loyal to God his future loyalty to him was not guaranteed, and the deacon was beheaded.[192] John Malalas knew another story,

[187] For the 6th-cent. attitude to Trajan as a military figure, see Procopius, *build.* 4.6.11 and John Lydus, *On powers*, 2.28. But Claudian represented Theodosius I as suggesting to his son Honorius that Trajan would be a good exemplar, not so much because of his military successes as because of his mildness: *pan. de quarto consolatu Honorii*, 315–20 (*MGH AA* 10.162).

[188] 'quidam purpuratus videretur esse philosophus': Cassiodorus, *var.* 9.24.8. 'sapientia': *var.* 11.1.19. His court may have been a centre for geographical study: F. Staab, 'Ostrogothic Geographers at the Court of Theoderic the Great: A Study of Some Sources of the Anonymous Cosmographer of Ravenna', *Viator*, 7 (1976), 27–64, esp. 54–8.

[189] Procopius, *BG* 1.1.29.

[190] Anon. Vales. 61.

[191] Anon. Vales. 62, perhaps to be read in conjunction with Suetonius, *Claudius*, 15.2, although the judgement of Solomon is a possible source: 1 Kings (= 3 Kings) 3: 16–18. But the concluding words, 'sunt eius et multa alia', invite comparison with John 21: 25 (Vulgate: 'sunt autem et alia multa quae fecit Iesus').

[192] *Kirchengeschichte*, 131.16ff., cf. Theophanes, *chron.* AM 5991.

according to which the widow Juvenalia, who had been involved for thirty years in a lawsuit with the patrician Firmus, asked Theoderic to speed matters up. The king summoned the lawyers and ordered them to finalize the case in two days. When this had been done he asked them why a case capable of being settled so quickly had dragged on for so long, and had them beheaded.[193] The three tales are all in sources of the sixth century, and while we may note uneasily that two of them involve beheadings, they are ample testimony to the esteem in which the Romans held Theoderic. Apparently disinterested sources compare him with the best of the emperors,[194] and a ruler who treated the Church and the Jews in the way Theoderic did could easily have inspired love.

It is more difficult to establish what the Goths made of their king. The period of barbarian settlements saw a rise in the power of the rulers of the various peoples, but it was claimed, almost certainly spuriously, that Theoderic belonged to a royal family which extended back for sixteen generations. Just as the Franks had the Merovingians, the Visigoths the Balts, and the Vandals the Hasdings, so the Ostrogoths, we are told, had the Amals.[195] The ancient Goths are said to have called their leaders kings, priests, and semi-gods; one of their kings was worshipped among the gods when he died.[196] The Amals made their appearance at a later stage. Jordanes, claiming to be using Gothic stories which were still told in his day, reported that the first of the line was Gapt, a name which in other contexts may have been applied to the god Odin; four generations later came Amal, from whom the line took its name.[197] The coming of Christianity must have changed the way the Goths looked

[193] John Malalas, *chron.* 384. 7th-cent. sources shorten the period of the case to three years: *chron. pasc.*, ed. Dindorf, 604f., John of Nikiu, 88.52ff. Note as well the comment of the last-named: Theoderic 'possessed the respect of the magistrates and senate' (88.51).

[194] Anon. Vales. 60; Procopius, *BG* 1.1.29.

[195] See in particular Wolfram, *History*, 29–32 and appendix 3; H. Moisl, 'Anglo-Saxon Royal Genealogies and Germanic Oral Tradition', *Journal of Medieval History*, 7 (1981), 215–48, at 219–22. The key sources are Cassiodorus, *var.* 11.1.19 and Jordanes, *get.* See now the important study of P. Heather, 'Cassiodorus and the Rise of the Amals: Genealogy and the Goths under Hun Domination', *Journal of Roman Studies*, 79 (1989), 103–28.

[196] Jordanes, *get.* 40, 48, 71, 78.

[197] Jordanes, *get.* 81. On the name Gapt: O. Holfer, 'Theoderich der Grosse und sein Bild in der Sage', *Anzeiger der Österreichischen Akademie der Wissenschaften, phil.-hist. Klasse*, 111 (1974), 349–72 at 354f. and, in a wider context, id., 'Der Sakralcharacter des germanischen Königtums', *Vorträge und Forschungen*, 3 (1954), 75–104 at 78. Wolfram, *History*, 110f., is cautious.

at their leaders, but in the sixth century members of the royal family took Holy Communion from a special chalice, a practice apparently unknown among the Franks, and it is just possible that a description of Theoderic as a *princeps* with respect to his strength, watchfulness, and success but as a priest with respect to his mildness reflects a pattern of thought which is Gothic as well as Roman.[198] When Cassiodorus thought of the Amals the quality which came first to mind was their brightness.[199] The precise weight of this in the sixth century is difficult to establish, but something of the mystique of barbarian kingship is suggested by the importance of long hair for the Merovingian Franks, most graphically illustrated by the decision of Clovis' distraught widow Chlotilde, confronted with the choice of her sons' having their hair cut or being killed, to opt for the latter; by the circumstance that the word Hasdingi may be cognate with the Gothic for 'long women's hair'; and by the only representation of Theoderic which may be accurate, that on the Senigalla medallion, in which he is depicted with long hair.[200] Further, something of the specific importance which could be held to attach to membership of the Amali at the time of Theoderic is suggested by the names of his sister Amalafrida, his daughter Amalasuintha, his niece Amalaberga, and his grandson Amalaric. This last name is particularly interesting, for whereas the second component of his name must derive from the name of his father Alaric, the first has nothing in common with his mother's name, Theodegotha, but reflects the family of which she was a member. Less obviously, another of Theoderic's daughters, Ostrogotho, may have been named after Ostrogotha, the purported grandson of Amal himself. We may also note Jordanes' concern to claim, falsely, that Theoderic's early rival Theoderic Strabo was not of the Amal line whereas Eutharic, the husband Theoderic brought from Spain for Amalasuintha, was, and Cassiodorus' comment that Tuluin, by

[198] Special chalice: Gregory of Tours, *hist. franc.* 3.31. *Princeps* and *sacerdos*: Ennodius, *pan.* 80, but cf. e.g. Leo, *ep.* 115.1, 'regia potentia et sacerdotalis ... industria' (*ACO* 2.4.67.17).

[199] *Claritas*: Cassiodorus, *var.* 4.1.1, 8.2.3, 9.25.4, 10.3.3; cf. Ennodius, *ep.* 4.10.1, 'haec claritas dominorum inserta natalibus'.

[200] Chlotilde: Gregory of Tours, *hist. franc.* 3.18. Hasdingi: Schönfeld, *Wörterbuch*, 129, with which cf. another story in Agathias, *hist.* 1.3.3–5, with Holfer, 'Sakralcharacter', 83. The attempt of S. Fuchs, *Kunst der Ostgotenzeit* (Berlin, 1944), 61 ff., to demonstrate that a mosaic purportedly depicting Justinian in the church of S Apollinare Nuovo in fact represents Theoderic is not convincing; cf. Ensslin, *Theoderich*, 258.

means of a marriage otherwise unrecorded, had been joined to the Amal family.[201]

Yet caution is called for, for it would be easy to overestimate the importance of family, at least in the case of Theoderic. Ennodius' panegyric displays little interest in his lineage. True, Theoderic's father had seemed to stand before his eyes prior to the battle of Verona. But Ennodius was concerned to show that although Theoderic was by origin a lord, it was his *virtus* which declared this; his rule came because of the splendour of his *genus*, but if such distinction had been lacking he would have been elected *princeps* because of his mind.[202] More significantly, in the 235 letters Cassiodorus wrote on Theoderic's behalf as *quaestor* and *magister officiorum* there are only three references to the Amal line: king Herminifrid the Thoringian is told that, although he himself is of royal stock, because of his marriage to Amalaberga he will shine all the more with the brightness of Amal blood (*var.* 4.1.1); Theoderic's wayward nephew Theodahad is reminded that he is of Amal blood (*var.* 4.39.2), and Thrasamund is told that his wife Amalafrida is the singular glory of the Amal family (*var.* 5.43.1). It comes as a surprise to discover there is no direct reference to Theoderic himself being an Amal. Yet, following his death, one of the arguments which Cassiodorus could urge in support of his weak successors Athalaric and Theodahad was their membership of the Amal family. Oddly enough these references to the Amal line tend to occur in letters to Romans rather than Goths, so that, for example, we find Cassiodorus helpfully explaining to the members of the senate that anyone who proceeded from the Amals was thereby proved to be most worthy of a kingdom, just as a senatorial origin could be an attribute of their own children.[203] This is not to suggest that membership of the family was not important to Theoderic, for the Gothic History, in which Cassiodorus seems to have had much to say about the Amals, was written at his behest,[204]

[201] Jordanes, *get.* 270 (Theoderic Strabo, but see Wolfram, *History*, 32); Cassiodorus, *var.* 8.9.7 (Tuluin).

[202] Ennodius, *pan.* 43, 88. Counterparts in imperial panegyric are discussed by R. Ficarra, 'Fonti letterarie e motivi topici nel panegirico a Teodorico di magno Felice Ennodio', in *Scritti in onore di Salvatore Pugliatti*, 5 (Milan, 1978), 233–54 at 239.

[203] *Var.* 8.2.3 (the passage referred to), 8.5.2, 10.3.3, 11.13.4. The reference to Tuluin's marrying into the Amal line (*var.* 8.9.7) similarly occurs after the death of Theoderic.

[204] *Var.* 9.25.4f., although this dates from after Theoderic's death; *Ordo generis Cassiodororum*, ed. Mommsen, 21f.

and in Jordanes' *Getica*, a work which proclaims itself to have been based on Cassiodorus' history, although it is not clear when Cassiodorus finished working the text of the history available to Jordanes, the Amals play a key role.[205] But weak kings need more ideological support than strong ones, and as the fate of Theodahad, replaced as king by a man 'not of a conspicuous house'[206] was to make clear, it need not have been enough to compensate for incompetence; for the Goths Theoderic's success in leading them from the poverty of a wandering life in the Balkans to the wealth of Italy was presumably much more important than his membership of a family to which none of our sources written in his lifetime directly refer. Still less than the Romans, the Goths had no reason to find Theoderic's rule objectionable. Doubtless some Goths were hostile to their king, for just as the beginnings of the reign of Odovacer were marked by the murder of Count Brachila in 477 and of the noble Adaric, together with his mother and brother, in 478, so Theoderic found it necessary to dispose of Odoin, perhaps in 504, and Count 'Petia' in 514.[207] But from the time he led his people from Moesia in 488 until his death Theoderic is not known to have lost a battle, and it is hard to imagine a more satisfactory situation for a barbarian people than having been led by their king from hunger in the Balkans to the mastery of Italy. One may conclude that this, rather than his belonging to the Amal line, was important in the attitude of the Ostrogoths to their sovereign.

BY WAY OF SUMMARY

Relations between the peoples who lived in the Italy of Theoderic were complex and are difficult to summarize. Doubtless the Romans had reason to dislike the Goths. The Goths, a people of *virtus*, did not always trouble themselves to behave *civiliter*, and no matter what compact Theoderic came to with Constantinople the fact re-

[205] Jordanes describes his work as dealing with 'the origin of the Goths and the nobility of the Amals and the deeds of brave heroes': *get*. 314f. On the relationship between Cassiodorus and Jordanes, A. Momigliano, 'Cassiodorus and Italian Culture of his Time', *Proceedings of the British Academy*, 41 (1955), 207–45 is learned and thought-provoking, if not ultimately persuasive.

[206] Procopius, *BG* 1.11.5.

[207] Brachila: *PLRE* 2.241. Adaric: ibid. 7. Odoin: ibid. 791 (although the date is questionable). 'Petia': *auct. haun. s.a.* 514.

mained that Italy was still what it had been in the time of Odovacer, a land under the control of barbarians. In the Gothic war unleashed by Justinian, most of its inhabitants supported the armies of the Byzantines.[208] But by the time of Theoderic the people who lived in Italy were used to barbarians in positions of authority, and power always attracts people to itself. As has already been hinted at in our discussion, some Romans were pleased to come to an accommodation with Theoderic's administration, and as we shall see the king was adept at flattery and judicious deployment of the considerable means of patronage at his disposal. The care Theoderic generally took to avoid interfering in the affairs of the Church must also have been attractive, as must the peace his regime provided, after the disasters of the fifth century. We may conclude that, by and large, most Romans were content to live with the Goths, without wishing to be like them. Apart from the members of one family who learned Gothic and Ennodius' acquaintance who grew a beard there is no evidence for Romans adopting Gothic ways. And the circumstance that, relative to the Romans, the Goths constituted a tiny minority of the population of Italy, a minority, furthermore, restricted to a part of Italy, must have helped make their rule tolerable, for many Romans must have been able to go about their daily lives rarely encountering Goths.

The situation of the Goths was different. A military élite with political power, they nevertheless found themselves a tiny minority of the population of Italy. The Arab conquerors of the seventh century in Syria, Egypt, and Africa were in a broadly similar position; ultimately the bulk of the native peoples of these areas came to be Arabic-speaking Muslims. But the experience of the Visigoths in Spain was to show that the tide there flowed in the opposite direction, and such evidence as we have suggests that in Theoderic's reign Ostrogoths were beginning to take Roman names and speak Latin, just as some were becoming Catholics and so, in religious terms, *Romani*, while those who remained Arians conducted baptisms at Ravenna under a cupola with a mosaic which imitated that of the nearby Catholic baptistery. Of course, such tendencies were only beginning in the time of Theoderic, for in terms of exposure to Roman ways the Ostrogoths were at a much earlier stage than their Visigoth and Vandal contemporaries. While the Os-

[208] J. Moorhead, 'Italian Loyalties during Justinian's Gothic War', *Byzantion*, 53 (1983), 575–96; Thompson, *Romans and Barbarians*, 92–104.

trogoths were still keeping company with the Huns a generation of Visigoths was growing up who had never left Gaul, and the Ostrogoths' entry into Italy took place sixty years after that of the Vandals into Africa. With regard to the process of Romanization to which all the barbarian peoples were subjected, both before and after their settlement in once Roman lands, the Ostrogoths were a young people. But just as Romans responded in different ways to Gothic political power, so did the Goths respond differently to the challenge presented by Roman civilization. At a time of increasing social differentiation, the wealthy Goths took most quickly to Roman ways. Doubtless there was the danger of a backlash, and in the weak rule of Theoderic's daughter Amalasuintha tensions among the Goths were to come out in the open.[209] But in general, wealth and Romanization went together, a conjunction perhaps most clearly to be seen in members of Theoderic's family.

What of Theoderic himself? He seems to have been content to keep the two peoples apart and allow them to play complementary roles. Goths were to fight and Romans continue with their traditional activities; the differing excellences of *virtus* and *prudentia* were to be maintained; the two peoples would be judged according to different systems; Ranilda suffered for converting to Catholicism. At one level this would have been perfectly straightforward: as we have seen the Goths tended to live in particular parts of Italy, and those who lived in Ravenna and Rome apparently had their own quarters.[210] Put like this Theoderic's system may seem to have been so straightforward as to have been almost inevitable, but Theoderic's own comment on how the poor Roman imitated the Goth and the rich Goth the Roman may be taken as implying an awareness of difficulties even in the first generation of Gothic settlement in Italy. Theoderic's policy was very different to that Jordanes

[209] Procopius, *BG* 1.2.

[210] At Ravenna, the *civitas barbarica* was outside the walls of the town (B. M. Felletti-Maj, 'Una carta di Ravenna romana e bizantina', *Rendiconti della pontifica accademia romana di archeologia*, 41 (1968–9), 85–120 at 115). In Rome the Arian church of S Agatha in Subura, already in use in the time of Ricimer, was located in an area of ill-fame and colourful nightlife (*Lib. pont.* 312.10f., with Duchesne's n. 8; Gregory of Rome, *dial.* 3.30, *reg.* 4.19), while another is described as 'iuxta domum Merulanum', i.e. near the Lateran (Gregory of Rome, *reg.* 3.19). The location of Arian churches in Rome and Ravenna is discussed by C. Cecchelli, 'L'Arianesimo e le chiese Ariane d'Italia', in *Le chiese nei regni dell'Europa occidentale e i loro rapporti con Roma sino all'800* (Settimane di studio, 3; Spoleto, 1960), 742–74 at 765–8.

attributed to the Visigoths in Italy at the beginning of the century, and had more in common with the second plan which Orosius believed Athaulf, a Visigothic king of the early fifth century, had entertained. Initially Athaulf had planned to obliterate the Roman name and make the Roman empire an empire of the Goths so that, to speak in vulgar language, Gothia would be what Romania had been, and he would be what Caesar Augustus had been. But he learned from experience that because of their barbarism the Goths would not obey the laws. Athaulf felt he could hardly abrogate the laws of the state (*res publica*), without which the state would not be the state. So he sought glory for himself by restoring it in its fullness and augmenting the Roman name by the strength (*vires*) of the Goths.[211] Theoderic's rule was of such a kind as to attract the support of some Romans well equipped to voice their support of it, and a reference in the *De anima*, which Cassiodorus wrote during the early stages of the Gothic war, to the Devil looking spitefully on two great peoples may reflect his unhappy awareness that the policies of the king had ultimately failed.[212] But as we have seen Theoderic's Italy was not free of tensions, and we shall have to consider later the extent to which failure was inevitable.

[211] Orosius, *adv. pag.* 7.43. On Athaulf and Theoderic, see further Ensslin, *Theoderich*, 220.

[212] 'Invidit, pro dolor, tam magnis populis, cum duo essent': Cassiodorus, *de anim.* 18 (ed. CCSL 96.574.10f.). For discussion, see Halporn's comments in the introduction to his edn., 505f., together with L. Schmidt, 'Theoderich, römischer Patricius und König der Goten', *Zeitschrift für Schweizerische Geschichte*, 19 (1939), 404–14 at 411 (in the context of an argument I cannot accept, but to query it is not to endorse the *Tendenzliteratur* against which Schmidt argued) and H. Löwe, *Von Cassiodor zu Dante* (Berlin, 1973), 24.

4

Schism in Rome

THE LAURENTIAN SCHISM

Theoderic's finding in favour of Pope Symmachus and his subsequent visit to Rome must have seemed to have resolved the schism which had broken out on the election of that Pope in 498. But the issues which underlay the opposition to Symmachus refused to go away, and before long the vehemence so often characteristic of Church politics had combined with the urban unrest which was common in late antiquity to produce in Rome a situation on which Theoderic looked with extreme distaste. Traditionally, it had been the task of emperors to deal with ecclesiastical controversies, but Theoderic was hampered from intervening as forcefully as he may have wished by his not being a Catholic, and the schism persisted for years. Our task in the following pages will be twofold: to provide a narrative of the schism, with particular attention to the role of Theoderic, and to attempt to answer some questions suggested by the events, those of the general nature of the support enjoyed by the rival claimants to the see of Rome, the supporters of either side it may be possible to name, the issues which generated the schism, and whether Theoderic inclined to support one side or the other.

In 501, or perhaps 502, trouble broke out again in the Roman Church.[1] Various charges against Symmachus were laid before Theoderic, and the Pope was summoned to Ravenna. The accusations seem to have fallen under three heads.[2] Firstly, Symmachus was accused of not having celebrated Easter 'cum universitate'. The phrase is awkward, but 501 was a year for which different systems of computation yielded different dates for Easter, the old Western cycle of 84 years yielding 25 March and both the compilation of

[1] 'Post annos vero IIII' (*lib. pont.* 260.10), but it is not clear whether the author counts from Symmachus' accession (Nov. 498) or the synod of Mar. 499, nor whether he counts inclusively or exclusively; cf. the more vague 'post aliquod autem annos' of the *frag. laur.* (*lib. pont.* 44). In any case, Duchesne's 'au commencement de l'année 501' (*lib. pont.* 264 n. 7) is hard to sustain.

[2] They are known from *frag. laur.*, ed. *lib. pont.* 44.

Victurius and the Alexandrian tables yielding 22 April, and is probably to be taken to mean that the pope had celebrated the festival at the earlier date, in opposition to the *universitas*, presumably that of the city of Rome, which implies that there must have been something close to a state of schism in Rome by Easter 501.[3] Secondly, Symmachus was accused of having sinned with various women. His enemies felt that he was something of a bon vivant, comparable to the biblical Esau who sold his birthright for a mess of potage. He was alleged to have been followed by crowds of worldly women, and rumours about his relationship with one of them, who was commonly known as 'Spicy', were to persist.[4] Finally, the pope's enemies among the Roman clergy claimed that, contrary to a decree of his predecessors, he had squandered the wealth of the Church and so bound himself with anathema; the decree they had in mind was presumably that issued not by a pope but by the praetorian prefect Basilius in 483, which will require our attention later. It was a mixed bag of accusations. For some reason Theoderic ordered Symmachus, *en route* to Ravenna, to remain for a while at Rimini, whither he had presumably made his way along the Via Flaminia, and strolling along the beach there one morning he caught sight of the women with whom he had been accused of illicit relations making their way to Theoderic's court. That night, while all slept, Symmachus left for Rome with one follower, and on arrival he enclosed himself in St Peter's. The priests, deacons, and other clergy who had been travelling with him shamefacedly proceeded to Ravenna to tell Theoderic that Symmachus had slipped away without their knowing it, whereupon Theoderic, according to our source, which is hostile to Symmachus, issued a condemnation of him.[5]

As the Easter of 502 approached the enemies of Symmachus struck again and, doubtless with the events of the preceding Easter in mind, asked Theoderic to appoint a visitor who would officiate at the liturgical ceremonies appropriate to the festival. The king

[3] The possible dates for Easter in 501 are discussed by B. Krusch, 'Die Einführing des griechischen Paschalritus in Abendlande', *Neues Archiv*, 9 (1883), 100–69 at 104–6.

[4] Compared to Esau: Ennodius, *lib. pro syn.* 29. Worldly women: ibid. 65. The woman called Conditaria: *frag. laur.* 46; I borrow the translation 'Spicy' from Chadwick, *Boethius*, 32. Criticism of Symmachus may lie behind Ennodius' 'Praeceptum quando iussi sunt omnes episcopi cellulanos habere' (*opusc.* 7).

[5] *Frag. laur.* 44.

obliged by appointing Bishop Peter of Altinum. There was an excellent precedent for this, as the emperor Honorius had sent the bishop of Spoleto to Rome to celebrate Easter in 419 during the controversy between Boniface and Eulalius over occupancy of the see. Pope Gelasius had recently reasserted the traditional position of the Church that baptisms were to be confined, if possible, to Easter and Pentecost, and there can be no doubt that the celebration of Easter was one of the high points in the civic year of the towns and cities of late antiquity, perhaps of Rome in particular, for Paulinus of Nola had made a practice of visiting the city every Easter to venerate the apostles and martyrs.[6] But the appointment of Peter is particularly difficult to interpret. The degree of support the move to have a visitor appointed enjoyed is not as clear as we would like it to be. The *Fragmentum Laurentianum*, an account of Symmachus' pontificate written by one hostile to the pope, asserts that the appointment was sought 'by all', while the account provided in the *Liber pontificalis*, written from the opposite perspective, simply mentions the senators Festus and Probinus as having sought it. Another account was to attribute the demand for the appointment to 'a faction of the clergy and some laity',[7] and is probably to be preferred, in which case both our narrative sources for the pontificate of Symmachus seem to have been guilty of distortion. Furthermore, the year in which Peter received his commission to go to Rome is not certain, and it could be argued that the Easter for which Symmachus' enemies sought the appointment of a visitor was not that of 502 but the one they proposed to observe on 22 April 501, following Symmachus' earlier celebration. However, this does not seem likely. After Easter had passed Theoderic ordered that a synod be held in Rome to give judgement concerning Symmachus.[8]

Theoderic's command produced tension in the Church. Some felt that, in principle, charges against a Roman pontiff could not be heard, no matter how grave the accusations; others believed that the king's order should be obeyed. On their way to Rome the

[6] Appointment of Peter: *frag. laur.* 44f.; *lib. pont.* 260.13–15. Appointment of visitor in 419: *coll. avel.* 21 ff., CSEL 35.68 ff. Gelasius' decree: *ep.* 10 (ed. Thiel); cf. Cassiodorus, *var.* 8.33.7. Paulinus of Nola: Augustine, *ep.* 94.1.

[7] 'ab omnibus': *frag. laur.* 44. Festus and Probinus: *lib. pont.* 260.13. 'a parte cleri vel aliquibus laicis': *MGH AA* 12.427.18f.

[8] *Frag. laur.* 45, claiming that Theoderic acted in accordance with the wishes of the senate and clergy; *MGH AA* 12.420.19–23, 425.3–8, 426.9–11. For the year see Stein, *Histoire*, 793f.

bishops of Liguria, Aemilia, and Venetia intimated to Theoderic that, because of the power which pertained to the see of Peter, Symmachus should have convoked the synod, but Theoderic was able to assure them that Symmachus had written to him agreeing that a synod be held.[9] Indeed the Pope thanked Theoderic for summoning the synod, and subject to only two conditions was prepared to appear before it: his demands were the removal of Peter of Altinum and the return of all he had lost through the accusations of his enemies. It is easy to understand the difficulty Peter's presence in Rome caused Symmachus, for it implied a situation of *sede vacante*, particularly as Peter did not confine himself to celebrating Easter but also performed ordinations.[10] Most of the bishops who had come together considered that Symmachus' conditions were reasonable, but felt unable to act without Theoderic's consent. But this was not forthcoming, the king decreeing that before Church property was restored to Symmachus he would have to answer the charges which had been made against him.[11] The bishops, some of whom were already drifting away from Rome, must have found it hard to see their way ahead, and proposed that the council be transferred to Ravenna, but Theoderic, in a *Praeceptio* issued on 8 August, turned down their suggestion.[12] On 27 August Theoderic was in touch again, pressing the bishops to find Symmachus innocent or guilty, and announcing the dispatch of his household officers Gudila, Bedeulf, and Arigern to expedite the presence of Symmachus at their deliberations, a step dictated by the disturbed conditions which were prevailing in Rome.[13] Five days later the bishops came together in synod.

The meeting of the bishops on 1 September took place in the

[9] *MGH AA* 12.426f.; see more generally *frag. laur.* 45; Ennodius, *lib. pro syn.* 19. It is curious that the bishops of Liguria and Aemilia made their way to Rome by way of Theoderic's court.

[10] *MGH AA* 12.423.1–3, 427f.; see too Ennodius, *lib. pro syn.* 48 on Symmachus' preparedness to accept the authority of the council. Peter's ordinations: *MGH AA* 12.427.19f.

[11] Ibid. 423.4, 428.6–8.

[12] Ibid. 419f. The document, sent via the bishops Germanus and Carosus, was addressed to the bishops Laurentius (of Milan), Marcellianus (of Aquileia), and Peter (of Ravenna) and all the bishops in Rome, and concludes with the mysterious words 'Reḡ relt senatus vel Marcellini epsc cum ceteris'. Pfeilschifter interprets it as indicating a report the supporters of Laurentius, namely the senate and Bishop Marcellianus, made to Theoderic; but see E. Caspar, *Geschichte des Papsttums*, 2 (Tübingen, 1933), 95 n. 4.

[13] *MGH AA* 12.420–2.

basilica of Sta Croce in Ierusalem. The church had been a part of the imperial patrimony and was therefore at Theoderic's disposal;[14] the holding of a synod there is thus yet another sign of the extent to which the bishops were executing Theoderic's will. On this occasion there was tabled a *libellus* prepared by the accusers of Symmachus, which made two points against him: his crimes had come to the notice of the king, and he could be convicted on the evidence of his slaves. The bishops did not find these points impressive.[15] It is not entirely clear what happened next, but the journey from St Peter's to Sta Croce would have taken Symmachus from one side of Rome to the other, and it seems that, on his way across the city to appear before the council, in the company of crowds of followers, the pope was attacked. Priests were killed and Theoderic's officers Arigern, Gudila, and Bedeulf were witnesses of Symmachus' wounds.[16] The *Liber pontificalis* describes a situation in which blood was flowing daily in the streets of Rome, and it may be that these turbulent conditions, most forcibly expressed in the attack on his own person, as much as failure to regain control of the property of the Church, led Symmachus to write to the bishops: 'Henceforth I will not submit myself to your examination. It is in the power of God and the lord king, what he decides to ordain concerning me.'[17]

The bishops, compelled to judge someone who refused to be judged, found themselves in an impossible situation. They wrote to Theoderic explaining that it was beyond their power to summon Symmachus against his will or pronounce in his absence, and asked him to be considerate of their infirmities and weakness. Doubtless we are dealing with a measure of rhetoric here, for this is not the only occasion when the physical weakness of the bishops was mentioned, but even the youthful have been known to find the Roman summer trying, and as autumn began the attractions of the city must

[14] C. Pietri, *Roma Christiana: Recherches sur l'église de Rome, son organisation, sa politique, son idéologie de Miltiade à Sixte III (311–440)* (Rome, 1976), 14 with n. 2, 92.

[15] *MGH AA* 12.428.10–21, well discussed by Caspar, *Geschichte*, 96f. The *Edictum Theodorici* provides for the torture of slaves who give evidence against their masters (cap. 101).

[16] *MGH AA* 12.423.4f., 429.1–6; Ennodius, *lib. pro syn.* 60ff.; Symmachus, *ep.* 12.7 (ed. Thiel; this letter to the emperor Anastasius mentions stoning); *lib. pont.* 261.4–6, where the priests Dignissimus and Gordianus are mentioned as havig een killed, although not specifically on this occasion. Gordianus was present at the synod of 499 (*MGH AA* 12.411 no. 4, cf. 401 no. 4) but not that of 6 Nov. 502, and so was probably dead by the latter date. Dignissimus attended neither synod.

[17] *MGH AA* 12.423.5f. *Lib. pont.* 261.1– dscribes blood in the streets of Rome.

have begun to pall. The bishops argued that, as Symmachus refused to co-operate, there was no more they could do, and asked Theoderic's permission to return to their respective churches.[18] Theoderic's reply, written on 1 October, was blunt: the bishops were to complete the task they had undertaken. Doubtless he could have come to a just decision by himself, but he did not consider it his task to give judgement in matters related to the Church. He forwarded to the bishops a communication which argued on the basis of the Gospel and the apostle Paul for their coming to a decision, and pointed out that if they failed to do this they would be setting an example of bad conduct to priests.[19] More than ever the bishops must have felt trapped between the Devil and the deep blue sea, and on 23 October they finally came to a decision which allowed them to go home. At the *Synodus palmaris*[20] the bishops confessed how difficult it was to pass judgement and how desirable mercy was; to these considerations, which one would have thought were of a fairly general nature, they added the authority pertaining to the see formerly occupied by Peter. So it was that the bishops decreed that the judgement of Symmachus was to be left entirely to God. The property of the Church of Rome was to be restored to the pope, and all were urged to receive communion from him and keep the peace. Clergy who had been in schism from Symmachus were to be restored to their offices when they had given satisfaction, but any who persisted in saying Mass without his knowledge were to be driven forth from the Church as schismatics in accordance with the canons. Seventy-five bishops subscribed to the proceedings of the synod.[21]

By declining to pass judgement the bishops had, in effect, decided for Symmachus, and this was made more clear in another synod held a fortnight later. Some of the bishops had already left Rome,

[18] *MGH AA* 12.422f. With the 'infirmitates et debilitas' of 423.23, cf. ibid. 426.13f. and Ennodius, *lib. pro syn.* 12.

[19] *MGH AA* 12.424–6.

[20] Mommsen was reluctant to see any geographical content in this term, suggesting that the synod was called 'palmaris' because it finally came to a decision (*MGH AA* 12.417f.). The theory that the bishops met in the porticus ad Palmaria in St Peter's (proposed by A. Lumpe, 'Die konziliengeschichtliche Bedeutung des Ennodius', *Annuarium historiae conciliorum*, 1 (1969), 15–36 at 20) may be rejected; we have no reason to believe that they joined Symmachus.

[21] *MGH AA* 12.426–32 (meeting of 23 Oct.), 432–7 (list of bishops). The list of subscriptions runs to 76, but Bishop Hilary of Temesa signed twice (nos. 15, 43). It is hard to recognize this synod in the coloured and inaccurate account of *lib. pont.* 260.15–19.

for which one can hardly blame them, but fifty-two were still on hand at the later synod, and their number was augmented by the arrival of new bishops. But in other respects the synod of 6 November recalled that of 499, for it was held in St Peter's basilica with Symmachus presiding. As we have seen, Symmachus' enemies had accused him of squandering the wealth of the Church, contrary to a decree of his predecessors. Against what this wording of the accusation, known only from a source hostile to Symmachus, would lead one to believe, the decree had in fact been issued by a lay person, the praetorian prefect Basilius, acting on behalf of (*agens etiam vices*) Odovacer in 483. Professing to be following the instructions of Pope Simplicius, Basilius had insisted on the need to consult with the laity in papal elections and forbidden the alienation of land or goods given to the Church by the laity. The decree was read to the synod by the deacon Hormisdas, who was ultimately to succeed Symmachus as pope, just as the deacon Anastasius, who read material before the Roman council of 495, was probably the man who succeeded Gelasius in 496. But Hormisdas' reading was punctuated by the comments of bishops irate that lay people, however religious, had sought to legislate on Church affairs. It is clear that the enemies of Symmachus had attacked him on this score, and the pope proceeded to enact detailed legislation concerning the uses to which church property could be put.[22]

This, however, was not to be the end of the story. One of the most important bishops among those who gathered in Rome in 502 was Marcellianus of Aquileia. Theoderic's *Praeceptio* to the bishops of 8 August had been addressed to him, together with the bishops of Milan and Ravenna, and it can be argued from some curious words at the foot of the *Praeceptio* that Marcellianus was an enemy of Symmachus.[23] But the list of the bishops who put their names to the synod of 23 October, while beginning with the bishops of Milan and Ravenna, does not contain the name of Marcellianus; nor was he present at the synod of 6 November, at which bishop Eulalius of Syracuse, who does not seem to have been in Rome earlier in the year, suddenly emerged to play a leading role, together with the bishops of Milan and Ravenna.[24] It may not be coincidental that

[22] *MGH AA* 12.444–51; the attacks on Symmachus are revealed by 448.15–20. For Anastasius in 495, *coll. avel.* 103.

[23] Ibid. 419f. with n. 12 above.

[24] Signatory list of 23 Oct.: ibid. 432. Eulalius on 6 Nov.: ibid. 447f. together with signatory list, p. 451.

Peter of Altinum was one of the suffragans of Marcellianus, and when the bishop of Aquileia died a few years later the supporters of Symmachus exerted themselves to see that he was replaced by a sound man.[25] But the opponents of the decision reached on 23 October felt able to claim that Theoderic had not summoned all the bishops to the council and that, more importantly, not all the bishops had agreed with its decision.[26] Further, Symmachus had failed to carry much of the Roman Church with him, for the synod of 6 November was signed by only thirty-six of his priests and four deacons, whereas the synod of 499 had been attended by seventy-four priests and the full college of seven deacons, while sixty-seven priests and six deacons signed its proceedings.[27] The synod of 23 October had manifestly failed to pronounce on the charges made against Symmachus, and while its reluctance found expression in language which is important in the development of notions of papal power,[28] Theoderic, who had been so insistent in seeking a decision as to the guilt or innocence of Symmachus, can scarcely have looked on it with approval.

The pope's enemies lost little time. Before long the supporters of Laurentius had petitioned Theoderic to allow him to return from Ravenna to Rome, and Theoderic acceded to their request. It is not clear for how long Laurentius had been in Ravenna, nor what he had been doing there, but his presence near Theoderic's court can have brought little joy to Symmachus.[29] A pamphlet, 'Against the Synod of the Incongruous Absolution' (that is, the synod of 23 October) was quickly circulated, and in 503 Ennodius, a deacon in the church of Symmachus' supporter Bishop Laurentius of Milan, wrote a refutation.[30] Ennodius had no doubt that his opponents were 'slaves of hell, obviously servants of Satan', and measured even by the standards of ecclesiastical polemic his language was frequently intemperate. Against the arguments put forward by such people Ennodius propounded a high doctrine of papal

[25] Ennodius, *ep.* 4.29 (note the strong language), 4.31, 5.1 (from which it is apparent that Marcellianus' successor was, confusingly, Marcellinus).

[26] Ennodius, *lib. pro syn.* 9.

[27] *MGH AA* 12.441–3 (6 Nov.), 401 f. and 410–15 (499).

[28] *MGH AA* 12.430.

[29] *Frag. laur.* 45. According to *lib. pont.* 260.12 f., Laurentius had been summoned back to Rome by stealth before the appointment of Peter of Altinum; if this is true, Laurentius presumably left Rome during 502.

[30] Title of the pamphlet known from Ennodius, *lib. pro syn.* 7. Date of Ennodius' response: Sundwall, *Abhandlungen*, 15.

power.[31] His *libellus* concludes with orations by Peter, Paul, and finally Roma. The patron saints of the city were obviously figures of its unity, and the address of Roma was clearly aimed at the noble families of the city: rejoicing at their progress from paganism to Christianity and current importance, she urges them to put away discord and enmities.[32]

The battle to influence public opinion was also waged in a remarkable series of forged documents produced in the interests of Symmachus.[33] The *Synodi sinuessanae gesta* purport to describe a synod which was held after Pope Marcellinus was witnessed offering incense to the gods during the reign of Diocletian. Marcellinus finally admitted his guilt, but was saved by the invocation of the principle 'the first see will not be judged by anyone'.[34] The principle would obviously have been useful to Symmachus, and a similar formulation occurs in the forged *Constitutio Silvestri*.[35] In the *Gesta Liberii* we learn how Pope Liberius was forced out of Rome by the heretical emperor Constantius but was able to conduct the Easter baptisms at the Ostrian cemetery where St Peter himself had baptized and those of Pentecost at the basilica of the apostle, that is, St Peter's. Points of contact with the situation of Symmachus are clear: Peter of Altinum had been brought to Rome to celebrate Easter, which he would presumably have done in the Lateran, leaving Symmachus to conduct such baptisms as he could elsewhere, quite possibly at St Peter's, for it was there that he took residence following his flight from Rimini; and any suspicions that the heretical emperor was meant to suggest Theoderic can only be strengthened by the anti-Arian flourish at the end of the work.[36] But allusions to the questions raised by the Laurentian schism are

[31] Strong language: 'mancipia tartari et liquido satanae ministri', *lib. pro syn.* 9; cf. 'sibilantium ... venena linguarum' (4), 'dementissimi hominum' (7), 'nescio verbis an latratibus' (29), 'lymphatici more ... insanissimi' (80), etc. The theme of papal power is discussed by Caspar, *Geschichte*, 103f.

[32] Ennodius' tract has been frequently misinterpreted, as by J. Haller, *Das Papsttum: Idee und Wirklichkeit*, I (Basle, 1951), 238 (it was intended for the ear of Theoderic) and Richards, *Popes and Papacy*, 74 (the synod of 6 Nov. began with a reading of this document). We lack a good detailed study of Ennodius.

[33] Best discussed by Caspar, *Geschichte*, 107–10.

[34] 'prima sedes non iudicabitur a quoquam': PL 6.11–20; the words quoted are at 20A.

[35] PL 8.829–40, c. 20: 'nemo enim dijudicet primam sedam'. Contemporary relevance may also be detected in the provision that 72 witnesses were needed when charges were made against bishops (c. 3).

[36] PL 8.1387–93.

clearer in the *Gesta de Xysti purgatione*, according to which Pope Xystus was troubled by the issue of control over property and the testimony given by a slave that he had sinned with a nun. The parallels with the accusations made against Symmachus are evident, for two of these accusations were of squandering property and misdemeanours with women; the similarities become closer if we accept that the evidence of slaves his enemies proposed to bring against him on 1 September concerned sexual sin. The *Gesta* go on to affirm that Valentinian called a synod which met in Rome in the 'basilica Heleniana quod dicitur Sessorium', but that the ex-consul Maximus observed that judgement could not be given against a pontiff.[37] When we recall that the bishops had been meeting in this church which had formed part of the imperial patrimony, and that Valentinian is one of the emperors the Romans felt Theoderic resembled (Anon. Vales. 60), the correspondence between the *Gesta* and the events of 502 becomes uncannily close. Indeed, some modern commentators have tried to assign names from the time of Theoderic to various actors in the drama of Xystus. Such attempts are not entirely convincing,[38] but the parallels are clear, and they become still more clear in the *Gesta Polychronii* appended to the tale of Xystus. We are told that Polychronius, bishop of Jerusalem, was found guilty of simony by a party dispatched from Rome. A vicar was imposed and Polychronius, suspended from his see, lived outside the city. Some months later famine broke out in Jerusalem, and such was the generosity of Polychronius that Pope Xystus reinstated him. One recalls that Symmachus seems to have lived outside Rome when a visitor was appointed, and as we shall see the pope was famed for his generosity towards the poor; but, as do other forged documents in this series, the *Gesta Polychronii* make their way towards the affirmation of a principle which here takes the form 'it is not right for anyone to accuse his bishop (*pontifex*), since a judge may not be judged.'[39] If this were true of the bishop of Jerusalem, *a fortiori* it would hold for the bishop of Rome. Taken together, these forgeries offer remarkable testimony to the liveliness of the debate between the supporters and opponents of Symmachus, and confirm the nature of the issues which were the subjects of controversy.

[37] PLS 3.1249–52.
[38] Pietri, 'Aristocratie et société clericale', 457f. is properly cautious.
[39] PLS 3.1252–5; the quoted words appear at 1254.

There followed a period of some years for which we have hardly any evidence, but as we shall shortly see, an analysis of the correspondence of Symmachus' ally the deacon Ennodius suggests that there were quite clear divisions in Rome between the supporters of Symmachus and those of Laurentius, and the years before the settlement of the schism were marked by a good deal of fighting in the streets of Rome. Symmachus is known to have made major improvements and extensions to St Peter's basilica,[40] and it is certainly possible that they were carried out in these years when that basilica constituted his headquarters and many if not most of the other churches of Rome were outside his control. It would be reasonable to believe that both Symmachus and Laurentius maintained contact with Theoderic's court at Ravenna, although the means by which this was done are obscure.[41] But finally, some four years after the return of Laurentius to Rome, Theoderic acted decisively, in a manner which won the admiration of Ennodius: by one short letter, Theoderic achieved something which earlier princes had scarcely been able to bring about with much effort.[42] In accordance with a petition of Symmachus made through the deacon Dioscorus he ordered that the churches of Rome be handed over to him. Laurentius was compelled to withdraw to the estates of his patron Festus, and spent the rest of his life in abstinence.[43] The location of the estates of Festus is unknown, but one possibility is of interest. On one occasion Festus was instructed by Theoderic to keep an eye on the *domus* of the patrician Agnellus, who was to spend some time in Africa. We know from a letter written in 527 that at that time Agnellus possessed a *domus* in the castrum Lucullanum, almost certainly to be identified with the place to which Odovacer had sent the emperor Romulus Augustulus.[44] Later we shall see that this may have been significant.

One would like to be able to assign a precise date to Theoderic's *fiat* ending the schism. If it occurred about (*circiter*) four years after

[40] *Lib. pont.* 261.11 to 262.8.

[41] Tjäder, *Papyri*, 2.255 n. 4 suggests that one Rusticus, 'acolytus sanctae ecclesiae catholicae Romanae', who received some land near Ravenna in 504, was *apocrisarius* of the Roman Church. But this seems clutching at straws.

[42] Ennodius, *ep*. 9.30.6.

[43] *Frag. laur.* 46. Dioscorus is otherwise known to have been a confidant of Symmachus; cf. Ennodius, *ep*. 6.33 (addressed to Hormisdas and Dioscorus), and references in *ep*. 7.28.4 (where Hormisdas is again mentioned) and 9.16.1.

[44] Cassiodorus, *var*. 1.15 (*tuitio* of Agnellus' *domus*), 8.25.3 (*domus* in castrum Lucullanum).

Laurentius returned to Rome, and if we date the return to not long after the synod of 23 October 502, this would allow Theoderic's decree to be dated to late 506 or early 507, which is consonant with both Theoderic's surviving *praeceptum* to the senate ordering the return of property to Symmachus, which seems to have been issued 11 March 507,[45] and the submission made to Symmachus, perhaps with an eye to the way the wind was blowing, by a certain John the deacon on 18 September 506.[46] The panegyric of Ennodius in honour of Theoderic makes no mention of the end of the schism, but clearly reflects support for the king in circles around Symmachus, and is probably to be dated to early 507; immediately afterwards Ennodius commenced his correspondence with Boethius who, we shall suggest, may be presumed to have been a supporter of Laurentius.[47] It is possible that Theoderic's decision in favour of Symmachus was prompted by war with the Byzantines in 505–6 and Byzantine involvement in the West at about the time of Clovis' victory over the Visigoths in 507,[48] but we are very badly informed about events in Italy from 502 to 507 and it is at least as likely that Theoderic was influenced by domestic considerations of which we are ignorant. It is possible that a Catholic emperor would have intervened at an earlier stage, but one wonders whether even this would have been effective, for the schism acquired a powerful momentum at an early date, and Theoderic's command that the churches be handed over to Symmachus was not enough to reconcile his most determined opponents. We know of a Roman deacon, Paschasius, almost certainly raised to the diaconate by Laurentius,[49] who remained in schism from Symmachus till the end of his

[45] *MGH AA* 12.392, dated 11 Mar. in the consulship of Venantius. The Western consulship was held by Venantii in 507 and 508; Mommsen's dating of the *praeceptum* to 'a. 507 (vel 508)', p. 392 *incip*. ignores the evidence from other sources in favour of the earlier year (see below, Appendix 1). Laurentius held the Roman Church 'per annos circiter quattuor', according to *frag. laur*. 45.

[46] Symmachus, *ep*. 8 (ed. Thiel).

[47] Date of panegyric: Sundwall, *Abhandlungen*, 42–4. Correspondence with Boethius: *ep*. 6.6, dated by Sundwall to mid-507 (*Abhandlungen*, 44); it seems from this letter that Boethius had already written to Ennodius. This author's *ep*. 9.30 'In Christi signo' is obscure, but seems to relate to affairs in Gaul (Sundwall, *Abhandlungen*, 69), rather than the ending of the schism.

[48] So Sundwall, *Abhandlungen*, 212; Stein, *Histoire*, 138f.; Chadwick, *Boethius*, 37.

[49] As M. Büdinger saw, we have no reason to believe there was a Roman deacon named Paschasius in 499, and Paschasius may therefore have been made deacon by Laurentius: 'Eugipius [*sic*]: Eine Untersuchung', *Sitzungsberichte der kaiserlichen Akademie der Wissenschaften* (*Vienna*) (Phil.-hist. Klasse, 91; 1878), 793–814 at 810.

life, which was certainly subsequent to the occasion when the abbot Eugippius submitted his *Vita Severini* to him for his approval in about 511.[50] Trouble which may have represented a direct continuation of that which broke out in 502 was still being experienced in the streets of Rome in 509 (below, Ch. 7), and only the death of Symmachus in 514 brought the schism to an end.[51] However, that the Laurentian fragment was written subsequent to the death of Symmachus reveals that feelings continued to run high.

THE *PLEBS* AND THE SCHISM

Unfortunately the schism was more than an occasion for pamphlet warfare and intrigue at Ravenna. The *Liber pontificalis* paints a grim picture of fighting in Rome between the supporters of Laurentius, led by the senators Festus and Probinus, and those of Symmachus, led by Faustus *niger*. It claims that clergy in communion with Symmachus were put to the sword, that nuns and virgins were taken from their convents and clubbed, and that fighting occurred in the middle of the city on a daily basis. Many priests were killed, although the only ones named are Dignissimus and Gordianus of the churches of St Peter ad Vincula and SS John and Paul; it was dangerous for the clergy to walk in the city by day or night.[52] On one occasion Pope Gelasius had been confronted with a situation in which two successive bishops had been murdered in Squillace, which furnishes a useful reminder of the violent turn which ecclesiastical politics could take.[53] The situation in Rome did not deteriorate to this level, but it was bad enough to cause Theoderic concern. Writing on 27 August 502 to the bishops assembled in Rome, the king stated that the lack of tranquillity in Rome at a time when, thanks to God, all things were peaceful, was intolerable and something his

[50] Gregory the Great, *dial.* 4.42.1 f. records his persistence in schism. Eugippius wrote to Paschasius some two years after the consulship of Inportunus (509): see his letter to Paschasius prefacing the *Vita Severini*, 1 (*MGH AA* 1.2.1.4). On Eugippius and Paschasius, consult F. Lotter, *Severinus von Noricum: Legende und historische Wirklichkeit* (Stuttgart, 1976), 21 ff. I. Bóna, 'Severiana', *Acta antiqua Academiae scientiarum Hungaricae*, 21 (1973), 281–338, seems to me unsound on Paschasius (at 285–7).

[51] *Frag. laur.* 46; cf. Cassiodorus, *chron. s.a.* 514, and the epitaph of Symmachus' successor Hormisdas, which records the healing of a schism (*lib. pont.* 274 n. 25 with 272 n. 4).

[52] *Lib. pont.* 260.19 to 261.7.

[53] Gelasius, *ep.* 37 (Thiel), placing the see under the control of two other bishops.

love for the royal city would not allow; he implored the bishops to give judgement so that no disorder nor any discord would remain in the city.[54] Theoderic's strong language may have been prompted by an awareness that urban riots could easily take an anti-government turn.[55] The *Fragmentum Laurentianum* uses the strong expression *bella civilia* to describe the tumults in Rome,[56] and as it happens Marcellinus *comes* uses the same expression to describe riots which broke out in Constantinople, during which statues of the emperor and empress were bound with ropes and dragged through the city (*s.a.* 493). The chronicle of Marcellinus is punctuated with references to an uprising in the theatre which saw more than 3,000 killed (*s.a.* 501), *seditio popularis* in the circus (*s.a.* 507), and an attempt by the people of Constantinople to depose the emperor after a minor change had been made to the liturgy (*s.a.* 512), while Antioch saw major riots in 494–5 and Ravenna, as we have seen, was troubled by violence between Christians and Jews in about 519.[57] The popular discontent which was endemic in the cities of late antiquity could easily turn into demonstrations against the government, as some of these Eastern examples show; the Nika riots of 532, which nearly brought about the fall of Justinian, is the most famous case. In 468 Sidonius Apollinaris, the prefect of the city, had professed that he was afraid of hearing in the theatre the cry 'the hunger of the Roman people'.[58] When Cassiodorus held office he found himself 'harassed with fear lest the cities should lack their supplies of food—food which the common people insist upon more than anything else, caring more for their bellies than for the gratification of their ears by eloquence', and it is clear that the supply of food to the people of Rome was a major concern for Theoderic. We may presume it was dictated by prudence at least as much as benevolence.[59]

[54] *MGH AA* 12.422.2–5 and 421.24; cf. 424.21f. and 419.5–8, 420.7f., 421.9, 14–16.

[55] Note that the *Edictum Theodorici* prescribes that one responsible for *seditio* among the people or in the army is to be burned (c. 107), the penalty apparently a new one.

[56] *Frag. laur.* 46.

[57] On Antioch, John Malalas, *chron.* 392f. See in general Stein, *Histoire*, 81f.

[58] Nika riots: Procopius, *BP* 1.24, with e.g. Stein, *Histoire*, 449–56. 'fames populi Romani': Sidonius Apollinaris, *ep.* 1.10.2.

[59] Cassiodorus is quoted from Hodgkin's rather free translation of *var. praef.* 5. But see too the formula 'praefecturae annonae' (*var.* 6.18, esp. 'plebs quam industria tua satiat', 2, 'nescit plebs tacere', 7) and *var.* 11.2.1 for 'fames popularis' as a concern of the state. Theoderic's generosity to the poor is indicated by Anon.

But the government was not the only organization concerned with the feeding of the poor. Rome, in common with all cities in late antiquity, witnessed the steadily rising power of its bishop, and the Church was committed to charity. As early as the middle of the third century the Roman Church was supporting over 1,500 widows and poor people; at the end of the fourth century Paulinus of Nola was able to congratulate Pammachius, the son-in-law of St Paula, for having fed 'the poor, the patrons of our souls . . . the religious throngs of the miserable plebs' at a banquet in St Peter's basilica, and a century later Pope Gelasius was 'a lover of the poor. He freed the city of Rome from the danger of famine.'[60] In his tract against the Lupercalia, Gelasius enquired as to why the Castores had not intervened in Rome at a time of want during the winter; writing to the *inlustris femina* Firmina he sought the restoration of property to the see of Peter so the needy could receive sustenance.[61] Such generosity was also a characteristic of Symmachus. We have already noted the largesse which Bishop Polychronius of Jerusalem, clearly meant to represent Symmachus, was alleged to have bestowed on his people in time of famine, and to this indirect evidence may be added the clear testimony of the *Liber pontificalis*.[62] It is difficult to see in what perspective Symmachus' benefactions should be located: he may simply have been following in the footsteps of Gelasius, he may have been following traditions associated with St Peter's basilica, which constituted his headquarters in Rome for some years, or his generosity towards the poor may have been nothing more than an attempt to win support in Rome while his stocks were low. But, however it is to be interpreted, Symmachus did enjoy the support of the Roman *plebs*, 'that multitude of the common people devoted to God' who, as the bishops assembled on

Vales. 67 and Procopius, *anec.* 26.29. See in general M. Lecce, 'La vita economica dell'Italia durante la dominazione dei Goti nelle "Variae" di Cassiodoro', *Economia e storia*, 3 (1956), 354–408 at 367f., and, on the threat of famine in Rome, Chastagnol, *Préfecture urbaine*, 296.

[60] Mid-3rd cent.: Eusebius, *HE* 6.43. Pammachius: Paulinus, *ep.* 13.11 (cf. Ammianus Marcellinus, 27.3.6 and Procopius, *anec.* 26.29 on beggars at the Vatican). Gelasius: *lib. pont.* 255.5f. On the general background, Jones, *Later Roman Empire*, 904f.

[61] Castores: *coll. avel.* 100.18. Firmina: Gelasius, *frag.* 35f., cf. 28 (ed. Thiel).

[62] *Lib. pont.* 263.2f. on his building of shelters for the poor by the basilicas of SS Peter, Paul, and Laurence. Note Duchesne's restitution of the first edition: 'Hic amavit clerum et pauperes, bonus, prudens, humanus, gratiosus' (p. 97); he gave clothing and *alimonia* to the poor (99).

23 October 502 observed, nearly all remained inseparably in his communion.[63] Again, evidence from the Symmachan forgeries corresponds with reality, for according to the *Gesta de Xysti purgatione* Pope Xystus enjoyed widespread popular support in the city.[64] The *Liber pontificalis* observed that after Peter of Altinum and Laurentius had been condemned, 'all the bishops, priests and deacons, all the *plebs* and clergy' sided with Symmachus.[65] The estimate offered of Symmachus' support is indeed generous, but the senate is conspicuous by its absence, and whenever senators are mentioned in the account of Symmachus provided by this source at least some of them are characterized as supporters of Laurentius.[66] There is nothing to be surprised at in senatorial involvement in papal affairs, for as we have seen Felix III came to the papal throne in 483 under the auspices of Basilius; what is of interest is that the senate seems to have been so one-sided. When one of the Symmachan forgeries claimed that one of the words for 'senate' was derived from a word for 'blood' (*curia a cruore dicitur*) it expressed the hostility of Symmachus' supporters for that body.[67]

SENATORS AND THE SCHISM

We therefore seem to be confronted with a situation in which, by and large, the Roman *plebs* supported Symmachus, and the senators Laurentius. It will be worth our while to investigate whether we can be more specific about individual senators. The *Liber pontificalis* states that Festus and Probinus were Laurentius' chief supporters, while the principal patron of Symmachus was Faustus *niger*.[68] They were three distinguished men. Festus, whose support for Laurentius is also known from his role in his protegé's election after his return from Constantinople in 498 and the circumstance

[63] Ennodius, *lib. pro syn.* 61; *MGH AA* 12.431.7f.

[64] 'multus populus Romanus, omnis plebs urbana' (3); cf. 'pauperorum senecta sublevare desiderat' (2), PLS 3.1250.

[65] *Lib. pont.* 260.17f.

[66] Ibid. 260.4, 10, 13, 20; so too *frag. laur.*, in particular 'clerus ergo et senatus electior qui consortium vitaverat Symmachi', 45.

[67] *Const. Silv.*, c. 16. The popular nature of Symmachus' support is well brought out by C. Pietri, 'Le Sénat, le peuple chrétien et les partis du cirque à Rome sous le pape Symmaque (498–514)', *Mélanges d'archéologie et d'histoire*, 78 (1966), 123–39 at 129f.

[68] *Lib. pont.* 260.10, 13, 19f., 261.7f.

that Laurentius finally retired from Rome to his estates, was 'head of the senate', that is, the most senior of the living consuls.[69] He had held the consulship as long ago as 472, and as no further consul was appointed in the West until Odovacer came to power, when he became 'head of the senate' he must have been the last surviving consul to have been appointed during the days of the empire in the West. He had also conducted embassies to Constantinople on Theoderic's behalf, journeying there in 490 and again in 497, on the latter occasion making peace with Anastasius concerning Theoderic's taking of the kingdom and securing the return to Italy of the palace ornaments Odovacer had sent to Constantinople.

Probinus was a younger man. Consul in 489, he was the son of Rufius Achilius Maecius Placidus (consul 481) and the father of Blesilla and Rufius Petronius Nicomachus Cethegus (consul 504).[70] Inscriptions on the Flavian amphitheatre dating from the time of Odovacer record that the neighbours of Festus when he attended the games were Rufius Achilius Sividius and Rufius Valerius Messala;[71] the name of the first in particular is suggestive of kinship with the father of Probinus. The names of Probinus' family also put one in mind of Jerome's friend Paula, who claimed descent from the Gracchi and the Scipiones of the republican period,[72] for both her mother and one of her daughters were named Blesilla.[73] Jerome is also our source for the information that Paula enjoyed a connection of some kind with the *stirps* of Furia,[74] and we know of a fourth-century senator, Cethegus, the father of Furius Maechius Gracchus.[75] In terms of familial status, therefore, we may presume that Probinus ranked very highly, enjoying descent from lofty Christian and, at least by repute, pagan forebears.

Faustus, for his part, was consul in 490. His father, Gennadius Avienus, had been consul in 450, and was one of the group of lay people who accompanied Pope Leo when he went forth to treat with Attila in 452.[76] When Sidonius Apollinaris visited Rome in 467–8 he

[69] He is described as 'caput senati' by Anon. Vales. 53 and *lib. pont.* 260.19. Cassiodorus pays him the pleasant compliment of saying that he deserved to be 'senatus prior' (*var.* 1.15.1).

[70] I should like to take this opportunity to record my debt to *PLRE* for data of this kind.

[71] Chastagnol, *Sénat romain*, 74, 80f.

[72] Jerome, *ep.* 108. Ennodius mentions Scipio as among those condemned to hell because he did not know Christ: *lib. pro syn.* 131.

[73] *PLRE* 1.674f.

[74] Jerome, *ep.* 54.2.

[75] *PLRE* 1.199f.

[76] Prosper Tiro, *chron. sa* 452.

found in the city two men who were easily its leading men, excluding the emperor and military men: Avienus, whom he described as having come to the consulship because of *felicitas*, and a Decius, Basilius, who owed his consulship to *virtus*. Each man was to be seen surrounded by clients. Sidonius, who had come to the city on business, needed a *patronus*. Observing that Avienus saw to the interests of his family, while Basilius was more interested in other people, Sidonius attached himself to the latter, who was responsible for his subsequent appointment as *Praefectus urbis Romae*.[77] Faustus' father, then, had been in his day one of the two leading senators of the city. Faustus had been entrusted by Theoderic with an embassy to Constantinople, and his sons Rufus Magnus Faustus Avienus and Ennodius Messala were to hold the consulship, in 502 and 506 respectively. In short, the affairs of the schism engaged the impassioned attention of some of the most distinguished of the senators.

Passing beyond these three men, it becomes difficult to establish the positions taken by other leading Romans. Symmachus dedicated a church erected in honour of St Martin by one Palatinus, but it is not clear whether this took place during or after the schism;[78] so too with the church built by the praetorian prefect Albinus and his wife Glaphyra,[79] although it is interesting that Ennodius found Albinus a bad correspondent during the years of the schism.[80] Attempts made by modern scholars to establish the position of Cassiodorus have reached contradictory conclusions, but given that his family lived in Brutium and that he was in Ravenna during most of the schism he need not have taken a pos-

[77] Sidonius Apollinaris, *ep*. 1.9.2–6.

[78] *Frag. laur*. 46. I cannot begin to explain why this is the only piece of building work recorded in the *Fragmentum*, and it is tempting to assign its dedication to Laurentius (so Pietri, 'Aristocratie et société cléricale', 246 no. 41), but the 'hic' who is described as dedicating the church seems from the context to have been Symmachus. A date of 499 is suggested by W. Buchowiecki, *Handbuch der Kirchen Roms*, 3 (Vienna, 1974), 882; cf. R. Krautheimer *et al.*, *Corpus Basilicarum Christianarum Romae*, 3 (Rome, 1967), 89.

[79] *Lib. pont*. 263.1f. This is our only source for Albinus' holding the office of praetorian prefect, dated to ?500/3 by *PLRE* 2.51f. and 512/14 by Sundwall, *Abhandlungen*, 88 and Chastagnol, *Sénat romain*, 83f. The *Gesta de purgatione Xysti* lists an Albinus as having been among the accusers of Xystus (PLS 3.1251), but on the status of this list I accept T. D. Barnes, '*Patricii* under Valentinian III', *Phoenix*, 29 (1975), 155–70 at 163f.

[80] Ennodius, *ep*. 2.21 reveals that Albinus had failed to answer four letters; cf. *ep*. 2.22. Only in 508 was he the joint addressee of *ep*. 6.12.

ition.[81] Liberius, to judge from the trust Ennodius placed in him when he sought to ensure that the successor of Marcellianus in the see of Aquileia was a sound man, was one of the adherents of Symmachus.[82] Our most tantalizing piece of evidence is a letter from Bishop Avitus of Vienne to Faustus (presumably Faustus *albus*, holder at about that time of the sensitive office of prefect of the city of Rome, who is to be distinguished from Symmachus' supporter Faustus *niger*) and Aurelius Memmius Symmachus, supporting the decision made in favour of Pope Symmachus by the bishops meeting in Rome, from the tone of which it appears that Avitus did not consider these senators to have been supporters of Symmachus and may have feared that they were not.[83] As Ennodius, a great supporter of Pope Symmachus, only wrote to Faustus *albus* and the senator Symmachus subsequent to Theoderic's decision against Laurentius, we may suspect that both these men were among the supporters of Laurentius.[84] In the same way, Ennodius' only known letter to Probinus (*ep.* 9.4) was written in 511, while Festus is not known to have figured among his correspondents at all. It has been widely assumed that Boethius, who was close to the senator Symmachus and either was or was to become his son-in-law, was an adherent of Pope Symmachus, apparently on no more evidence than an interpretation of the letter of Bishop Avitus to Faustus *albus* and Symmachus as being to supporters of the pope.[85] Once this piece of evidence is dislodged

[81] Contrasting assessments are offered by Caspar, *Geschichte*, 114 n. 7 and P. A. B. Llewellyn, *Rome in the Dark Ages* (London, 1971), 40f., although the latter really has in mind the father of Cassiodorus.

[82] Ennodius, *ep.* 5.1 and perhaps 2.26; cf. 9.23.

[83] Avitus of Vienne, *ep.* 34. Interpreting the Faustus as Faustus *albus* is plausible as his consular date (483) was prior to that of Symmachus (485), whereas that of Faustus *niger* (490) was subsequent. The letter is well interpreted by Chadwick, *Boethius*, 9 and n. 7 ('an essay in persuasion, not congratulation'), against e.g. Caspar, *Geschichte*, 104, Wes, *Das Ende*, 100. I am not persuaded by the comments of G. Zecchini, 'I "Gesta de Xysti purgatione" e le fazioni aristocratiche a Roma alla metà del V secolo', *Rivista di storia della chiesa in Italia*, 34 (1984), 60–74.

[84] Ennodius, *ep.* 6.34 to Faustus *albus*, dated by Sundwall to 508 (*Abhandlungen*, 47; note the reference to 'longum silentium') and 7.25 to Symmachus, dated to 509 by Sundwall (*Abhandlungen*, 55f.).

[85] e.g. E. Demougeot, 'La Carrière politique de Boèce', in Obertello, ed., *Congresso internazionale*, 97–108 at 100; V. Recchia, 'San Benedetto e la politica religiosa', *Romanobarbarica*, 7 (1982–3), 201–52 at 210, in the context of a generally unpersuasive argument. V. Schurr goes so far as to see in the 'summorum virorum cura' (Boethius, *phil. cons.* 2.3.5) a reference to Faustus *niger* as well as the senator Symmachus: *Die Trinitätslehre des Boethius im Lichte der 'Skythischen Kontroversen'* (Paderborn, 1935), 113–15; Bertolini, *Roma*, 73, on the basis of no discernible evidence, sees the senator Symmachus as a member of a moderate party.

there are no grounds for numbering Boethius among the supporters of Pope Symmachus, and the fact that, as was also the case with the two addressees of the letter of Avitus, Ennodius only entered into correspondence with Boethius after Theoderic's *fiat* forcing Laurentius to withdraw from Rome should raise suspicions that his sympathies lay with Laurentius.[86]

Another circumstance makes it highly likely that Boethius was a supporter of Laurentius. The adherents of Laurentius among the Roman clergy included the deacon John, whose submission to the pope was made on 18 September 506. Some years later there arrived in Rome a proposal from some Eastern bishops for restoring peace between the Churches of Rome and Constantinople. Pope Symmachus rejected their proposal, but Boethius felt it was meritorious, and he expressed his feelings in a tractate, *Contra Eutychen et Nestorium*, addressed to his 'holy lord and venerable father John the deacon'.[87] Again, at a still later time of controversy on other issues, Boethius wrote two further tractates addressed to John.[88] Later we will have to consider further evidence which suggests very close links between Boethius and John (below, Ch. 6); for the time being we may simply note that, as the Roman Church only had seven deacons at one time,[89] it is likely that Boethius' friend the deacon John had thrown in his lot with Laurentius. As the senator Symmachus seems not to have supported his namesake, and as Boethius' friend John was, presumably for years, an adherent of Laurentius, we may accept what the corpus of the correspondence of Ennodius implies, that Boethius supported Laurentius as well. We know that most of the senatorial class were Laurentians,[90] and it seems likely Boethius was among them.

[86] Ennodius, *ep*. 6.6, written around the autumn of 507.

[87] Ed. Stewart and Rand, 72–128.

[88] *Utrum pater et filius et spiritus sanctus de divinitate substantialiter praedicentur*, ed. Stewart and Rand, 32–6; *Quomodo substantiae in eo quod sint bonae cum non sint substantialiter bona*, ibid. 38–50.

[89] So already in the 3rd cent. (Eusebius, *HE* 6.43.6), a principle perhaps based on Acts 6: 3. The argument advanced here falls short of absolute certainty: Symmachus is credited with ordaining 16 deacons during a pontificate of 16 years (*lib. pont.* 263.11), an astonishingly high figure, and we know that Peter of Altinum performed ordinations (*MGH AA* 12.427.19f.); it is possible Laurentius did as well. Confusingly, a deacon called John was present at the synod of 6 Nov. 502 (*MGH AA* 12.443 no. 2) and hence presumably a supporter of Symmachus. But more evidence linking the Laurentian John with Boethius will be considered below.

[90] *Lib. pont.* 260.10, 261.7f. in particular.

THE ISSUES AT STAKE

Is it possible to go beyond this and establish the issues which lay behind the schism? The existence of a schism within the Roman Church during the pontificate immediately preceding that of Symmachus, the certainty that this had to do with relations with the Eastern Churches, and the circumstances that just prior to the double election of November 498, the chief supporter of Laurentius had been involved in negotiations in Constantinople during which he was believed to have told the emperor that he could obtain papal assent to Zeno's *Henotikon*, combine to suggest that the schism was, *au fond*, to do with relations with the East. Particularly significant is the portion of an account of the pontificate of Anastasius surviving at the commencement of the Laurentian fragment, which indicates that Pope Anastasius had indicated to the emperor Anastasius by means of Bishops Cresconius and Germanus in authoritative writings that the evil schism between the Churches of the East and Italy had lasted quite pointlessly.[91] But the account of Pope Anastasius preserved in the official *Liber pontificalis* interprets his endeavours to end the schism in a very different light: 'Many clergy and priests separated themselves from communion with him because, without consulting the priests, bishops and clergy of the whole Catholic Church he had entered into communion with a deacon of Thessalonica named Fotinus [*sic*] who had been in communion with Acacius, and because he secretly wished to call back Acacius and was not able. This man was struck down by the will of God.'[92] Laurentius is known to have had forerunners as well as followers,[93] and these accounts provide clear evidence that the eirenic policies of Anastasius towards the East were supported in Laurentian circles but bitterly opposed by the adherents of Symmachus, who stood rather in the tradition of Gelasius. Just as Gelasius' generosity towards the poor was continued by Symmachus, while no evidence suggests that this was a concern for Anastasius or Laurentius, so Anastasius' friendly attitude towards the East was continued by Laurentius, while the evidence suggests that Gelasius and Symmachus were most definitely of another opinion. Relations between Symmachus and the emperor Anastasius were nothing if not frosty: the latter charged the pope with being a Manichean, and

[91] *Frag. laur.* 44.
[92] *Lib. pont.* 258.
[93] Ennodius, *lib. pro syn.* 125: 'adulteri Laurenti aut sequaces aut praevii'.

having not been properly ordained, and with having conspired with the senate to excommunicate him, to which accusations the pope replied with 'a strange mixture of unctuous phrases and gross invective'.[94] Following the death of Symmachus, Anastasius wrote to his successor complaining that he had been a harsh man.[95]

It must be said that while Anastasius had little time for Symmachus, there is no direct evidence that he supported Laurentius,[96] and as far as we know, the issue of relations with the East was not directly raised in the propaganda war which broke out in Rome after the acquittal of Symmachus in 502. But this need not invalidate the argument that the issue of relations with the East was uppermost in the minds of the Roman aristocrats and clergy who created and sustained the schism, as the need to find ammunition against a rival candidate and to focus on issues capable of influencing public opinion necessarily caused other considerations to be raised. But early in 502 Symmachus had been accused of three things: the celebration of Easter on the wrong date, sinning with women, and squandering the wealth of the Church. Accusations about sex and money have been used throughout the centuries to mask more important areas of controversy, and we may well believe that Symmachus' observance of Easter on a date which happened to be contrary to that of the East, together with what that may have been held to imply concerning his relations with the East, was the point genuinely at issue.

THE WEALTH OF OSTROGOTHIC ITALY

An interpretation of the schism along the lines of a continuation of an immediately pre-existing schism, which broke out over relations with the East, seems to fit the data best, but this is not to deny that other issues may have been involved. An impressive case has been made for locating the schism against the background of the decree

[94] Symmachus' letter to Anastasius is printed in Schwartz, *Publizistische Sammlungen*, 153–7; Anastasius' accusations are known from 154.15, 21; 155.16. I borrow the characterization of Symmachus' letter from Caspar, *Geschichte*, 119.

[95] *Coll. avel.* 107.2.

[96] Despite frequent assertions, e.g. by A. Gaudenzi, *Sui rapporti tra l'Italia e l'impero d'oriente fra gli anni 476 e 554 D.C.* (Bologna, 1888), 42; F. X. Seppelt, *Der Aufstieg des Papsttums* (Leipzig, 1931), 259 (which remains an interesting account from a confessional perspective); Richards, *Popes and Papacy*, 80.

issued by Basilius in 483 forbidding the alienation of goods left to the Church, which was overturned by the Roman synod of 6 November 502, and so arguing that the issue at stake was the economic power of the Roman Church.[97] It will be worth our while to consider the economic situation of Italy during the time of Theoderic.

The economy of late antique Italy is an area concerning which it is difficult to speak with confidence.[98] In general terms, however, it is clear that for some centuries it had been in decline, with respect to both the role Italy played in long-distance trade and its internal situation, and this for reasons not directly connected with the political turmoil of the times. The population was decreasing, with fewer sites being occupied, and Rome, among other cities, becoming noticeably over-catered for by the scale of its monumental buildings; it may not be accidental that the Goths could be represented as making the population numerous and creating peaceful conditions which would allow the Roman population to increase (Cassiodorus, *var.* 7.3.3, 8.3.4), while it is certainly significant that attention could be drawn to the scale of the walls, amphitheatre, and baths of Rome, by implication inappropriate to contemporary needs (*var.* 11.39.2). By the time of the Ostrogoths the forum at Luni was covered by sand.[99]

Against this gloomy background the reign of Theoderic offers relief, and it may well be that his enlightened policies played some part in an economic revival.[100] The state prospered, for according to

[97] C. Pietri, 'Le Sénat' (building on Picotti, 'Osservazioni', 188 ff., although the earlier parts of his article should be read in the light of Alan Cameron, *Circus factions* (Oxford, 1976)); id., 'Evergétisme et richesses ecclésiastiques dans l'Italie du IVe à la fin du Ve siècle: L'Exemple romain', *Ktema*, 3 (1978), 317–37; id., 'Aristocratie et société'.

[98] There is a useful account by P. Jones, in *Storia d'Italia*, 2 (Turin, 1974); see too R. Soraci, *Aspetti di storia economica italiana nell'età di Cassiodoro* (Catania, 1974). A major monograph is that by L. C. Ruggini, *Economia a società nell 'Italia annonaria'* (Milan, 1961), largely summarized in her 'Vicende rurali dell'Italia antica dall'età tetrarchia ai Longobardi', *Rivista storica italiana*, 76 (1964), 261–86. But very important work is now being done by archaeologists, concerning which synthesis is premature: A. Giardina, ed., *Società romana e impero tardoantico*, 3. *Le merci gli insediamenti* (Rome, 1986). Much of the current work is being published in two journals: *Archeologia medievale* and *Papers of the British School at Rome*.

[99] B. Ward-Perkins, 'Luni: The Decline and Abandonment of a Roman Town', in H. Blake, T. Potter, and D. Whitehouse, eds., *Papers in Italian Archaeology*, 2 (Oxford, 1978), 313–21.

[100] Theoderic's economic policies are discussed approvingly by F. Giunta, in *Magistra barbaritas*, 76 ff.

one source, although he found the treasury full of straw, thanks to his efforts it came to abound in wealth (Anon. Vales. 60). Theoderic's daughter Amalasuintha had at her disposal 400 *centenaria* when she thought of fleeing to imperial territory (Procopius, *BG* 1.2.26), and Procopius found the wealth of Theoderic 'a notable sight' (*BG* 3.1.3). Some of this was doubtless the treasure of the Goths, but the state certainly had great assets, and the same was true of churches; shortly after the death of Theoderic the Catholic church of Ravenna, for example, enjoyed an annual income of 12,000 solidi.[101] Similarly, we have no reason to believe that the aristocracy of Theoderic's time, whose members were able to invest in income-producing activities such as the draining of marshes (Cassiodorus, *var.* 2.32f.), could have pleaded poverty: from 493 to 526 approximately thirty Westerners were appointed to consulships, and this expensive office seems to have been coveted, while such evidence as we have, admittedly impressionistic in kind, suggests that the aristocracy was well-off.[102] It is an area concerning which it is difficult to speak with confidence; nevertheless we are surely justified in assuming that a decline in imported foodstuffs must have been beneficial to the holders of estates in Italy, given the need to feed the non-productive populace of Rome,[103] and that the conditions of peace Italy enjoyed under Theoderic must have been favourable to production and consumption.[104] In the light of this we may accept the comment made by Totila to the Roman senate in 546 that under Theoderic and Athalaric its members had 'amassed vast wealth',[105] despite one reference to the

[101] Procopius, *BG* 3.1.3 on the large treasure left by Theoderic, on which see D. Claude, 'Zur Geschichte der frühmittelalterlichen Königsschatz', *Early Medieval Studies*, 7 (*Antikvariskt Arkiv*, 54; 1973), 5–24 at 5ff. Agnellus of Ravenna, *Codex* 60.

[102] Note Ennodius' comment, in a discussion of Liberius' work assigning the *tertiae*: 'tuta enim tunc est subiectorum opulentia, quando non indiget imperator' (*ep.* 9.23.5).

[103] Rougé, 'Quelques aspects', 140–2; Ruggini, *Economia e società*, 349–59, 514–20; K. Hannestad, *L'Évolution des ressources agricoles de l'Italie du IV*ᵉ au VIᵉ *siècle de notre ère* (Copenhagen, 1962). Note scattered references to grain being exported to Gaul (Cassiodorus, *var.* 4.5, 4.7) and imported from Spain (ibid. 5.35).

[104] Anon. Vales. 72 (an achievement the more remarkable in that late antiquity was a time of insecurity in the Italian countryside: A. Kahane, L. M. Threipland, and B. Ward-Perkins, 'The Ager Veientanus, North and East of Rome', *Papers of the British School at Rome*, 36 (n.s. 23) (1968), 1–218 at 153); Ennodius, *V. Epiph.* 120; Cassiodorus, *var.* 9.10.2 'longa quies et culturam agris praestitit et populos ampliavit', with reference to Sicily.

[105] Procopius, *BG* 3.21.12; cf. 3.21.23 for a claim of the Goths that Anastasius and Theoderic 'filled their whole reigns with peace and prosperity' and Anon. Vales. 60.

sale of wheat at the extraordinarily low rate of 60 modii for one solidus.[106] Shortly after the death of Theoderic economic decline was to resume at an accelerated rate, and in retrospect his reign could have taken on the quality of a golden age. It was a period during which one could imagine an aristocracy playing ecclesiastical politics for reasons not connected with their economic situation.

But one point concerning the aristocracy and the schism deserves emphasis. As we have seen, when Sidonius Apollinaris visited Rome in 467–8 he found two men pre-eminent in the city, Basilius and Avienus. Basilius was subsequently to impose his decree on the papacy in 483, so providing grounds for the Laurentians attacking Symmachus; Faustus *niger*, the son of Avienus, was the most notable supporter of Symmachus. This suggests that Basilius' decree may have been an issue, if not the chief one, during the schism; more importantly, it indicates the ability of aristocratic factions to survive across the decades. Their capacity to do this is a theme we will encounter later.

THEODERIC AND THE SCHISM

Before leaving the schism, we must examine Theoderic's attitudes to the parties in it. Scholars have not hesitated to pronounce that he supported one side or the other, but their assertions have usually been based on *a priori* considerations such as Theoderic's presumed response to one of the parties being pro-Gothic, in response to the other's being allegedly pro-Byzantine in political as well as theological matters, an assumption for which there is no support in the sources.[107] I am not certain that we are in a position to be definite. Theoderic had, of course, decided for Symmachus shortly after the double-election of 498, but was in no hurry to intervene when trouble broke out again in 502. Ennodius told the supporters of Peter and Laurentius that they had no support in the authority of

[106] Anon. Vales. 73. But we know elsewhere of grain from the state stores being sold at 25 modii to the solidus (Cassiodorus, *var.* 10.27.2), so the rate quoted by Anonymous Valesianus must have been exceptional, if indeed it ever applied: Pferschy, 'Das Probleme'.

[107] As Wes points out, a theological view did not imply a political stance (*Das Ende*, 100f.). Theoderic's support for Symmachus has been asserted, in varying ways, by Gaudenzi, *Rapporti*, 42; Seppelt, *Der Aufstieg*, 254; O'Donnell, 'Liberius', 40. Support for Laurentius is urged by Duchesne, *L'Église*, 121; Picotti, 'Osservazioni', 196–200; D. M. Hope, *The Leonine Sacramentary* (Oxford, 1971), 77.

the king, without, however, claiming that Theoderic supported Symmachus.[108] He was writing in 503, presumably after Theoderic allowed Laurentius to return to Rome from Ravenna, which could have been interpreted as an act in favour of the latter. Theoderic may have been cast as Valentinian in the *Gesta de Xysti purgatione*, but it is difficult to argue on the basis of this. Similarly, appointments to high office yield no clear pattern. Avienus (consul 502) was a son of Symmachus' backer Faustus *niger*, who was himself appointed *quaestor* in 503, and Messala (consul 506) was another of his sons; but on the other hand Avienus (consul 501) was a son of Basilius, to whose decree Symmachus' supporters took violent exception, and Cethegus (consul 504) was a son of Probinus, one of Laurentius' chief supporters, while Faustus *albus*, who, we have suggested, may have been inclined towards Laurentius, seems to have held the office of *praefectus urbis Romae*, a sensitive one in a time of civil strife, in 502–3.[109] Such evidence as we have cannot be used to sustain an argument that Theoderic supported one side or the other; its implication that he preserved neutrality until 506 or 507 may well be correct; and, as suggested by his impatience with the dilatory bishops assembled in Rome in 502, the king may have been more concerned with the peace of Rome than the issues raised by a schism within a Church of which he was not a member.[110]

Our analysis of the Laurentian schism has been lengthy, and may seem to have taken us some distance away from Theoderic. Nothing in the detail of the foregoing account should be taken to imply that Theoderic followed papal affairs with breathless attention, and it would have been possible for us to have dealt with them more expeditiously.[111] But they do have the merit of presenting a body of material which is susceptible of analysis. Our discussion of the Laurentian schism has not only suggested ways in which the affairs of the Church of Rome can be understood, but has also offered ways of coming to grips with the role played by the Roman aristocracy during the time of Theoderic, a theme which, we shall suggest, became of great importance at the end of his reign. But the issues involved here are complex, and will require our further attention in the following chapter.

[108] *Lib. pro syn.* 19. [109] *PLRE* 2.451 f.

[110] See further along these lines Pfeilschifter, *Theoderich*, 91–103; Richards, *Popes and Papacy*, 78–80.

[111] See e.g. the concise summary in Udal'tsova, *Italiia*, 219.

5

Rome and Ravenna

Ostrogothic Italy was a land dominated by two cities. The elder, Rome, remained what she had been for centuries, the head and mistress of the world who gathered her children to her breasts.[1] She was the friend of liberal studies whither boys would be sent for their education, such as the young Benedict, sent there from Norcia; it was only natural for the government to be concerned about the payment of the salaries of the teachers, and Theoderic is known to have ensured that its resident teachers of grammar, rhetoric, medicine, and law were paid.[2] As the former centre of what could still be called the Roman Empire,[3] the place from which *libertas Romana* and the *nomen Romanum* took their origin, and the place of residence of the senators, whose assembly could be styled *curia Romana*,[4] the city enjoyed an unequalled prestige. Coins issued during this period by the senate showed on the reverse the Roman eagle, or Romulus and Remus being suckled by the she-wolf, or the fig-tree beneath which Livy reported that this scene took place.[5] The senators, by this time a small body of men,[6] included among their number scholars who, as we shall see, were interested in the transcription of such authors as Virgil, Macrobius, and Sedulius. But we have already suggested that the high degree of interest

[1] 'urbs illa caput orbis et domina': Jordanes, *get.* 291. Children at breasts: Cassiodorus, *var.* 2.1.2; cf. Corippus, *laud.* 1.288–90 with Cameron's commentary, 142, and the development of these images by Ennodius, *lib. pro syn.* 120. See on traditional epithets for Rome, Schneider, *Rom und Romgedanke*, 57f.

[2] 'urbs amica liberalibus studiis': Ennodius, *ep.* 6.23.1, cf. 5.9.2, 6.15.2. Boys sent to Rome: Gregory of Rome, *dial.* 2, prol. (Benedict); Cassiodorus, *var.* 1.39, 4.6; Ennodius, *ep.* 5.9.2, 9.2–4. Payment of salaries: Cassiodorus, *var.* 9.21; Justinian's *Constitutio pragmatica*, 22 (in Kroll, ed., *Corpus iuris civilis*, 3.802). On Rome as the source of eloquence, Cassiodorus, *var.* 10.7.2; as the unrivalled centre of education, R. Luiselli, 'La società dell'Italia romano-gotica', in *Atti del 7° Congresso internazionale di studi sull'alto medioevo, 1980* (Spoleto, 1982), 49–116.

[3] 'Romanum imperium': Cassiodorus, *var.* 3.18.2.

[4] Ibid. 9.22.4.

[5] Grierson and Blackburn, *Coinage*, 32.

[6] Chastagnol, *Sénat*, 45–7, with S. J. B. Barnish, 'Transformation and Survival in the Western Senatorial Aristocracy, *c.*AD 400–700', *Papers of the British School at Rome*, 56 (1988), 120–55.

displayed by senators in papal politics is testimony to the increasing power of the bishop in the city of Rome, and there can be no doubt that the processes which transformed Rome into what a contemporary author seems to have referred to as 'urbs ecclesiae' (Anon. Vales. 65) were well under way by the time of Theoderic. The see of Rome had come to dispose of enormous wealth,[7] the expenditure of which could enable factions seeking power within it to win the support of the volatile *plebs* of the city. Another source of its power was the political leadership demonstrated above all by Pope Leo in the troubled 450s; it is scarcely surprising that the attitude of Pope Silverius was a matter of concern to both Witigis and Belisarius as the Byzantine conquest of Italy began.[8] The growing strength of the Church was reflected in the city's changing skyline in the fifth century, as it erected buildings 'meant to compete with palaces and public buildings and the temples of the gods', and by the second third of the century the building of churches in Rome was controlled by the papacy.[9] Of itself this may not seem surprising, but episcopal control was certainly not exercised in Constantinople, where Justinian and Anicia Juliana were later engaged in the building of churches on a massive scale, or in Ravenna or other Italian cities. Furthermore, it was a time when the figure of St Peter was becoming increasingly important in the consciousness of the city—as demonstrated by the confidence of its citizens during a siege early in the Gothic war that the Apostle would see to the defence of a part of the wall against the Goths—and despite the devotions Theoderic performed at Peter's tomb in 500 we are surely justified in seeing the saint's cult as being, at least implicitly, anti-Arian.[10] Rome was the seat of the vicar of Peter, to which bishops from a wide area repaired in 495, 499, and 502 to attend synods, whose bishop claimed the right to give permission when other bishops sought to visit the king, and whose jurisdictional rights Theoderic maintained,[11] and which was attracting such devout visitors as the African monk Fulgentius, Bishop Caesarius of Arles, and the Burgundian prince Sigismund. Given the possible political bearings of the cult of the Apostle, we may suspect that some of the pilgrims to his shrine were allying

[7] Jones, *Later Roman Empire*, esp. 781 f.

[8] Witigis: Procopius, *BG* 1.11.26; Belisarius: Caspar, *Geschichte*, 230 ff.

[9] R. Krautheimer, *Rome: Profile of a City, 312–1308* (Princeton, NJ, 1980), 34 f.

[10] Procopius, *BG* 1.23.5; see further Arator, *De actibus apostolorum*, 1.1070–6.

[11] Gelasius, *frag.* 7 (ed. Thiel) to the bishop of Volterra, in which Gelasius expresses himself with some heat; on papal rights, *MGH AA* 12.391.8, cf. 6.

themselves with a cult which was not entirely neutral with respect to the state. One circumstance which may have given Theoderic pause was the enormous success of the papacy in attracting donations from Catholic rulers.[12] Rome may not have seemed an entirely loyal city.

Quite different was the situation in Ravenna. Perched in marshes by the Adriatic Sea, almost due north of the politically marginalized Rome, it was a government town: an imperial residence from the beginning of the fifth century, it was the seat of government for Odovacer and the Ostrogothic sovereigns, and following the Byzantine conquest it became the headquarters of the imperial administration in Italy.[13] It was at Ravenna that an inscription recorded the achievement of Theoderic, 'with God supporting him, glorious in both war and leisure', in draining an unproductive swamp and planting an orchard.[14] If the secular life of Rome was dominated by senators, Ravenna was a town of Goths and bureaucrats. The former would have lived in the *civitas barbara* which grew up along the *platea maior*, the modern via Roma, a zone which included the palace, the court church of S Apollinare Nuovo, and the Arian episcopal complex,[15] and they could have patronized the *Codex argenteus*. The civil servants would have been Romans, but aware, perhaps uneasily, that their fortunes depended on the Goths, and if we accept that the workshop which produced the *Codex argenteus* also produced a manuscript of Orosius, we have evidence for either Goths reading a Catholic Latin author or Romans and Goths utilizing the one source of codices.[16] But the contrast between Rome and Ravenna can most easily be brought out with reference to the building of churches. Many of the wonderful ecclesiastical buildings erected at Ravenna in the fifth and sixth centuries were the fruit of royal or imperial rather than episcopal patronage, and the famous mosaics depicting Theoderic's palace in the church of S Apollinare Nuovo, and Justinian and Theodora in S Vitale, would have been unthinkable in a church in Rome. Indeed, the choice of buildings in

[12] *Lib. pont.* 271 (gifts from Clovis, hard to interpret in view of the date, and Justin as well as Theoderic), 276 (gifts from Justin).

[13] It may be worth noting, however, that Valentinian III seems to have had a penchant for living in Rome.

[14] Fiebiger/Schmidt, *Inschriftensammlung*, no. 179 (*CIL* 11.10).

[15] Lusuardi Siena, 'Tracce', in *Magistra barbaritas*, 526ff.

[16] G. Cavallo, 'La cultura a Ravenna tra corte e chiesa', in *Le sedi della cultura nell'Emilia Romagna: L'alto medioevo* (Milan, 1983), 29–51, provides an excellent discussion of literary life at the Gothic court at pp. 29–36.

the background to the palace in the surviving mosaic at S Apollinare Nuovo can be held to represent Ravenna as Theoderic's creation.[17] Moreover, Ravenna was noteworthy for its number of Arian churches. When the Byzantines captured the town there were six such churches which had to be 'reconciled', and they were well-endowed, although the Catholic churches were better off.[18] Such considerations incline one to believe that Theoderic may have felt more at home in Ravenna than Rome.

Theoderic's panegyrist Ennodius proclaimed that during his reign Rome, the mother of cities, regained her youth, and according to the evidence of stamps on bricks Theoderic claimed to govern for the good of Rome; he wished the city to gleam as buildings rose.[19] But his visit in 500 was the only occasion in a reign of thirty-three years when he is known to have set foot in Rome. While in the city he must have lived on the Palatine, and he is known to have carried out some building activity there.[20] But this was nothing to his works at Ravenna; as has been well observed, whereas Theoderic restored buildings already in existence in Rome, in Ravenna he built new ones.[21] In the ninth century Agnellus of Ravenna was still able to see a mosaic with important implications for the status of Rome *vis-à-vis* Theoderic's capital of Ravenna. He describes it as showing Theoderic on horseback clad in mail, holding in his right hand a lance and in his left a shield. On the side of the shield stood Rome, with spear and helmet, but on the side of the lance, a weapon which may have been perceived as having explicitly royal associations, was Ravenna, hastening towards the king, her right foot in the water and her left on the land.[22] The suggestion of Ravenna being of both

[17] See in general Deichmann, *Ravenna*, but one misses a study of Ravenna which integrates various classes of material, as does that of Krautheimer on Rome. On Ravenna as Theoderic's creation, Johnson, 'Theoderic's Building Program', 88.

[18] Six churches: Agnellus, *Codex*, 228f. Well endowed: Jones, *Later Roman Empire*, 821. Catholic churches: ibid. 908.

[19] 'illa ipsa mater civitatum Roma iuveniscit': Ennodius, *pan*. 56, cf. 48. Stamps on bricks: H. Bloch, 'Ein datierter Ziegelstempel Theoderichs des Grossen', *Mitteilungen des Deutschen Archäologischen Instituts: Röemische Abt*. 66 (1959), 196–203. *Urbs* to gleam: Cassiodorus, *var*. 4.30.3. See in general G. Della Valle, 'Teoderico e Roma', *Rendiconti della Accademia di archeologia lettere e belli arti* (Naples), 34 (1959), 119–76, and Lusuardi Siena, 'Trace', 525 in *Magister barbaritas*.

[20] *CIL* 15.1665 (1–29), 1666, 1669.3.

[21] V. Righini, 'I bolli laterzi di Teoderico e l'attività edilizia teodericana in Ravenna', *XXXIII Corso di cultura sull'arte ravennate e bizantina* (Ravenna, 1986), 371–98.

[22] Agnellus, *Codex*, 228f. On the interpretation of this, Johnson, 'Theoderic's Building Program', 86f.

sea and land is also present in another mosaic of the period in S Apollinare Nuovo which depicts the port of Classis.[23] Ravenna standing on Theoderic's right in the mosaic described by Agnellus occupies the position of honour, but standing as they did on either side of the king the two cities must have seemed to belong to roughly the same category. Thus Ravenna had been promoted.[24] The evidence of copper coinage points in the same direction, for if Theoderic issued 40 and 20 nummi pieces depicting *invicta Roma*, he also issued 10 nummi pieces showing *felix Ravenna*, and this despite *felix Roma* being an expression in common use.[25] In 502 Theoderic described Rome as the *urbs regia*, but later in his reign and in the reigns of his successors the expression *regia civitas* was applied to Ravenna, while Jordanes goes so far as to style Ravenna *urbs regia*.[26] Undoubtedly, Ravenna had gained in power at the expense of Rome.

THE SENATE

But the senate continued to meet in Rome, and it, or perhaps more accurately its members, remained important to Theoderic. The emperor Anastasius, as he himself was reminded in a letter, had frequently exhorted Theoderic to love the senate (Cassiodorus, *var.* 1.1.3) and it is reported that, as he lay dying, love of the senate was one of the things Theoderic enjoined on the leaders of the Goths (Jordanes, *get.* 304). The death of Theoderic and the accession of Athalaric in 526 were announced in a series of eight letters which Cassiodorus preserved in his *Variae*; of these the letter to the senate comes second in the sequence in which he arranged his correspondence (*var.* 8.2), immediately after the letter to the emperor Justin. It could well be that the tendency which can be observed during Theoderic's reign for coins to be minted in Rome rather than Milan

[23] Deichmann, *Ravenna*, 3, fig. 111, with 2.1.145f.

[24] Reydellet, *Royauté*, 236.

[25] Hahn, *Moneta*, 77, 79, with types 70–6; see too E. Ercolani Cocchi, 'Osservazioni sull'origine del tipo monetale Ostrogoto "Felix Ravenna"', *Studi Romagnoli*, 31 (1980), 21–44. For *felix Roma*, see Cassiodorus, *var.* 6.1.2, 6.18.4, and Bloch, 'Ziegelstempel', 198.

[26] Rome as *urbs regia*: *MGH AA* 12.419.7, 420.7; cf. *regia civitas*, 422.3. Ravenna as *regia civitas*: Cassiodorus, *var.* 8.5.1. Ravenna as *urbs regia*: Cassiodorus, *var.* 12.22.5; Jordanes, *get.* 147, 293.

or Ravenna arose from a desire to conciliate the senate,[27] and the numerous letters Theoderic dispatched to the senate are frequently full of flattering sentiments and testify to the same desire. One recalls that early in his reign the emperor Marjorian had asked the senate to share in the care of affairs.[28]

Theoderic's letters to the senate are interesting from more than one point of view. When Cassiodorus organized his official correspondence into the twelve books of the *Variae* he arranged the letters so that each book containing letters written on behalf of a monarch begins with a letter or group of letters to royalty, whether Byzantine emperors or barbarian kings. But in most cases the correspondence with sovereigns is immediately followed by a letter to the senate, or by a doublet of letters on the same topic of which the second is to the senate.[29] This, of course, is testimony to the attitude of Cassiodorus at the time he came to edit the *Variae*, and not to any feeling Theoderic may have had, although it hints at editorial practices on the part of Cassiodorus which would repay more detailed study. These letters concern a multitude of topics, but the largest group contains notifications of appointments to office for which the assent of the senate was sought; the strategic locations assigned to letters dealing with Cassiodorus' father and himself witness to a certain amount of vanity (*var.* 1.4, 11.1). But it is clear that the *Variae* fail to document every appointment for which senatorial approval may have been sought. To take the most obvious example, of the consuls appointed while Cassiodorus was *quaestor* (507–11), only Felix, who held office in 511, is commemorated (*var.* 2.3). It follows from this either that the practice of the government in seeking senatorial approval was inconsistent, or that only an incom-

[27] Hendy, *Studies*, 396f., and 'From Public to Private', 44. 'SC' still appeared on some coins: Hahn, *Moneta*, 79. Cf. an inscription: 'Ex praecepto domini nostri Theoderici et senatus consulto' (Fiebiger/Schmidt, *Inschriftensammlung*, no. 194). The same duality in an oration of Cassiodorus: 'nos gloriamur de sententia boni principis; laetamur de consensu senatus' (*MGH AA* 12.468.19–21).

[28] Note in particular the addressing of a letter, 'Domitori orbis, praesuli et reparatori libertatis senatui urbis Romae Flavius Theodericus rex': *MGH AA* 12.392.2f. Cassiodorus' letters to the senate are usually addressed 'Senatui urbis Romae', the exception being 'Senatui romano' (*var.* 3.12). Traube's index itemizes the fulsome terms applied to it (p. 583f.). Marjorian's request is expressed in his *nov.* 1 of 458.

[29] This policy is clearly to be seen in books 2, 3, 4, 5, 8, and 10. It is not immediately obvious at the beginning of book 1 because of Cassiodorus' tendency to organize his letters into groups. The opening letter to the emperor Anastasius is fittingly followed by a letter on the *sacra vestis*, then a pair of letters on the appointment of Cassiodorus' father as *patricius*, of which the second is to the senate.

plete run of material was available to Cassiodorus when he edited the *Variae*, or that Cassiodorus chose to publish only a small selection of the material available; later we shall see that the third of these explanations may be the true one.[30]

Another large group of letters concerns senatorial misdeeds. Writing to the senate after Rome had been troubled by civil disturbances, Theoderic encouraged its members not to overreact to the empty words of the people, and to bring any complaints to the *praefectus urbis Romae*, so that the guilty would be punished in accordance with the laws.[31] On another occasion the king was forced to bring to the attention of the senate a complaint made by the officials of the praetorian prefect, quite possibly Faustus *niger*, that senators were evading the payment of tax, causing a heavier burden to fall on the less wealthy (*var.* 2.24); they were also accused, among other things, of diverting public water to turn mills and irrigate their gardens, and were instructed to co-operate with the *vir spectabilis* John, whom Theoderic sent to Rome to investigate the matter (*var.* 3.31). One has the feeling that the senate was not necessarily the easiest body to deal with. Its members occupied themselves with ecclesiastical politics, and quite apart from the cases when individual members are known to have participated in such activities during the Laurentian schism and the termination of the Acacian schism, we may note that Theoderic's *fiat* ending the Laurentian schism in 507 was addressed to the senate rather than ecclesiastics, that in 516 the emperor Anastasius wrote to the senate seeking its approval for his ecclesiastical policies, that shortly after the death of Theoderic his successor Athalaric wrote to the senate advising its members to accept a pope imposed by Theoderic, and that in 530 the senate intervened in the election of another pope.[32] Doubtless the concentration of material bearing on the ecclesiastical interests of the senate is a pointer towards the growing power of the Church and a desire by senators to accommodate themselves

[30] This despite the claim that the *Variae* constitute 'quod . . . a me dictatum in diversis publicis actibus potui reperire' (*praef.* 13).

[31] *Var.* 1.30; cf. 1.31 on the same subject, 1.32 (showing some senators had been taking matters into their own hands), 1.44.

[32] Theoderic's *fiat*: *MGH AA* 12.392. Anastasius' letter: *coll. avel.* 113 (but the letter is addressed in full to the proconsuls, consuls, praetors, tribunes, plebs, and his senate, and begins with the splendid words 'Si vos liberique vestri valetis, bene est; ego exercitusque meus valemus'); the reply was sent in the name of the senate (*coll. avel.* 114). Athalaric's letter: Cassiodorus, *var.* 8.15. Intervention in 530: *lib. pont.* 282.

to this and, if possible, exercise control over it,[33] but the Roman aristocracy remained a powerful and refractory body, by no means innocent of self-interest. It will be worth our while examining some detailed evidence in an attempt to determine whether Theoderic, ruling from Ravenna, had any fixed policies in the appointment of its members to high office.

THEODERIC'S APPOINTMENTS

An enquiry into the men appointed to office by Theoderic must reflect the limitations of the evidence. One would love to know more about Senarius, who, according to his epitaph, discharged twenty-five missions for the state, in one year visiting both the far west and Constantinople twice. It is interesting that Ennodius was able to describe him as a relative—although as we shall see Ennodius was generous with such descriptions—and that he received correspondence from both Avitus of Vienne and the deacon John, but unfortunately we lack the material which would allow the pieces of information we have which bear on him to be placed in a useful context. Nevertheless, it is obvious that early in his reign Theoderic maintained the practice which had been followed in the days of Odovacer and the later empire in the West of appointing aristocrats to high office.[34] Admittedly, Odovacer's last consular nominee, Faustus *niger*, who may have traced his ancestry back to M. Valerius Messala Corvinus (consul AD 59), was a hard act to follow, but one is struck by the weightiness of Theoderic's nominees.[35] The first of these, Albinus (493) was a Decius, the son of Odovacer's first nominee to this office, Caecina Decius Maximus Basilius (480), who is probably to be seen as the brother of Decius Marius Venantus Basilius, oddly enough the colleague of Theoderic when he held his consulship (484) and of Caecina Mavortius Basilius Decius (consul

[33] See in general, Pietri, 'Aristocratie et société'.

[34] Matthews, *Western Aristocracies*, studies 'the manner in which the government of the Western empire seems progressively in these years [364–425] to fall from public into private hands' (his summary, p. xi; see in particular pp. 355, 386–8). Senarius is discussed in *PLRE* 2.988f.

[35] Technically, Theoderic, like Odovacer, did not appoint but nominated consuls: Chrysos, 'Amaler-Herrschaft', 456–60. On the ancestry of Faustus' father Gennadius Avienus, *PLRE* 2.193.

486), and the son of Caecina Decius Basilius (consul 463).[36] Both Odovacer and Theoderic nominated a Decius as their first consul, and as the first nominee of Theoderic's successor Athalaric, Vettius Agorius Basilius Mavortius (consul 527) was, to judge by his names, a member of the same family, it is clear we are dealing with a tradition which may owe something to a desire to maintain an alliance with at least part of the Roman aristocracy. Generally Theoderic nominated only one consul annually, but the consuls of 494 were both nominated by him. Turcius Rufius Apronius Asterius must have been a descendant of the Turcius Apronius who had been converted to Christianity in the late fourth century by the famous ascetic Melania and married her niece Avita; their children were Asterius and Melania.[37] A literary man, during the year of his consulate Asterius co-operated with his friend, the otherwise unknown Macharius, in revising a text of Virgil's Eclogues; a subscription to a manuscript of Virgil indicates that he was already a *patricius* and had held the offices of *comes domesticorum protectorum*, *comes privatarum largitionum*, and *praefectus urbis*, presumably under Odovacer.[38] During or after his consulship he produced what may have been the first edition of Sedulius' *Carmen paschale*,[39] and it is a fascinating possibility that Ennodius may have had access to his text.[40] Nothing definite is known of the other consul in 494, Praesidius, who was also a

[36] On Albinus, *PLRE* 2.51, with stemma 26 (p. 1324). Again, I must record my debt to this work. The early date of Albinus' consulship relative to that of his father already implies what was to become a feature of Theoderic's reign, the appointment of young men: A. Cameron and D. Schauer, 'The Last Consul: Basilius and his Diptych', *Journal of Roman Studies*, 72 (1982), 126–45 at 129; cf., for earlier centuries, T. Mommsen, *Römisches Staatsrecht*, 1 (Leipzig, 1887), 574. For consular *fasti*, see too A. Degrassi, *I fasti consolari dell'impero Romano* (Rome, 1952).

[37] *PLRE* 1.87; cf. Paulinus of Nola, *carm*. 21.210ff., 283ff.; Palladius, *hist. laus*. 41, 54.

[38] Virgil, *Bucoliques*, ed. E. de Saint-Denis (Paris, 1967), pp. xxivf.; note the verses penned by Asterius referring to the consular games with races and animal fights.

[39] CSEL 10, p. vii. The subscription refers to him as consul; see M. Schanz, C. Hosius, and G. Krueger, *Geschichte der römischen Literatur*, 4/2 (Munich, 1920), 312f.

[40] In a poem written on the occasion of the thirtieth anniversary of the ordination of bishop Epiphanius of Pavia, therefore in 495–6 (Sundwall, *Abhandlungen*, 53f.), Ennodius shows awareness of Sedulius (*carm*. 1.9.20; see *MGH AA* 7.332 for more references). Sedulius' work was approved in a work which has been attributed to Pope Gelasius, *De libris recipiendis et non recipiendis* (ed. E. von Dobschutz (1912), 47), but Gelasius was not its author (Ullmann, *Gelasius*, 256–9).

Westerner,[41] nor of Theoderic's nominee for 495, Viator,[42] but the Speciosus who was nominated for 496 may have been associated with the family of the senator Olybrius. An inscription implies that he may have enjoyed his third term as *praefectus urbis* in *c*.493–6, and even given the short tenure customary for this office we can hardly be mistaken in seeing in him a man first appointed by Odovacer.[43] Speciosus' consulship was not widely published, and it is possible to interpret this, together with the lack of Western consuls in 497, 499, and 500, as a sign that Theoderic's relations with Constantinople during this period were cool, although the peace made through Festus in about 497 suggests that an explanation along these lines may be simplistic. Perhaps there were simply no seekers after the office in those years.[44] Theoderic's nominee for 498, Paulinus, is a fairly obscure figure, but to judge from a reference in Boethius' *Consolation of Philosophy* to the dogs of the palace who sought to devour his wealth he was a man of substance.[45]

Theoderic's consular nominees for the 490s were moderately distinguished, but they do not compare with those for the period 501–10. The Western consul for 501, Avienus, was a member of the Decii and a brother of Albinus (consul 493), hence the brother, son, and grandson of consuls. By 509 he was a *patricius*.[46] In 502 the office was held by Rufius Magnus Faustus Avienus, the son of Anicius Probus Faustus *niger* who, as consul in 490, would have been Odovacer's last nominee to this office, and the grandson of Gennadius Avienus (consul 450). Little is known of Volusianus, consul in 503, although in 510–11 he is referred to as *patricius*, and his name had certainly been one to conjure with in the third and fourth centuries.[47] Rufius Petronius Nicomachus Cethegus, on the other hand, sole consul in 504, was the son of Probinus, who was Odovacer's second last consular nominee (489) and one of the

[41] He was probably not the Presidius, 'a man of no mean station', who lived in Ravenna and was on bad terms with the Goths in the time of Vitigis: Procopius, *BG* 2.8.2.

[42] Wolfram, *History*, 284 misleads on the consulship in 495.

[43] *PLRE* 2.1024f., where a connection with Olybrius is suggested on the basis of the similarity of his name to that of Speciosa, who was related to Olybrius (Ennodius, *ep*. 2.13.6); and for his prefecture, *CIL* 6.37130 (= 15.7119).

[44] Stein, *Histoire*, 113f. n. 3 is sensible.

[45] He is probably the Paulinus 'consularis vir' of Boethius, *phil. cons*. 1.4.13. See too Cassiodorus, *var*. 1.23, 2.3, 3.29.

[46] Cassiodorus, *var*. 1.33.

[47] Ibid. 4.22f. Earlier centuries: *PLRE* 1.976–81.

strongest supporters of Laurentius in the schism which still divided Rome, and the grandson of Odovacer's second nominee to the consulship, Rufius Achilius Maecius Placidus (consul 481); by 512 he was *patricius*, and at an uncertain date Cassiodorus wrote his *Ordo generis Cassiodororum* for him.[48] The office was held in 505 by Theodorus, a brother of Albinus (consul 493) and hence a well-connected member of the Decii, who was a *patricius* by 509,[49] while in 506 it passed to Ennodius Messala, the brother of Avienus (consul 502) and hence an Anicius. The Western consul for 507, Venantius, was the son of Petrus Marcellinus Felix Liberius, who had not held a consulship but was a patrician who had held the office of praetorian prefect of Italy and was soon to become praetorian prefect of Gaul, and who had some connection with Avienus, the consul of 502.[50] Basilius Venantius (consul 508) was presumably the son of Decius Marius Venantius Basilius (consul 484), who may have been a brother of the consul of 480; at about the time of his consulate he was *patricius*.[51] In 509 the office passed to Inportunus, the brother of Albinus (consul 493), and so the fourth son of Caecina Decius Maximus Basilius (consul 480) to be appointed to the office during the reign of Theoderic. It is clear that for two if not three years the office had been held by members of the Decii, but in 510 it passed to Anicius Manlius Severinus Boethius, the son of Marius Manlius Boethius (consul 487) and perhaps a relative of the Severini who were consuls in 461 and 482.[52]

We are obviously dealing with an extraordinary group. There were ten Western consuls in the period: eight seem to have been the

[48] Ennodius, *MGH AA* 7.314.38 refers to him as *patricius*. The *Ordo generis Cassiodororum*, also known as the *Anecdoton Holderi*, is most conveniently edited by Mommsen, *MGH AA* 12. v. It may also be consulted in the editions of Krautschick (*Cassiodor*, 84) and O'Donnell (*Cassiodorus*, 260f.); the edition of Fridh (CCSL 96A.1, pp. vf.) is too free in its emendations.

[49] Cassiodorus, *var*. 1.27.

[50] On Liberius, L. Cantarelli, 'Il patrizio Liberio e l'imperatore Giustiniano', in his *Studi Romani e Bizantini* (Rome, 1915), 289–303, and O'Donnell, 'Liberius'. His family connections are elusive: Ennodius writes of him as the *parens* of Avienus (consul 502, *ep*. 9.7.2; the word is not suggestive of a close relationship) and of Venantius as having a blood relationship with himself (ep. 5.22.1). But 'Venantius' was a name in common use among the Decii, the consul for the year immediately following providing an example.

[51] Cassiodorus, *var*. 3.36.

[52] Lotter, *Severinus von Noricum*, esp. 246–53, has suggested, but I do not think demonstrated, that the holy man Severinus of Noricum is identical with the Severinus who was consul in 461. Cf. M. van Vytfanghe, 'Les Avatars contemporains de l'"hagiologie"', *Francia*, 5 (1977), 639–71, esp. 667f.

sons of consuls, seven are known to have held at some stage the dignity of *patricius*, and the consular *fasti* for the period following the death of Theoderic are replete with men whom we may see as their connections. Vettius Agorius Basilius Mavortius (consul 527) was presumably related to Caecina Mavortius Basilius Decius (consul 486), perhaps to be seen as the brother of the prolific Caecina Decius Maximus Basilius (consul 480) and Decius Marius Venantius Basilius (consul 484). The Western consul for 529, Decius, and that for 534, Paulinus, were sons of Basilius Venantius (consul 508), who seems to have had at least one other consular son, while the names of Anicius Faustus Albinus Basilius (consul 541) suggest close ties with earlier consuls, in particular that of 493.[53]

The consuls appointed in the period 511–21, on the other hand, were men of a different stamp. Felix, the first of them, was a provincial from Gaul, whose appointment may plausibly be associated with Ostrogothic occupation of territory formerly held by the Visigoths.[54] There was no Western consul in 512, and in 513 the office passed to one Probus, a scholarly man whose family does not seem to have been lofty.[55] Magnus Aurelius Cassiodorus Senator (consul 514) was obviously a very considerable figure, who had been appointed *patricius* by 537, but his roots were in Brutium rather than Rome, and he was certainly the first consul in the family.[56] Florentius, the Western consul for 515, is an almost totally obscure figure (although he may have been the man of this name who was the father of Constantiolus, a later *magister militum per Moesiam*);[57] so too was Peter, the sole consul for 516, although Cassiodorus purported to find his family impressive.[58] In 517 the Western consul was Agapitus. He may have been a *patricius* as early as 509–10, he may have been related to Faustus *niger*, and he may for that matter have been a Ligurian, but there were two contemporary Agapiti, and none of these pieces of information

[53] Cameron and Schauer, 'The Last Consul'.

[54] But note that Cassiodorus draws attention to his ancestry: *var*. 2.1.2, 2.2.5.

[55] His family was known for its 'studiis et probitate' (Ennodius, *ep*. 7.27.1); commendable virtues no doubt, but Ennodius was capable of higher flights of praise.

[56] *Patricius*: *var*., MGH edn., p. 1. Roots in Brutium: *var*. 1.4.13. I accept the characterization of Cassiodorus and his family given by J. Matthews, 'Anicius Manlius Severinus Boethius', in M. Gibson, ed., *Boethius: His Life, Thought and Influence* (Oxford, 1981), 15–43, at 28f.

[57] John Malalas, *chron*. 438; Theophanes, *chron*. AM 6031.

[58] 'parentum luce conspicuus': *var*. 4.25.3.

need relate to the consul of 517.[59] There was no Western consul in 518; in 519 the colleague of the emperor Justin in this office was a Goth, Eutharic Cilliga, husband of Theoderic's daughter Amalasuintha, following in the footsteps of his father-in-law, who had himself been consul in 484. The Western consul for 520, Rusticius, and that for 521, Valerius, appear totally obscure. With the joint consulship of Boethius' sons Symmachus and Boethius in 522 we enter a new period, and it will be best to defer consideration of it.

It must be said that there are circumstances which militate against the consuls of 511–21 being amply documented: Cassiodorus laid down his pen as Theoderic's *quaestor* in 511 and only took it up again as his *magister officiorum* in 523, and Ennodius seems to have stopped writing in 513. But when all allowances are made one cannot help noticing that several of the consuls are mere names, not attested otherwise; that one was a Goth and another a Gaul; and that only two are known to have been advanced to the patriciate. None can be shown to have been the son of a consul.

Theoderic's consular nominations, opening up this most coveted office to a wider spectrum of society as they did, can have brought no joy to the Roman aristocrats in the second decade of the sixth century. The consulship was by then the only surviving magistracy of the republican period with any stature,[60] and was particularly sought after because the consul still 'gave his name to the year'.[61] But families also desired the 'fama publica' won by the distribution of largesse and the holding of games; the favourable impression made by consuls was commensurate with how much they spent.[62] Contemporary accounts emphasize the public nature of consular

[59] *Patricius*: Cassiodorus, *var*. 1.23, 1.26. Faustus as *frater* of Agapitus: Ennodius, *ep*. 5.26.3, obviously not to be taken literally. Liguria: ibid. 1.13.3. On the Agapiti, *PLRE* 2.30–2; Tjäder, *Papyri*, 2.298 n. 27.

[60] Jones, *Later Roman Empire*, 532; see e.g. Boethius, *phil. cons*. 3.4.15: 'praetura magna olim potestas, nunc inane nomen et senatorii census gravis sarcina'.

[61] 'anno nomen imposuit', Ennodius, *ep*. 1 5.1 to Faustus *niger* on the consulship of his son Avienus in 502; see too *lib. pro syn*. 133. The same point occurs in Cassiodorus' *formula consulatus, var*. 6.1.1, 7, and *chron*., *praef*. This system of dating can create difficulties for the historian; see e.g. below, Appendix 1. With Western assessments of the office of consul we may compare Ioannes Lydus, *Powers*, 2.8 who notes the expenditure it entailed and its use in chronology before observing that it was 'the mother, as it were of the Romans' freedom. For it stands in opposition to tyranny; and, when the former prevails, the latter ceases to exist' (trans. Bandy). Could this have been written in Ostrogothic Italy?

[62] 'fama publica': Cassiodorus, *var*. 3.39.1, in a letter to Felix (consul 511) encouraging *liberalitas* and *munificentia*. 'Opinio' and expenditure: ibid. 2.2.6, 6.1.7 (*gratia publica* to be had by giving).

office and the possibilities it offered for gaining popularity among the people: Asterius (consul 494) was aware of the impression he made, Eutharic (consul 519) displayed different kinds of wild animals, some sent from Africa, and distributed largesse, while Boethius recalled that when his sons were joint consuls in 522 he had seen them being carried amid a throng of the *patres* and of the *plebs*. As Boethius delivered an oration in praise of Theoderic they had sat on their curule seats, and in the circus the proud father, between his sons, had satisfied the expectations of the crowd which had gathered about with triumphal largesse.[63] Consular ivories depict holders of the office sitting on the *sella curulis* presiding over the games, and bags of money for distribution to the people are frequently in evidence.[64] In 507 Ennodius contemplated the large number of consuls who held office in Theoderic's reign with satisfaction;[65] a few years earlier he had seen the consulship of Avienus as the answer to his father's prayers, and we have no reason to see the consulship as having been other than actively sought.[66] During Avienus' tenure of office (502) street fighting associated with the Laurentian schism broke out in Rome, and any goodwill won by consular largesse would have done no harm to the side supported by the donor. As it happens Avienus' father Faustus *niger* was, as we have seen, the only senator whom we know to have supported Pope Symmachus, and Symmachus enjoyed the support of the *plebs*. It may not be accidental that the two men whom one of the circus factions, the Greens, accused of perpetrating plots against it in 509 were Inportunus, the consul for that year, and his brother Theodorus, who had held the office in 505.[67] One cannot

[63] Asterius: above, n. 38. Eutharic: Cassiodorus, *chron. s.a.* 519. Boethius' sons: *phil. cons.* 2.3.8; for the *sellae* cf. Cassiodorus, *var.* 6.1.6: 'sellam curulem pro sua magnitudine multis gradibus enisus ascende'.

[64] e.g. those of Boethius, Clementinus, and Orestes, respectively figs. 6, 15, and 31 in W. F. Volbach, *Elfenbeiarbeiten der Spätantike und des frühen Mittelalters* (Mainz, 1952).

[65] 'hic actum est, ut plures habeas consules, quam ante videris candidatos' (*pan.* 48). Note however that consuls nominated by Odovacer held office annually from 480 to 490.

[66] Ennodius, *ep.* 1.5.12: 'fidelium orationum vestrarum retributio est circa suboles dignitas inpetrata'. In general, I agree with Cameron and Schauer, 'Last Consul', 139.

[67] Cassiodorus, *var.* 1.27.2. Cassiodorus was philosophical: 'ad circum nesciunt convenire Catones' (1.27.5), nicely translated by Hodgkin: 'It is not exactly a congregation of Catos that comes together at the circus.' There may be a reminiscence of the preface to the first book of Martial's epigrams:

> Non intret Cato in theatrum meum, aut si intraverit, spectet . . .
>
> Cur in theatrum, Cato severe, venisti?

imagine the Anicii and Decii taking pleasure in relinquishing consular office in the decade following 510.

Appointments to the office of *praefectus urbis Romae* seem to have followed a similar pattern. Initially Theoderic appears to have maintained the general policy of Odovacer. Thus, the Speciosus who became consul in 496 had served three terms as *praefectus urbis*, presumably under both Odovacer and Theoderic.[68] If the Catulinus who seems to have held the office in about 500 remains almost totally obscure,[69] Anicius Acilius Aginantius Faustus *albus*, whose office is probably to be dated to 502–3, had held it previously towards the beginning of Odovacer's regime and had subsequently gone on to become consul (483).[70] But Constantius, who may have held office 506–7, was a Ligurian,[71] and if the Agapitus who held office in 508–9(?) was a senator and is to be identified with the Agapitus who was consul in 517,[72] Artemidorus, *praefectus urbis* 509–10, was an Easterner who preferred to live under Theoderic;[73] the father of Argolicus (*praefectus urbis* 510–11) had been *comes privatorum* and his grandfather *comes sacrarum largitionum* and *magister officiorum*, but neither attained the office of prefect,[74] and if we choose to see the Florianus whose tenure of the office of *praefectus urbis* can be dated securely only to the period in which Anastasius and Theoderic were both reigning (i.e. before 518) as holding office after 507, which seems reasonable, we are confronted with a

[68] *PLRE* 2.1024f.

[69] On the date, Sundwall, *Abhandlungen*, 107.

[70] *PLRE* 2.451f.

[71] Ennodius, *ep*. 2.19.1, a letter of some theological interest.

[72] *PLRE* 2.30–2; S. Krautschick, 'Bemerkungen zur PLRE II', *Historia*, 35 (1986), 121–4 at 121. His tenure of this office is only known from an undated inscription, Fiebiger/Schmidt, *Inschriftensammlung*, 185 (*CIL* 6.1665); Ennodius, *ep*. 5.26, of 506–7, implies he held high office in Ravenna then.

[73] Cassiodorus, *var*. 1.43.2, which indicates that he was related to Zeno; earlier he had participated in an embassy from Zeno to Theoderic (Malchus, *frag*. 18 (Blockley, 20)). A problem is caused by a reference to 'Arthemidori v.c.' in an inscription on the Colosseum; the possibilities are discussed by Chastagnol, *Sénat romain*, 81. It can be argued on the basis of Cassiodorus, *var*. 1.43.3 that he had held the office of *tribunus voluptatum* (so Sundwall, *Abhandlungen*, 94), in which connection it is interesting to recall a charge later made by the Goths that the only Greeks they had seen in Italy were 'actors of tragedy and mimes and thieving sailors' (Procopius, *BG* 1.18.40).

[74] Cassiodorus, *var*. 3.12.

successful career bureaucrat.[75] Of these five men whose tenure of the office of *praefectus urbis* seems to have been subsequent to 506 not one became consul, and not one can be securely located in any of the great families of Rome. It is possible that similar changes occurred in the kind of men being appointed to other high offices, but the evidence does not permit generalization, as the urban prefecture, being subject to a high turnover, is inevitably better documented.[76]

How are we to assess this evidence? It would certainly be possible to draw too sharp a distinction between court *apparatchiks* and the members of senatorial families: as we shall see, Boethius received commissions from Theoderic long before he took up office in Ravenna, and Ennodius was undiscriminating in badgering prospective patrons. Further, one of the characteristics of noble classes is their capacity to survive by recruiting new members, and we need not be surprised if new fish were finding their way into the tank in the time of Theoderic. Nevertheless, it is hard to avoid the impression that Theoderic turned decisively towards *novi homines* in the 510s. Hitherto, his consular appointments had echoed those of Odovacer, at least half of whose appointees had been sons of consuls.[77] We may take it, then, that towards the end of the first decade of the sixth century Theoderic's policies concerning recruitment to the highest offices

[75] *PLRE* 2.480; the case for identifying Florianus with Valerius Florianus seems strong.

[76] *Fasti* for the various offices may be consulted in *PLRE* 2.1242 ff. and Sundwall, *Abhandlungen*, 171 ff.

[77] I take Fl. Caecina Decius Maximus Basilius iunior (consul 480), Decius Marius Venantius Basilius (consul 484) and Caecina Mavortius Basilius Decius (consul 486) to have been sons of Fl. Caecina Decius Basilius (consul 463), Anicius Acilius Aginatius Faustus iunior (consul 483) to have been son of Anicius Acilius Glabrio Faustus (consul 438), Fl. Anicius Probus Faustus iunior (consul 490) to have been son of Gennadius Avienus (consul 450) and, perhaps unfairly in the context of the present argument, (Petronius) Probinus (consul 489) to have been son of the consul of 481. It seems to me likely that Severinus iunior (consul 482) was son of Fl. Severinus (consul 461) and that Q. Aurelius Memmius Symmachus iunior (consul 485) was son of Symmachus (consul 446). It is hard to detect consular antecedents for Placidius (consul 481), Boethius (consul 487), or Claudius Iulius Eclesius Dynamius (consul 488); however, the colleague of the last named, Rufius Achilius Sividius, was probably a descendant of Acilius Glabrio Sibidius Spedius, whose son, Anicius Acilius Glabrio Faustus, was consul in 438.

changed.[78] It is difficult to see why he turned away from the aristocrats at that particular time, but it is possible that his change of policy was connected with his final decision against Laurentius, who enjoyed widespread senatorial support, in 507; perhaps a degree of punishment, and conceivably fear, were involved. But we may accept that there was such a change, and that this affected the way the Roman nobility looked at their king. For whatever reason, relations between Theoderic and the lay holders of power in Rome cannot have been good in the second decade of the century.

It need hardly be said that the holding of office permitted the exercise of power in an environment where the restraint of someone who placed integrity above gold could cause astonishment,[79] and the correspondence of Ennodius allows fascinating insight into how patronage could be exercised at Ravenna. A sequence of letters addressed to one Agapitus is instructive. The first letter, written at the beginning of 503, complains that Agapitus had not been in touch with Ennodius when he entered into office (*ep.* 1.13); unfortunately we do not know what the office was. Ennodius dispatched another letter in 505, again complaining of Agapitus' silence and asking him to carry out the request (*suggestio*) made by the deliverer of the letter.[80] Writing towards the end of 505 or early in 506 Ennodius commented that Agapitus seemed to have lapsed into silence, but asked him to help Panfronius (*ep.* 4.16), on whose behalf he had also written to Faustus *niger* (*ep.* 4.14); within the year Panfronius had entered into an office at Ravenna (*ep.* 5.16), and we may plausibly associate the appointment with Ennodius' efforts on his behalf. Meanwhile, a *sublimis et magnificus vir* was labouring under the plots of his enemies, and Ennodius wrote to Agapitus seeking his assistance (*ep.* 4.28); and at the beginning of 507 he wrote again requesting the intervention of Agapitus in church matters (*ep.* 5.26).[81] When the deacon Stephen travelled to Ravenna in 508 Ennodius

[78] The classic discussion is that of Sundwall, *Abhandlungen*, 215 f. (but see already Hartmann, *Geschichte*, 180), followed e.g. by Bury, *History*, 466 and Bertolini, *Roma*, 76–80.

[79] Augustine, *conf.* 6.10.16. See too Cassiodorus, *var.* 9.24.4.

[80] *Ep.* 4.6. For the phrase 'Ravennates excubiae' (*ep.* 4.6.1) cf. *ep.* 2.17.2, 2.27.1, 4.24.3, 6.21.1.

[81] See further on this last letter Sundwall, *Abhandlungen*, 41.

was able to provide him with a letter of introduction to his powerful friends Liberius, Eugenes, Agapitus, Senarius, and Albinus (*ep.* 6.12).

But Ennodius' letters to Agapitus are as nothing when compared to the flood of letters of introduction or recommendation addressed to Faustus *niger*.[82] It is noteworthy, however, that Ennodius' approaches to his powerful friend on behalf of people involved in disputes over property were made while Faustus held office at Ravenna,[83] and Faustus was by no means omnipotent. Shortly before he became *quaestor* in 503 he received from Ennodius a letter containing allusions to his enemies;[84] in 506 Ennodius wrote to him complaining about enemies, and very shortly afterwards Faustus had ceased to be *quaestor*;[85] before long Ennodius was writing to Senarius mentioning his distress at the various reports that were reaching him concerning Faustus.[86] Two adjacent letters in Cassiodorus' *Variae* are evidence for the trials to which Faustus was exposed: in one, referring to a dispute over property, Faustus, then praetorian prefect, was described in hostile terms (*var.* 3.20.4), and in the immediately following letter Faustus, addressed simply as *vir illustris* and not as prefect, was given permission to absent himself from Rome for four months (*var.* 3.21). It is difficult to know whether the leave of absence granted in the second letter is to be seen as a punishment, and, if it is, whether it was the consequence of the matter mentioned in the first letter, but even the phrase 'notus ille artifex' used by Cassiodorus gives point to Ennodius' fears concerning the enemies of Faustus. Eugenes, the

[82] *Ep.* 2.24, 3.3, 3.22, 4.9 (written in 505 on behalf of the *vir clarissimus* Venantius, perhaps to be identified with the Venantius, son of Liberius, who was consul in 507), 4.14, 6.15, 6.25, 7.30, 9.2. On one occasion Ennodius asked Faustus to write a letter of introduction for someone else (*ep.* 4.18); on another he suggested to an abbot the sending of *commendaticia* to Faustus at Ravenna (*ep.* 3.4).

[83] *Ep.* 2.23 on behalf of Lupicinus, written in late summer 503, and 4.5 on behalf of Dalmatius, written in early 505; for the dates of Faustus' period as *quaestor*, *PLRE* 2.455 and Krautschick, 'Bemerkungen', 122. See too *ep.* 3.20 seeking help for Julianus, troubled by the plots of Marcellinus.

[84] *Ep.* 2.10.4; Faustus was *quaestor* by the time Ennodius wrote *ep.* 2.25.

[85] *Ep.* 4.24.2; *ep.* 4.26 is addressed to Eugenes, the new *quaestor*. The first sentence contains a pun on 'Faustus'.

[86] *Ep.* 4.27; see too *ep.* 6.20 for Ennodius' anxieties concerning Faustus.

successor of Faustus as *quaestor*, was a friend of Ennodius, but he was in turn quickly succeeded by Cassiodorus, a man who, towards the end of his long career, was aware that there had been those who had spoken ill of him.[87] Cassiodorus is not addressed or even mentioned directly in Ennodius' voluminous correspondence. The atmosphere among the Romans at Theoderic's court must have been one of strenuous competitiveness between individuals and factions. But by the second decade of the sixth century the aristocrats who for decades had been able to enjoy the pleasures of high office found themselves excluded from the consulship and urban prefecture. We do not know how far down the hierarchy of positions Theoderic's new policy extended, but it is clear that the Romans of Rome, the men who had thrown their weight behind the candidacy of Laurentius for the papacy, were no longer as powerful as they had been.

THE ROMANS OF ROME

What were men such as Festus and Symmachus doing as the new men moved into positions of power? Each of these men had been engaged in some form of political activity early in Theoderic's reign, Festus on his two missions to Constantinople and Symmachus with his speech *pro allecticiis*,[88] but references to them in the *Variae* of Cassiodorus written during his period as *quaestor*, hence 507–11, allow them to be seen engaged in other pursuits. They were jointly involved in a legal action against the patrician Paulinus,[89] each was entrusted by Theoderic with the oversight of youths from the provinces studying in Rome, the only people known to have been so honoured,[90] and each was involved in public good works, Festus in

[87] *PLRE* 2.415, 266. Cassiodorus' awareness: *var.* 9.24.7.

[88] *PLRE* 2.1045 and Chadwick, *Boethius*, 6, suggest that Symmachus undertook a mission for Theoderic in Constantinople. While he travelled there at least once as we shall see later in this chapter, and L. Obertello, *Severino Boezio* (Genoa, 1974), 21f. has argued for a trip to Constantinople on the basis of *coll. avel.* 221.2 and 229.1, there is no reason to see Theoderic as having been involved.

[89] Cassiodorus, *var.* 1.23. The cause of the litigation is not clear, but the language of Cassiodorus ('quietus populus, concors senatus totaque res publica') makes it tempting to associate it with the troubles in which the Decii were involved. We shall return to this subject in Ch. 7.

[90] Cassiodorus, *va.* 1.39 to Festus (cf. 2.22), 4.6 to Symmachus.

seeing to the transport of marble from the domus Pinciana to Ravenna and Symmachus in undertaking repairs to the theatre of Pompey, for which he was reimbursed out of state funds.[91] To this extent it is possible to see them as something of a pair. But it is noteworthy that Cassiodorus' letters to Symmachus are longer and altogether more literary than those he wrote to Festus, and it will be worth our while examining Symmachus as an intellectual figure.

In his *Ordo generis Cassiodororum*, Cassiodorus reports that Symmachus wrote a history of Rome in seven books, imitating his forefathers (*parentes*).[92] Some scholars have argued that this lost work of Symmachus attributed great significance to the deposition of Romulus Augustulus by Odovacer in 476, and suggested on the strength of this that Symmachus must have been opposed to the government of the Goths.[93] However, the evidence for this position is decidedly weak,[94] and we may refrain from speculation as to the contents of the lost history beyond acknowledging that the use to which Jordanes put it in his *Romana* shows that Symmachus made use of the *Scriptores historiae augustae*.[95] This source is of interest, for the only possible known forefather of Symmachus whom he would have been in a position to imitate was the pagan historian Nicomachus Flavianus, who committed suicide following the defeat of Eugenius—a usurper who had agreed to restore the altar of Victory—by Theodosius I in 394.[96] Nicomachus' son married the daughter of the famous pagan intellectual Quintus Aurelius Symmachus, and quite apart from Cassiodorus' comment we have every reason

[91] Ibid. 3.10 to Festus (the palatium Pincianum became the headquarters of Belisarius following his capture of Rome: *lib. pont.* 291.18, 292.17); 4.51 to Symmachus, 'the last recorded case in Italy of private patronage of a traditional secular building': Ward-Perkins, *Urban Public Building*, 44.

[92] Ed. Mommsen, p. vi.

[93] See esp. Wes, *Das Ende*; as well, A. Momigliano, 'La caduta senza rumore di un impero nel 476 D.C.', *Annali della scuola normale superiore di Pisa*, 3rd ser. 3 (1973), 394–418.

[94] Well discussed by Croke, 'A.D. 467', but see too among others R. Luiselli, 'Sul De summa temporum di Iordanes', *Romanobarbarica*, 1 (1976), 82–134; R. Bratóz, *Severinus von Noricum und seine Zeit* (Österreichische Akademie der Wissenschaften, Phil./Hist. Klasse, Memorial vol. 165; Vienna, 1983), 12–14 (n. 35); Krautschick, 'Zwei Aspekte', 355 ff., and G. Zecchini, 'Il 476 nella storiografica tardoantica', *Aevum*, 59 (1985), 3–23. One cannot help but note that not all the contributors to this discussion have been aware of their predecessors.

[95] Jordanes, *get.* 83, 88.

[96] Rufinus, *hist. eccl.* 2.33; Sozomen, *Kirchengeschichte*, ed. J. Bidez and G. C. Hansen (Berlin, 1960), 7.22.

to believe that a strong tradition operated in the family, for the full name of the younger Symmachus was Q[uintus] Aurelius Memmius Symmachus. The elder Symmachus is most famous for his attempt, thwarted by St Ambrose of Milan, to restore the altar of Victory in Rome in 382, and memories of this lingered, for one of Ennodius' epigrams concerns this very subject.[97] That Ennodius sent this epigram to his patron Faustus *niger* is suggestive of tensions among the Roman aristocracy of the early sixth century, but we would perhaps be safer in seeing it as a product of the strong Ambrosian traditions of the Church of Milan, of which Ennodius was then a deacon. It is easy to contrast Symmachus with the author of another lost history, Cassiodorus, who undertook his Gothic history under the auspices of the government, 'by the order of king Theoderic' as he explains, perhaps not without pride.[98] It was a work which its author saw as assimilating the origin of the Goths into Roman history,[99] and what this meant in practice we can deduce partly from the *Getica* of Jordanes, which proclaims itself to be in essence an abbreviation of Cassiodorus' twelve volumes on the origin and deeds of the Goths,[100] and partly from another historical work Cassiodorus wrote for a Goth, his *Chronica*, which he states Eutharic directed him to undertake in the year of his consulship (519) and which reveals a persistent reworking of older material to highlight the Goths and smooth over inconvenient episodes in their past.[101] Whatever the lost history of Symmachus was like, it can scarcely have been like this. That the intellectual interests of Symmachus were antiquarian and hospitable to pagan thought is also suggested by a subscript to Macrobius' *Somnium Scipionis*, one which indicates

[97] Attempt of elder Symmachus: Ambrose, *ep.* 17; Symmachus, *rel.* 3. Ennodius' epigram: *carm.* 2.142 (p. 266).

[98] *Ordo generis Cassiodororum*, ed. Mommsen, p. vi. See most recently on the date of this work S. J. B. Barnish, 'The Genesis and Composition of Cassiodorus' "Gothic History"', *Latomus*, 43 (1984), 336–61; W. Goffart, *The Narrators of Barbarian History* (Princeton, NJ, 1988), 32–5.

[99] 'Originem Gothicam historiam fecit esse Romanam': Cassiodorus, *var.* 9.25.5; for the turn of phrase cf. 'ut Graecorum dogmata doctrinam feceris esse Romanam', *var.* 1.45.3.

[100] Jordanes, *get.* 1. The richest treatment of relations between Cassiodorus and Jordanes remains Momigliano, 'Cassiodorus', but both Momigliano's thesis in particular, and a close relation between the works of Cassiodorus and Jordanes in general, have been denied by Goffart, *Narrators*, 21–42, 58–62, 68–73.

[101] The propagandistic purpose of this work is brought out by O'Donnell, *Cassiodorus*, 38 ff.

that he revised the text of this work in co-operation with Macrobius Plotinus Eudoxius.[102] One has the feeling that the past weighed heavily on Symmachus; indeed, even his repairing the theatre of Pompey may be a sign of adherence to family tradition.[103] In a letter addressed to Symmachus, Cassiodorus described him as 'a most careful imitator of the ancients, a most noble teacher of the moderns',[104] and nothing we know tells against the accuracy of the description.

Symmachus, in short, was a considerable figure. When Cassiodorus came to organize his correspondence for publication as the *Variae*, of the books containing letters written while he was *quaestor* he concluded book 4 with a letter to Symmachus, and books 1, 2, and 5 with letters to kings; it is a minor puzzle why a letter to the *comes privatarum* Apronianus was chosen to conclude book 3. It is scarcely surprising that Symmachus attracted young men of scholarly bent. In 509 Ennodius sought to initiate an exchange of correspondence with him, wishing to be regarded as his 'cliens et famulus' (*ep.* 7.25), and in the following year he informed the student Beatus that a letter of admonition which he had addressed to him had been forwarded to Symmachus for the latter's emendation (*ep.* 8.28). This is presumably to be identified with *Opusculum* 6 of the works of Ennodius which, although it is addressed to Ambrosius and Beatus, concludes with a blatant appeal for the patronage of Symmachus.[105] We have no reason to believe that these endeavours, which occurred only after the official termination of the Laurentian schism, bore fruit. Quite different was the story of the relations of Symmachus with a still younger man, Anicius

[102] On the relationship of the colleague of Symmachus with the author of the text, consult J. Flamant, *Macrobe et le néo-Platonisme latin à la fin du IVe siècle* (Leiden, 1977), 131 f., and S. Panciera, 'Inscrizione senatorie di Roma e dintorni', *Tituli*, 4 (1982), 591–678 at 658–60.

[103] The 'auctores vestri' of Cassiodorus, *var.* 4.51.3 may refer to Aurelius Anicius Symmachus, *praefectus urbis* 418–20, who carried out works in the portico of the theatre of Pompey: Della Valle, 'Teoderico e Roma', 156f.

[104] 'antiquorum diligentissimus imitator, modernorum nobilissimus institutor' (*var.* 4.51.2); cf. 'antiqui Catonis fuit novellus imitator' (*Ordo generis Cassiodororum*, ed. Mommsen, p. v) for another reference by Cassiodorus to Symmachus as an imitator, clearly a word which came readily to mind with regard to him. Note too a similar turn of phrase in Jordanes, *get.* 58, 'Dio storicus et antiquitatum diligentissimus inquisitor.'

[105] Note in particular: 'Nil moror: en supplex venio, miserere precanti, Vilia divitibus commendans dicta patronus' (p. 315).

Manlius Severinus Boethius. Born in about 480, he was the son of Marius Manlius Boethius, who had served Odovacer as prefect of the city of Rome and praetorian prefect, holding a consulship in 487 and being appointed patrician.[106] Boethius later recalled that he had been orphaned but had been cared for by some of the most lofty men and subsequently been joined to some of them by marriage, so referring to his marriage to Symmachus' daughter Rusticiana.[107] It was presumably under the aegis of Symmachus that he received the education which was to equip him as such a formidable scholar. Eastern Christian theological method provides a background against which his thought can be located, and it is generally believed that he studied in the East, although there is no agreement as to where.[108] It is certainly possible that Boethius found himself at some time during his education among Christians who were out of communion with the see of Rome. But the strongest influence on Boethius seems to have been his father-in-law. It is clear that he retained an intense admiration for Symmachus until the end of his life.[109] In his last work, the *Consolation of Philosophy*, Boethius drew on the *Somnium Scipionis* of Macrobius, and it is pleasant to think that he may have availed himself of the text revised by Symmachus and Macrobius Plotinus Eudoxius.[110]

As his full name indicates, Boethius was a member of the *gens*

[106] On Boethius' father, *PLRE* 2.232f.; on his name, Alan Cameron, 'Boethius' Father's Name', *Zeitschrift für Papyrologie und Epigraphik*, 44 (1981), 181–3. Obertello puts the birth of Boethius back to 475/7: *Severino Boezio*, 17ff. On relations between Symmachus and Boethius, Wes, *Das Ende*, 94–6 is good.

[107] Boethius, *phil. cons*. 2.3.5; Procopius, *BG* 3.20.27; Anon. Vales. 92.

[108] As is well known, Cassiodorus, *var*. 1.45.3, 'Atheniensium scholas longe positus [*not* positas] introisti' need not be taken literally, and the most influential modern commentator has argued for Alexandria as the place of his education: P. Courcelle, *Late Latin Writers and their Greek Sources* (Cambridge, Mass., 1969), esp. 316–18. But Obertello is sceptical (*Severino Boezio*, 27–9), and Athens seems to be envisaged by C. J. de Vogel, 'Boethiana I', *Vivarium*, 9 (1971), 49–66 and Pietri, 'Aristocratie et société cléricale', 440. On the Eastern theological background, B. E. Daley, 'Boethius' Theological Tracts and Early Byzantine Scholasticism', *Mediaeval Studies*, 46 (1984), 158–91. It has been suggested, but not demonstrated, that the figure of a writer on a diptych of the period preserved at the cathedral at Monza represents Boethius: A. Effenberger, *Frühchristliche Kunst und Kultur* (Leipzig, 1986), 285.

[109] Boethius, *phil. cons*. 1.4.40, 2.3.6, 2.4.5; cf. 'paterna gratia' (*de arith.*, *praef.*, ed. Friedlein, 5.21) and the dedication to Symmachus of tractate 1.

[110] Boethius, *phil. cons*. 2.7.7f.

Anicia. Towards the end of the fourth century the Anicii, famed for their wealth and adherence to Christianity, had enjoyed a period of extraordinary renown in Rome.[111] In 535 Cassiodorus, writing on the occasion of the elevation of Maximus to the office of *primicerius domesticus*, recalled that 'an earlier age brought forth Anicii who were almost equal to emperors' (*var*. 10.11.2). There is more to this than meets the eye, for one of Maximus' ancestors, Petronius Maximus, had been Augustus briefly in 455 following the murder of Valentinian III.[112] In another letter concerning the appointment of Maximus, Cassiodorus noted that what an Anicius did should not be spoken of in a low style, for the family was famous throughout the world and could truly be called noble when (*quando*; perhaps 'since') proper conduct did not depart from it (*var*. 10.12.2). Again Cassiodorus may be proceeding by allusion rather than speaking openly, for the slightly unexpected comment about proper conduct may conceal a reference to the behaviour of Boethius which led to his fall towards the end of Theoderic's reign, and the fact that the letter expressing this sentiment was addressed to the senate may not be without significance. But it is difficult to argue from such data for the existence of a powerful clan whose members had values in common and acted in similar ways. On the evidence of names, Ennodius' patron Anicius Probus Faustus *niger* was as much an Anicius as Boethius, but he is reported to have been the only senator to have supported Pope Symmachus during the Laurentian schism, whereas Anicius Acilius Aginantius Faustus *albus* seems to have inclined towards the other side, and the animosities which rent the Decii early in the sixth century should warn us against attributing cohesion to the members of a family.[113] Ennodius could describe Albinus as a 'parens' of Faustus *niger* (*ep*. 2.22.1), yet as we have seen Albinus was indubitably a Decius. Most significant,

[111] Jerome, *ep*. 130.3.1, Prudentius, *contra Symm*. 1.553, and esp. Ammianus Marcellinus, 27.11.1 on Probus: 'claritudine generis et potentia et opum magnitudine cognitus orbi romano, per quem universum paene patrimonia sparsa possedit'. See in general A. Momigliano, 'Gli Anicii e la storiografia latina del VI sec. D.C.', *Rendiconti accademia dei Lincei*, 8th ser. 11 (1956), 279–97, most conveniently available in his *Secondo contributo alla storia degli studi classici* (Rome, 1960), 231–53, from which subsequent citations will be made.

[112] Procopius, *BG* 1.25.15.

[113] J. Moorhead, 'The Decii under Theoderic', *Historia*, 33 (1984), 107–15.

however, is the lack of interest displayed by Cassiodorus (at least in his correspondence) and Ennodius in the Anicii, and their apparently higher level of interest in the Decii;[114] further, as we have seen, the first consular appointees of Odovacer, Theoderic, and Athalaric were all Decii. Membership of the Anicii was doubtless not to be despised, but Boethius owed his fame to other factors.

During the years when Cassiodorus was serving Theoderic in Ravenna as *quaestor*, Boethius received three commissions from the king: he was to enquire into the payment of the soldiers of the palace, see to the construction of a water-clock and sundial for Gundobad, king of the Burgundians, and select a harpist to be sent to Clovis, king of the Franks. It is indicative of Boethius' standing that the second and third of these commissions involved matters of cultural prestige and were made in letters of quite extraordinary length, and when Cassiodorus edited the *Variae* he positioned them immediately before the letters to Gundobad and Clovis which he placed in positions of honour at the end of books 1 and 2.[115] The phrasing of these letters, all written in the period 507–11, makes it clear that Boethius was already a person of considerable scholarly reputation, and works of Ennodius which date from this period vouch for his youthful accomplishments and skill at Greek.[116] But Boethius was a thinker whose ability was fully extended by his intellectual ambition, and beyond his theological tractates and translations of works on the liberal arts he conceived the project of translating into Latin all the works of Aristotle and the dialogues of Plato, writing commentaries where necessary and showing their fundamental agreement.[117] It was a task of no small

[114] Cassiodorus, *var.* 3.6.2f., 8.22.3, 9.22.3, 9.23.5 on the Decii, and only 10.11.2, 10.12.2 on the Anicii. None of the letters addressed to Boethius (or Symmachus) mention membership of the Anicii. Lists of great families supplied by Ennodius include the Decii (*ep.* 1.5.5, *lib. pro syn.* 130) but do not mention the Anicii. I accept, however, that the *Ordo generis Cassiodororum* indicates an interest in the Anicii on the part of Cassiodorus.

[115] Cassiodorus, *var.* 1.10, 1.45, 2.40, the last being by a comfortable margin the longest letter written by Cassiodorus in Theoderic's name. The second longest was *var.* 1.4, announcing the appointment of Cassiodorus' father as *patricius* to the senate.

[116] Scholarly reputation: Cassiodorus, *var.* 1.45.3–5, 2.40.1. Youthful accomplishments: Ennodius, *ep.* 7.13.2 and *opusc.* 6.21 (pp. 314f.); cf. *phil. cons.* 1.3.2. Skill at Greek: Ennodius, *ep.* 8.1.4.

[117] Boethius, *De interpretatione*, 22 (ed. Meisser, 79f.).

dimension, made no easier by the lack of interest Romans had generally displayed in Greek thought, which was reflected even at the level of the inadequate Latin vocabulary Boethius had at his disposal as a translator. But he was a born scholar. Such was his enthusiasm that when he was consul in 510 he complained that his labours were being hindered by the burdens of office,[118] while at the end of his life he commented that he had become involved in affairs of state in accordance with Plato's doctrine that states would be happy if lovers of wisdom were to rule them, or if it should happen that their rulers loved wisdom.[119] The goal Boethius set himself was probably beyond the power of any one person to achieve, but the degree of his success may be measured by the constant use to which medieval scholars were to put his works.[120]

A popular approach to the history of ideas, to be found in countless introductory textbooks, sees Boethius as perhaps the last representative of rational classical thought before the onset of the benighted Middle Ages, and the last philosophical thinker of substance before St Anselm in the eleventh century. This caricature of reality is hard to take seriously, but for our purposes it may be more worthwhile to contrast Boethius with his contemporaries rather than with those who were to come later. Of course a figure such as Boethius would have been exceptional in any age, but we may doubt whether intellectuals among his contemporaries, such as Cassiodorus and Ennodius, were particularly sympathetic to his concerns, despite their professed admiration.[121] This is true of general intellectual issues and, in a more narrow way, of involvement in political life, for the contrast between Boethius' stated reason for assuming office and the fortune-seekers of Ennodius'

[118] *In categorias Aristotelis*, introduction to book 2, PL 64.201B.

[119] *Phil. cons.* 1.4.5; cf. Gruber, *Kommentar*, 116. While enemies of Boethius felt he had been guilty of 'ambitus dignitatis' (*phil. cons.* 1.4.37), his own judgement was the reverse (ibid. 2.7.1).

[120] Best approached by means of Gibson, *Boethius*.

[121] Note e.g. the judgement implied by *phil. cons.* 1.1.3. Cassiodorus' Greek may be better than has sometimes been thought (A. Garzya, 'Cassiodoro e la grecità', in *Atti della settimana di studi su Flavio Magno Aurelio Cassiodoro (1983)* (1986), 118–34), but Ennodius had little (J. Fontaine, 'Ennodius', *Reallexikon für Antike und Christentum*, 5 (1962), 398–421 at 406; L. Alfonsi, 'Ennodio letterato (nel XV centenario della nascita)', *Studi romani*, 23 (1975), 303–10 at 309). On Greek in Ostrogothic Italy, see in general W. Berschin, *Griechisch-Lateinisches Mittelalter* (Berne, 1950), esp. 99–105.

acquaintance is obvious. Further, while Ennodius doubtless had a genuine respect for the learning of Boethius, he was well aware of the philosopher's large patrimony and the claims poor members of the family had on it.[122] Relations between Boethius and Ennodius can be fairly well documented, and analysis of them is revealing.

BOETHIUS AND HIS CIRCLE

It is clear that Boethius and Ennodius were related, but Ennodius always expressed the connection in vague words:[123] Boethius was neither one of those whose precise relations with Ennodius can be determined from the latter's voluminous correspondence,[124] nor one of those with whom Ennodius merely claimed a tie by blood.[125] When his nephew Parthenius left Milan to study in Rome in 506, Ennodius was able to write letters on his behalf to Faustus *albus* (*ep.* 5.9), Pope Symmachus (*ep.* 5.10), his friend Luminosus, who is known to have moved in clerical circles (*ep.* 5.11; cf. 3.10, 4.11, 6.16), and Faustus *niger* (*ep.* 5.12); in 511 he wrote on behalf of Ambrosius to Faustus *niger* (*ep.* 9.2), Meribaudus (*ep.* 9.3), and Probinus (*ep.* 9.4). Similarly, he was able to write to Faustus on behalf of Simplicianus (*ep.* 6.15) and to Pope Symmachus on behalf of Beatus (*ep.* 8.38). The addressees provide an interesting picture of Ennodius' contacts in Rome, and they were certainly broad: to judge by his name, Meribaudus may have been a Goth,[126] and

[122] Ennodius, *ep.* 8.1.7; cf. Boethius, *phil. cons.* 3.3.5 for his wealth.

[123] *Proximitas* (*ep.* 6.6.3); Ennodius saw himself as the *propinquus* (*ep.* 7.13.3, 8.1.5) or *parens* (*ep.* 8.1.7) of Boethius, but see *pan.* 26 for *parens* used in a particularly loose way.

[124] Euprepia his sister (e.g. *ep.* 2.15.1), her son Lupicinus (ibid.), Parthenius the son of a sister of Ennodius (e.g. *ep.* 5.9.2).

[125] As Senarius (*ep.* 1.23.3), Armenius (*ep.* 2.1.1), Asturius (*ep.* 2.12.2), Promotus (*ep.* 3.14.1), Helisaea (*ep.* 5.4.1), Venantius (*ep.* 5.22.1), Dominica (*ep.* 6.18.1), Archotomia (*ep.* 6.24.1). It will be clear that I am unhappy with characterizations of Boethius and Symmachus as 'amici e congiunti' of Ennodius (Momigliano, 'Anicii', 234) and of Ennodius as 'un personaggio intimamente legato agli Anicii' (S. Roda, 'Alcune ipotesi sulla prima edizione dell'epistolario di Simmaco', *La parola del passato*, 34 (1979), 31–54 at 52).

[126] Schönfeld, *Wörterbuch*, 167, 284.

Probinus was an old enemy of Pope Symmachus, whom Ennodius supported. But neither Boethius nor his father-in-law, despite their familial prestige and intellectual eminence, figure among them, and it is clear that Boethius was not close to Ennodius. The latter, at a time when he was a deacon in the Church of Milan, deemed the consulship of Boethius in 510 a suitable occasion to ask for the gift of a house (*domus*) in that city which he felt was surplus to his wealthy relative's requirements, and when his initial approach proved futile he bombarded the consul with a series of shameless begging letters.[127] There is no reason to believe that this campaign succeeded, any more than that Ennodius waged in seeking the gift of a horse from one Agnellus when he was appointed to high office (*ep*. 7.26, 8.20). But a letter subsequent to those sent to Boethius was dispatched to the deacon Helpidius, known from other sources to have been Theoderic's physician and well connected at the court; it may be worth recalling that on one occasion when Gregory the Great wished to send a private message to the emperor Maurice he used the latter's physician, Theodore.[128] Oddly enough the name 'Hilpidius' may be Gothic in origin,[129] and while 'Helpidius' was a common enough name among Romans of the fourth and fifth centuries for us to have no reason to see the correspondent of Ennodius as anything but a Catholic and a Roman, there is no doubt that he was more accommodating to the Goths than some. Bishop Avitus of Vienne was mildly scandalized to receive from Helpidius a letter carried by Arian clergy, and commented in his reply that they had not made the message unclean, no more than the beaks of the horrid ravens which brought food to Elijah.[130] We may doubt whether Helpidius shared this concern; nor such people as Ennodius, who, as we have seen, corresponded with Goths, and furthermore found Helpidius well placed to gain the ear of Theoderic when the king's

[127] Ennodius, *ep*. 8.1.7, 8.31 (mentioning that Ennodius had already dispatched *crebrae litterae* concerning the house), 8.36 (no overt reference to the house, but Ennodius suspects Boethius of lacking in affection towards him), 8.37 (the end of this letter makes it clear that Ennodius envisaged the house as *consularis sportula*, 8.40; cf., perhaps, 8.12 to Florus).

[128] On Helpidius, *V. Caes. Arel*. 1.41, Procopius, *BG* 1.1.38, and Cassiodorus, *var*. 4.24 all imply he was a powerful figure. Gregory the Great: *reg*. 3.64.

[129] Schönfeld, *Wörterbuch*, 138.

[130] Avitus of Vienne, *ep*. 38; cf. 1 Kings (= 3 Kings) 17: 6.

support was sought.[131] Writing to Helpidius after the sequence of letters to Boethius concerning the house was finished, Ennodius reported that he had sent Helpidius' slave to tell the lord prefect and his master that he had received legal documents concerning a villa (*suburbanum*); Helpidius and his 'son' the lord Triggua were to do what they saw to be necessary (*ep.* 9.21). It is clear from this letter that Helpidius had been to Milan, and we may take it that the *suburbanum* Ennodius acquired there through the machinations of Helpidius and his agent took the place of the *domus* sought in vain from Boethius. It is an extraordinary commentary on Ennodius' range of contacts with respect to Boethius that Helpidius was the confidant to whom Theoderic was to confess his remorse following the executions of Symmachus and Boethius,[132] that the lord prefect involved in the transaction at Milan was almost certainly Ennodius' patron Faustus *niger*, quite possibly involved in a controversy with Boethius while he held this office,[133] and that Triggua must be the Goth Trigguilla mentioned by Boethius in his *Consolation of Philosophy* as being particularly villainous.[134]

One suspects that Boethius had little time for the company Ennodius kept, and his shady contacts in Ravenna. On the other hand, the feelings of Ennodius towards Boethius may have been less warm than his correspondence with him implies. In his letter to Boethius on the occasion of the latter's accession to the consulate Ennodius commented that any military triumphs which the new consul had won were of a strictly metaphorical kind,[135] and in an epigram presumably not intended for general consumption he launched a withering attack on Boethius' military and, by implication, sexual incapacity which concluded:

[131] 'Vere, domne Helpidi, si dignatur pius rex de servo suo esse sollicitus, tu fecisti' (Ennodius, *ep.* 9.14, writing in 511).

[132] Procopius, *BG* 1.1.38.

[133] The letter to Helpidius may be dated to 511 (Sundwall, *Abhandlungen*, 67), comfortably within Faustus' tenure of the praetorian prefecture (ibid. 119; *PLRE* 2.455). Note too Ennodius, *ep.* 9.7 to Faustus' son Avienus, asking him to act with Liberius 'de suburbano illo'. That the praetorian prefect whom Boethius accused of acting badly at a time of want in Campania (*phil. cons.* 1.4.12) was Faustus has been widely asserted (e.g. by Chadwick, *Boethius*, 24) but cannot be proven (so Ruggini, in Obertello, ed., *Congresso internazionale*, 94 n. 118).

[134] Boethius, *phil. cons.* 1.4.10. On the various forms of the name, Wrede, *Sprache*, 78–80.

[135] Ennodius, *ep.* 8.1.3: 'noster candidatus post manifestam decertationem debitum triumphum, dum numquam viderit bella, sortitur . . . '.

Reprobate, a javelin borne by you turns into the staff of Bacchus.
Steadfast in the pursuit of Venus, abandon the tool of war![136]

A poem of Maximianus indicates that Boethius may have been regarded as an authority on love, if not lust,[137] and one wonders whether the impression Boethius made on his contemporaries was uniformly favourable.

But Boethius and his father-in-law cannot only be located against Italian backgrounds. It goes without saying that Symmachus was fluent in Greek,[138] and on a trip to Constantinople which cannot be dated he came in contact with the grammarian Priscian, who, enormously impressed by his visitor from Rome, dedicated three treatises to him.[139] A reference in one of his works suggests that Priscian sent a son to Rome for study.[140] Be this as it may, the early circulation of some of Boethius' works can be traced to a circle around Priscian.[141] A subscription to a manuscript which contains most of Boethius' writings on logic, of tenth-century provenance, indicates that a very early codex of Boethius, owned by Martius Novatus Renatus, had been copied by 'Theodorus antiquarius',[142] while another subscription to a work now lost refers to a codex which Theodorus wrote in 527.[143] Theodorus must be identical with the Flavius Lucius Theodorus who is known from subscriptions to Priscian's *Ars grammatica* to have copied this work, finishing his task on 29 May 527.[144] Renatus is known to have made at least one trip to Constantinople,

[136] Ennodius, *carm*. 2.132 (p. 249), *De Boetio spata cincto*. Ruggini's remarks on the date of this epigram (in Obertello, ed., *Congresso internazionale*, 91 n. 77) are not convincing; on its interpretation, D. Shanzer, 'Ennodius, Boethius and the Date and Interpretation of Maximianus' *Elegia III*', *Rivista di filòlogia e di instruzione classica*, 3 (1983), 183–95.

[137] Ed. Baehrens, *Poetae Latini minores*, 5.334.

[138] 'tu utrarumque peritissimus litterarum': Boethius, *de arith.*, *praef.*, ed. Friedlein, 4.25.

[139] The treatises were *De figuris numerorum, De metris fabularum Terentianis*, and *Praeexercitamina*, ed. H. Keil, *Grammatici Latini*, 3.

[140] Keil, 2.407.14. But see R. A. Kaster, *Guardians of Language: The Grammarian and Society in Late Antiquity* (Berkeley, 1988), 348.

[141] For what follows, Momigliano, 'Anicii', 240f.; M. Salamon, 'Priscianus und sein Schülerkreis in Konstantinopel', *Philologus*, 123 (1979), 91–6; Chadwick, *Boethius*, 255f.

[142] Schanz, Hosius, and Krueger, *Geschichte*, 4.2.152.

[143] Chadwick, *Boethius*, 256.

[144] Keil, 2.192, 451; 3.106 (where the name occurs in its fullest form), 208.

for while Severus, later patriarch of Antioch, was in that city, and so presumably within the period *c*.508–12, he had occasion to speak with two men from the West who spoke Greek, Petronius from Rome and Renatus from Ravenna.[145] Petronius was probably Rufius Petronius Nichomachus Cethegus, consul in 504 and the son of Probinus, one of the chief supporters of Laurentius during the Roman schism. During the Gothic war he was to flee to Constantinople where he participated in the affairs surrounding the second ecumenical council of Constantinople,[146] and Cassiodorus wrote his *Ordo generis Cassiodororum* for him.[147] This short work, as it survives, is devoted to brief discussions of the achievements of Symmachus, Boethius, and Cassiodorus, and provides evidence of an indirect kind for seeing Cethegus as belonging to a circle around Symmachus and Boethius; we may doubt whether Cassiodorus himself belonged to it.[148] Renatus, the other allocutor of Severus, is obviously to be identified with the owner of the manuscript of Boethius on logic.

One has the impression of a coterie of scholarly highbrows, and links between Italian Hellenists and Latinist circles in Constantinople were doubtless primarily formed on the basis of intellectual interests, but it does not follow from this that they were politically innocent. Priscian was probably an African, who may have owed some of his success in the capital to a feeling that Africans 'conversed more elegantly than Italians'.[149] We do not know how Priscian had come to be in Constantinople, but by the end of the fifth century the Mediterranean world was full of refugees from Vandal Africa, and he may well have been one of them. The phenomenon of the African diaspora has never been fully examined, and it may be that some of the Africans were driven not by persecution but by ambition, as the young Augustine had been. Nevertheless, the popes kept in touch with exiled bishops in

[145] Severus, *Liber contra impium grammaticum*, 3.2.29, trans. J. Lebon, *CSCO* 102 (*Script. Syr.* 51), 72.

[146] ACO 4.1, e.g. p. 27.22.

[147] *MGH AA* 12 p. v.

[148] Momigliano, 'Cassiodorus', 214f. discusses relations between Cassiodorus and the Anicii.

[149] Priscian's background: Schanz, Hosius, and Krueger, *Geschichte*, 4.2.221f.; R. Helm, 'Priscianus', PW 22.2328–46 at 2328. Elegant conversing of Africans: John the Lydian, *Powers*, 2.73.

Constantinople,[150] and we may doubt whether they were above applying political pressure. Just as an African bishop was among the very few witnesses Gregory of Tours could summon to illustrate his thesis that Catholics in territory controlled by the Visigoths wished to become subject to the Franks,[151] so African magnates later urged Justinian to make war on the Vandals.[152] One wonders to what extent Priscian was hostile to the non-Catholic Germanic kingdoms of the West. The panegyric he wrote in honour of Anastasius early in the sixth century would have given Theoderic little joy: it mentioned the warm welcome which people from the old Rome received at the emperor's court and the fortune and safety which they owed him, and describes them as praying for him day and night.[153] It would be interesting to know who the people so described were, but Zachariah of Mytilene knew of one Dominic, or Demonicus, one of the chief men of Theoderic's kingdom, who 'had a quarrel with the tyrant and took refuge with King Justinian, and gave him information about the country'.[154] As it stands this information is awkward, as Theoderic died before the accession of Justinian i527, but it may remind us that discontented Italians may well have found a welcome in Constantinople. But Priscian did not merely describe people from Rome being made welcome, for he went on to express the hope that both Romes, that is Rome and Constantinople, would come to obey the emperor alone.[155] Such thoughts may have been exhilarating, but in the light of substantial Byzantine interest in the West, which will occupy our attention

[150] Gelasius with Bishop Succonius (Schwartz, *Publizistische Sammlungen*, 56), Hormisdas with Bishop Possessor (*coll. avel.* 230). It is possible that Pope Symmachus, described in *lib. pont.* as 'natione Sardus' (260.1) fled because of the Vandal occupation of Sardinia: Caspar, *Geschichte*, 2.88 n. 1; note his generosity to exiled bishops in Sardinia (*lib. pont.* 263.8f.). P. A. B. Llewellyn has argued for Pope Symmachus having received support from Africans in Rome: 'The Roman Clergy during the Laurentian Schism (498–506): A Preliminary Analysis', *Ancient Society*, 8 (1977), 245–75.

[151] Gregory of Tours, *hist. franc.* 2.35 and *Lib. vit. pat.* 4.1 on Quintianus of Rodez.

[152] Zachariah, *hist. eccl.* 9.17.

[153] Priscian, *pan.* 242–7 (ed. Baehrens, *Poetae Latini minores*, 5.272). On the date of this work, the arguments for 503 advanced by Alan Cameron, 'The Date of Priscian's *De laude Anastasii*', *Greek, Roman and Byzantine Studies*, 15 (1974), 313–16, do not seem to be met by A. Chauvot, 'Observations sur la date de *l'Éloge d'Anastase* de Priscien de Césarée', *Latomus*, 36 (1977), 539–50, reasserting the traditional date of 512.

[154] Zachariah, *hist. eccl.* 9.8.

[155] *Pan.* 265 (ed. Baehrens, 273).

shortly, Theoderic could have found them alarming. If Theoderic's Italy was a land of two cities, one of them may have been considerably less loyal than the other; if the thought of two Romes obeying the emperor pleased some, their pleasure may not have escaped notice in Ravenna.

6

Foreign Affairs

By 500 Theoderic, having gained control of Italy and established peaceful relations with all the major barbarian kingdoms of the West and the emperor in Constantinople, had become unquestionably the most powerful ruler in western Europe. The fame which he enjoyed is indirectly reflected in stories which circulated among the Germanic peoples of the Middle Ages. That Theoderic fought Odovacer was a fact known, however obscurely, to the author of the *Hildebrandslied*, which in its present form dates from approximately the time of Charlemagne, and legends concerning Dietrich von Bern are common in the Germanic vernacular literatures of the early Middle Ages. The material they contain is of no direct value to the modern historian of Theoderic, but it does indicate the survival of traditions about him which were independent of Latin and Greek sources, and so indirectly suggests the impact made by Theoderic on the Germanic peoples of his day. Another indication of this is provided by a ninth-century Scandinavian inscription on the Rökstone, which refers to Theoderic as having been dead for nine generations. While the inscription is difficult to interpret, it provides further evidence for the great prestige Theoderic enjoyed among the Germanic peoples.[1]

Despite this, it would have been too much to have anticipated that he would have lived in peace with his neighbours of the early sixth century. Modern scholars have found it very difficult to establish the precise borders of his kingdom, discord being particularly acute as to whether it extended as far north as the Danube, and it may be that the problems that have been encountered reflect not merely unsatisfactory evidence but also differing concepts of 'frontier', for beyond the firm borders of his kingdom there remained

[1] J. Heinzle and E. E. Metzner, 'Dietrich v. Bern', *Lexikon des Mittel Alters*, 3 (1986), 1016–21, provide a recent introduction to the subject. On the Rökstone, J. M. Wallace-Hadrill, *Early Germanic Kingship in England and on the Continent* (Oxford, 1971), 9–11, although aspects of his interpretation seem to me doubtful.

areas strongly under his influence.[2] Further, the barbarian kingdoms of the time were still in a state of flux which rendered aggression easy. Moreover, the government of Constantinople was by no means reconciled to the loss of its Western provinces, and while direct military intervention was rare during the reign of Theoderic it was not averse to meddling in Western affairs, probably on a scale we cannot now detect. In short, Theoderic's diplomatic achievements of the 490s and the stature he had gained by 500 were not a guarantee against future troubles, and relations with other powers were to prove one of the chief concerns of the Ostrogothic kingdom throughout its existence. More than at any other time, a complex series of events which began in 504 and were to involve all the major powers of the West as well as Constantinople required Theoderic's attention. We shall suggest that the outcome of these events was mixed from his point of view.

In 504 Theoderic sent a force under the noble Goth Pitzias to the city of Sirmium, then controlled by the Gepids. These people were traditional enemies of the Goths, their king at the time, Traseric, being the son of the Trapstila killed by the Goths on their way to Italy, and our one detailed source for these events locates them in the context of relations between Goths and Gepids.[3] But Sirmium was part of the province of Pannonia Secunda and hence traditionally part of the Eastern Empire, and it is at least equally likely that Theoderic's expedition should be seen as an attempt to strengthen his kingdom at a time when the imperial armies were fighting the Persians.[4] Whatever Theoderic's motive, the expedition was a success: Sirmium was taken, Traseric and his ally Gunderith fled, neither of them to be heard of again, and the captives included Traseric's mother.[5] Shortly afterwards the leader of a gang of robbers, one Mundo, probably a Gepid, who operated from a

[2] See in particular Wolfram, *History*, 315 ff., with Bierbrauer, 'Zur ostgotische Geschichte', 2–10 (*Ostgotische Grab- und Schatzfunde*, 17–25). The attempts of scholars such as F. Beyerle, 'Suddeutschland in der politischen Konzeption Theoderichs des Grossen', in *Grundfragen der Alemannischen Geschichte* (*Vorträge und Forschungen*, 1; Lindau, 1955), 65–81, and B. Behr, *Alemannische Herzogtum*, 45–7, to extend Theoderic's territory to the Danube do not convince, but this is not to deny that both the Raetias could have been within his sphere of influence.

[3] Ennodius, *pan.* 61 f.; cf. Wolfram, *History*, 321.

[4] Ensslin, 'Beweise der Römverbundenheit', 518. On Anastasius and the Persians, Capizzi, *Anastasio*, 183–5. On Sirmium as part of the East, Cassiodorus, *var.* 11.1.9, and Stein, *Histoire*, 145 f.

[5] Cassiodorus, *chron. sa* 504; Ennodius, *pan.* 60–2; Jordanes, *get.* 300; Procopius, *BG* 1.11.5.

fortress named Herta on the right bank of the Danube, was attacked by Sabinianus, *magister militum per Illyricum*. Sabinianus' father, Sabinianus Magnus, had been in years past an enemy of Theoderic, and as Mundo was a *foederatus* of Theoderic we are surely to see the attack of Sabinianus as having been made indirectly against Theoderic. But Pitzias was still in the region and, responding to Mundo's plea for help, advanced with 2,000 infantry and 500 horse and defeated the imperial forces, in the main apparently Bulgars. Sabinianus retreated, the grateful Mundo became a subject (*subiectus*) of Theoderic, and the newly won territories were secured for the Goths.[6] Friends of Theoderic's government attributed the success to him. Ennodius, whose account of this campaign in his panegyric on Theoderic is perhaps surprisingly long given that the king played no part in the fighting, asserts that in his speech to the troops before the battle Pitzias advised against believing that Theoderic's eyes were not on them, and that when the fortunes of battle remained in doubt it was their memory of Theoderic which triumphed.[7] The entry in the chronicle of Cassiodorus, 'By the power of the lord king Theoderic the Bulgars were defeated and Italy regained Sirmium' (*s.a.* 504) evades the reality that the Goths were embroiled with Constantinople and by implication credits Theoderic with a success won by Pitzias, while Marcellinus goes to the other extreme and seems to deny any Ostrogothic involvement in Mundo's victory (*s.a.* 505). But the Ostrogoths had won a considerable success, and Theoderic subsequently minted coins in Sirmium.[8] Constantinople cannot have been pleased.

THE RISE OF THE FRANKS

Meanwhile there were developments to the west of Italy. In 500 Godegisel, who ruled part of the kingdom of the Burgundians from Geneva, had sought the help of Clovis against his brother Gun-

[6] Ennodius, *pan.* 63–9; Jordanes, *get.* 300f., *rom.* 387; Marcellinus, *chron. sa* 505. On Mundo's race, B. Croke, 'Mundo the Gepid: From Freebooter to Roman General', *Chiron*, 12 (1982), 125–35; see further on Mundo, W. Pohl, in H. Wolfram and F. Daim, *Die Fölker an der mittleren und unteren Donau im fünfsten und sechsten Jahrhundert* (Vienna, 1980), 292f. On Sabinianus Magnus and Theoderic, Malchus, *frag.* 18 (Blockley, 20); Marcellinus, *chron. sa* 479.

[7] *Pan.* 65, 67.

[8] I. Meixner, 'Three Unknown Coins of King Theoderic', *Numizmaticke vijesti*, 15 (1968), 53–5; Hahn, *Moneta*, 1.86f., 2.31.

dobad. Clovis marched on Gundobad who retreated to Avignon and promised to pay an annual tribute. Before long, however, Gundobad had refused to pay tribute and besieged Godegisel in Vienne. The city was captured, Godegisel murdered, the Franks found there were exiled to the Visigothic king Alaric at Toulouse, and those of the senators and Burgundians who had supported Godegisel were killed. The events of 500 cannot be shown to have had any direct impact on Italy, but Gundobad had been forced to deal with the Franks and the Visigoths, and henceforth the future of his kingdom was to be linked with his more powerful neighbours.[9]

Clovis was also involved with another barbarian people, the Alamanni, whom he defeated in about 497 and again in about 506.[10] After the second battle Theoderic wrote to Clovis congratulating him on his success but warning him not to prosecute further his war against the Alamanni who had fled in great fear to his own territory, perhaps to Raetia. Theoderic suggested that Clovis take the advice of an experienced man: the wars which had turned out happily for him were those which had been brought to completion with moderation.[11] In accordance with a wish Clovis had expressed Theoderic sent him a skilful harpist, who by his singing and playing would 'bring delight to the glory of your power'. But this was not the only language in which the mission of the harpist could be expressed, for the Franks were generally despised for their

[9] Gregory of Tours, *hist. franc.* 2.32f.; Marius of Aventicum, *chron. sa* 500. Binding's thesis of a religious background to this war (*Burgundisch-Romanische Königreich*, 145–54) is based on the *Collatio episcoporum*, a forgery of the notorious Jerome Vignier, but otherwise his account remains valuable. Judging by the presence of a representative of the bishop of Avignon at the Council of Agde in 506, this city may have been under Visigothic control by then: CCSL 148.214.42f.

[10] Gregory of Tours, *hist. franc.* 2.30 (the battle during which he invoked Christ) and Cassiodorus, *var.* 2.41, which I take to refer to different events, as suggested by Fredegarius, *chron.* 3.21. See e.g. the discussions of Zöllner, *Geschichte*, 56f.; K. F. Stroheker, 'Die Alemannen und das spätromische Reiche', in W. Müller, ed., *Zur Geschichte der Alemannen* (Darmstadt, 1975), 20–48 at 46f.; W. Hartung, *Süddeutschland in der frühen Merowingerzeit* (Wiesbaden, 1983), against e.g. L. Levillain, 'La Crise des années 507–508 et les rivalités d'influence en Gaule de 508 à 514', in Mélanges offerts à M. Nicholas Iorga (Paris, 1933), 537–67 at 541f.

[11] Cassiodorus, *var.* 2.41. On the settlement of the Alamanni in Theoderic's territory, Ennodius, *pan.* 72, but 'Alamanniae generalitas' cannot be pressed closely, and 'intra terminos Italiae' need not mean 'within Italy'; cf. *pan.* 60 for Sirmium as the frontier of Italy, with Clavadetscher, 'Churrätien', 163f., and Hartung, *Süddeutschland*, 103–9. Cassiodorus, *var.* 3.50 enjoins an exchange of the cattle of the men of Noricum with those of the Alamanni, 'more valuable on account of their size but worn out from the length of their journey'. Agathias, *hist.* 1.6.4 sees the Alamanni as tributaries of Theoderic.

barbarism, and the letter to Boethius asking him to select the man noted that he would be another Orpheus 'who will tame the fierce hearts of the gentiles with his sweet sound'.[12]

Hitherto relations between Theoderic and Clovis had been good, as far as we can judge. The two were brothers-in-law, and according to one report Clovis sent to Theoderic some of his sons.[13] But any hopes that the heart of Clovis would be tamed were destined to be disappointed, and Theoderic soon found himself trying to mediate between his son-in-law Alaric, king of the Visigoths, and the king of the Franks. Evidence for his endeavours is found in a series of letters preserved in the *Variae* of Cassiodorus. He reminded Alaric that even though the power of Attila had been laid low by the strength of the Visigoths, a long period of peace can soften the hearts of fierce peoples. Hence he advised him to be cautious, and not allow himself to be carried away by blind indignation.[14] Gundobad the Burgundian was approached in a letter which, a little fancifully, styled him and Theoderic as old men, and Alaric and Clovis as young kings, and asked him to co-operate in peaceful initiatives.[15] Writing to the kings of the Heruls, Warni, and Thoringians, Theoderic made Clovis out to be an aggressor. He asked them to recall the love the Visigoth Euric had for them, and how he had protected them from wars threatened by nearby peoples (*proximae gentes*), by which we are presumably to understand the Franks.[16] Most importantly,

[12] Cassiodorus, *var.* 2.40.17; the same Boethius attested to the power of Orpheus' music in *phil. cons.* 3 *met.* 12. Cf. a description of the impact of music on Theoderic the Visigoth, in Sidonius Apollinaris, *ep.* 1.2.9. The barbarism of the Franks: Ennodius, *opusc.* 4.13 (pp. 187.2f.), Jordanes, *get.* 176. This letter of Cassiodorus has been dated to 'verso 496' (C. Leonardi, in 'Boezio', *Dizionario biografico degli Italiani*, 11 (Rome, 1969), 142–65 at 142) and to 501 (Bierbrauer, 'Zur Ostgotische Geschichte', 28 (*Ostgotische Grab- und Schatzfunde*, 43), but no one has produced sound reasons for dating it other than shortly before the war which began in 507: Krautschick, *Cassiodor*, 53f.

[13] Jordanes, *get.* 296, a difficult passage: the names Celdebert and Heldebert both suggest Childebert, one of Clovis' sons by Chlotilde, and Theudebert was the son of Clovis' illegitimate son Theoderic. Unlike Weiss, *Clodwigs Tauf*, 56, and P. J. Geary, *Before France and Germany: The Creation and Transformation of the Merovingian World* (New York, 1988), 84, I see no significance in the name of this son.

[14] Cassiodorus, *var.* 3.1. Are we to see in the phrase 'longa pace' a reminiscence of Statius, *Theb.* 3.255?

[15] *Var.* 3.2.

[16] *Var.* 3.3; on the identity of the *proximae gentes*, Stroheker, *Euric*, 127f., n. 175. The location of the Heruls, Warni, and Thoringians is discussed by R. Wenskus in K.-U. Jaschke and R. Wenskus, eds., *Festschrift für Helmut Beumann* (Sigmaringen, 1977), at 128.

Theoderic wrote to Clovis urging the claims of the divine laws of family relationship (*affinitas*). He represented himself as addressing both Alaric and Clovis: 'Put away your iron, you who seek to shame me by fighting. I forbid you, by my right as a father and as a friend. But in the unlikely event that someone believes that such advice can be despised, he will have to deal with us and our friends as enemies.'[17]

Presumably Theoderic attempted by other means to keep the peace between Clovis and Alaric,[18] but when Clovis' attack came in 507 it was sudden; an appeal for help Alaric made to Theoderic failed to meet with a response.[19] Despite the fact that he had recently held discussions with Alaric on an island in the Loire near the village of Ambroise, a little upstream from Tours, where they had shared food and sworn friendship,[20] Clovis led his host south. They met Alaric's forces near Poitiers, perhaps at Vouillé. The Goths were put to flight, Alaric was killed—by Clovis himself, it was later believed—and many years later the remains of bodies could still be seen on the battlefield.[21]

When Jordanes came to write his Gothic history he claimed that the Goths never yielded to the Franks in Gaul while Theoderic lived, and twice referred to the death of Alaric without explaining how he had died.[22] Such deceit and evasiveness, which presumably reflects the attitude of Cassiodorus, is indicative of the defeat for Theoderic which the victory of Clovis represented. Nevertheless, the setback need not have been of the magnitude which some modern scholars believe it was. Some have seen the campaign of Clovis against the Arian Alaric as a religious war, indeed a kind of crusade, which would be the more plausible if his conversion to Catholic Christianity could be dated to 506, immediately prior to his

[17] *Var*. 3.4; the passage quoted is from 3.4.3f.

[18] See the odd tradition preserved by Fredegarius, *chron*. 2.58 (*MGH SRM* 2.82f.).

[19] Procopius, *BG* 1.12.34, 37.

[20] Gregory of Tours, *hist. franc*. 2.35; on the significance of this A. Lippold, 'Chlodovechus', PW supplementary vol. 13.139–74 at 160f. Note the description in a later account of a ceremony whereby Alaric touched the beard of Clovis and became his 'patrenus': Fredegarius, *chron*. 2.58 (*MGH SRM* 2.82), an account full of interest if not accuracy. For a meeting in mid-stream, cf. that of Valens and Athanaric in the middle of the Danube in 367: Ammianus Marcellinus, 27.5.9, 31.4.13.

[21] Gregory of Tours, *hist. franc*. 2.27; Procopius, *BG* 1.12.33–40; Venantius Fortunatus, *de virt. s. Hil*. 7.21.

[22] Jordanes, *get*. 296 (the widely used translation of C. C. Mierow is inaccurate at this point), 298, 302.

war with Alaric. But it is now generally accepted that Gregory of Tours was correct when he dated the conversion to the end of the fifth century,[23] and even if it had occurred shortly before the war with Alaric it is hard to see what practical difference it would have made: the letters in the *Variae* of Cassiodorus bearing on the subject make no allusion to Catholicism, and the narrative of Gregory of Tours provides scarcely any evidence to support that author's thesis that many people in lands controlled by the Visigoths wished to be ruled by the Franks; indeed, in so far as he mentions that Alaric's forces at the decisive battle included a contingent from the Auvergne which had come with Apollinaris, the son of Sidonius Apollinaris, he provides evidence to refute it.[24] It is true that Bishop Avitus of Vienne wrote to Clovis at the time of his baptism asserting that 'your faith is our victory', but this does not imply any change in the political balance of power.[25] Some scholars have seen Clovis' advance into Visigothic territory as part of a great swing towards the Mediterranean,[26] but, as his two battles with the Alamanni, his activities in the north of Gaul during the period 507–11, and the wars of his sons against a variety of enemies to the east indicate, the south was not the only direction in which the Franks were interested in expansion. The Frankish state was to remain firmly based in the north; following the death of Clovis his sons established their capitals at Orleans, Soissons, Paris, and Rheims. Only the first of these

[23] Gregory of Tours, *hist. franc.* 2.30f. The issues are complex, but against A. van de Vyver, 'La Victoire contre les Alamans et la conversion de Clovis', *Revue belge de philologie et d'histoire*, 15 (1936), 859–914 and 16 (1937), 35–94, followed by Stein, *Histoire*, 148, see P. Courcelle, *Histoire littéraire des grandes invasions germaniques*, 3rd edn. (Paris, 1964), 240; K. Hauck, 'Von einer spätantiken Randkultur zum karolingischen Europa', *Frühmittelalterliche Studien*, 1 (1967), 3–93 at 27f. (not entirely convincing); B. S. Bachrach, 'Procopius and the Chronology of Clovis' Reign', *Viator*, 1 (1970), 20–31; Zöllner, *Geschichte*, 61f., K. F. Werner, *Histoire de France*, 1. *Les Origines* (*avant l'an mil*) (Paris, 1984), 307. The thesis of Weiss, *Clodwigs Tauf*, has been refuted by K. Schäferdiek, 'Ein neues Bild der Geschichte Chlodwigs?', *Zeitschrift für Kirchengeschichte*, 84 (1973), 270–7. Recently, however, I. N. Wood has argued for a conversion, from Arianism, in 508: 'Gregory of Tours and Clovis', *Revue belge de philologie et d'histoire*, 63 (1985), 249–72. G. Tessier, *Le Baptême de Clovis* (Paris, 1964), summarizes the discussion up to that time.

[24] Gregory of Tours, *hist. franc.* 2.35 (desire for Franks), but cf. 2.37 (*MGH SRM* 1.88, on the role of Apollinaris, oddly misinterpreted by Jimenez Garnica, *Origenes y desarrollo*, 128, who sees the Gallo-Romans deserting during the battle). Against the interpretation of Clovis' Catholicism expressed in e.g. Hartmann, *Geschichte*, 1.155; Ensslin, *Theoderich*, 133; Tessier, *Baptême*, 96, see Grahn-Hoek, *Fränkische Obersicht*, 165; J. Moorhead, 'Clovis' Motives for Becoming a Catholic Christian', *Journal of Religious History*, 13 (1985), 329–39.

[25] Avitus of Vienne, *ep.* 46.

[26] e.g. Zöllner, *Geschichte der Franken*, 168.

could remotely be described as being southern in its location, and we know from Gregory of Tours that the Franks had fought there as early as 463 (*hist. franc.* 2.18).

Perhaps, then, Theoderic looked on Clovis in 507 not as a Catholic enemy of the Arian kingdoms, nor as the leader of a would-be Mediterranean power, but simply as someone who threatened to upset a stable situation, and whose ambition was the worse because it was directed against his kinsfolk the Visigoths. While sources which reflect the experience of Frankish involvement in Italy during Justinian's Gothic war are powerful testimony to the fears excited later by the rising power of the Franks,[27] what may have seemed the most important thing in 507 was that an attack had been launched, by whatever party, against the Visigoths. These were a people who had not only lent their support while Theoderic was fighting Odovacer[28] but who could be described as forming, with the Ostrogoths, two *populi* in one *gens* and who were generally, with them, simply styled 'Goths', the terms 'Ostrogoths' and 'Visigoths' only being used when it was necessary to contrast the two branches of the Gothic people.[29] Ostrogoths were happy to intermarry with them,[30] and they accepted an Ostrogoth, Theudis, as their king in 531. For Theoderic the identity of the defeated may have been more important than that of the victor.

Nevertheless, Theoderic had thereafter to deal with a radically changed situation. After the defeat of Alaric, Clovis' son Theoderic moved as far south as Albi before proceeding north to Rodez and Clermont-Ferrand, while Clovis spent the winter at Bordeaux and seems to have acquired some of the treasure of the Visigoths.[31]

[27] Jordanes, *get.* 296, with the ominous 'dum viveret Theodoricus'; Procopius, *BG* 1.12.33 (the chronology is confused, but Procopius seems to have believed that the Franks defeated the Visigoths after they had dismembered the territory of the Burgundians with the Ostrogoths in 524).

[28] But note already Sidonius Apollinaris, *carm.* 8.9.5.36ff. for Ostrogoths at Euric's court, perhaps to be identified with Vidimer and his followers (Jordanes, *get.* 284; see Stroheker, *Euric*, 74 n. 36, and further *PLRE* 2.1178 *s.v.* 'Vittamer').

[29] Jordanes, *get.* 98. On the terms 'Goths', 'Ostrogoths', and 'Visigoths', Teillet, *Goths*, 10 n. 44 with 309; on close ties between Ostrogoths and Visigoths, F. Giunta, *Jordanes e la cultura dell'alto medioevo* (Palermo, 1952), 105f. A generation earlier, a group of Ostrogoths had joined the Visigoths and formed 'unum corpus' with them: Jordanes, *get.* 284.

[30] Procopius, *BG* 1.12.49, whereas Rugians did not marry Ostrogoths (*BG* 3.2.3).

[31] Gregory of Tours, *hist. franc.* 2.37 (*MGH SRM* 1.88), although the claim that he took all of Alaric's treasure from Toulouse is awkward in the light of Procopius, *BG* 1.12.41, 47; see further D. Claude, *Geschichte der Westgoten* (Stuttgart, 1970), 135 n. 49, and id., 'Königsschatz', 13.

Meanwhile, the Burgundians began to intervene. Two letters in the *Variae* of Cassiodorus describe the sending of a water-clock and sundial to Gundobad the Burgundian, and while they cannot be precisely dated there seems no reason not to see in them an attempt by Theoderic to win the favour of his neighbour at that difficult time.[32] But the presents were to no avail. The Burgundians joined the Franks in mounting a siege of Arles and in sacking the old Visigothic capital of Toulouse, while Gundobad was able to go so far as to take Barcelona, and the evidence of inscriptions in Lyons suggests irregularities in the publication of the names of consuls in Burgundy in the period 507–10, and hence, by implication, bad relations between Theoderic and Gundobad.[33] An accusation of treason made against the bishop of Augusta in the period 507–11 probably reflects fears that the bishop of what is now Aosta, a town with easy access to the greater and lesser St Bernard Passes, had been intriguing with the Burgundians, whose territories were nearby; at about the same time the government was concerned that *annonae* should be forwarded to sixty soldiers stationed at Augusta with the purpose of denying entry to 'barbarians'.[34] A poem written in 507 describes one Bishop Honoratus as having built a *castellum*; he is presumably identical with the bishop of the north-eastern town of Novara of that name who built a basilica in honour of the apostles where there had been a pagan temple.[35] Gundobad's involvement in northern Italy during the war between Odovacer and Theoderic cannot have been forgotten, and it may be that rather than merely acting in concert with Clovis he was again seeking to

[32] Cassiodorus, *var.* 1.45f.; 1.45.1 makes the diplomatic purpose clear. P. Lamma, *Oriente e occidente nell'alto medioevo* (Padua, 1968), offers interesting speculative discussion. The attempt of L. M. de Rijk, 'On the Chronology of Boethius' Works on Logic', *Vivarium*, 2 (1964), 1–49, 125–62 at 142, following S. Brandt, 'Entstehungszeit und zeitliche Folge der Werke von Boethius', *Philologus*, 62 (1903), 141–54 at 146f., to date *var.* 1.45 to 515–16 runs aground on the circumstance that Cassiodorus was not in Theoderic's employ then. U. Pizzani, 'Boezio "consulente tecnico" al servizio dei re barbarici', *Romanobarbarica*, 3 (1978), 189–242 proposes a date late in the quaestorship of Cassiodorus, but it is not clear how closely *var.* 1.45.4 can be pressed as a guide to works already undertaken by Boethius.

[33] Arles: *V. Caes. Arel.* 1.20–4; Caesarius, *serm.* 298. Toulouse and Barcelona: *chron. gall.* 689f.; Isidore of Seville, *hist. goth.* 37. Inscriptions: Fiebiger/Schmidt, *Inschriftensammlung*, 1 no. 79 (of 508, = *CIL* 13.2373) and no. 80 (of 510, = *CIL* 13.2374).

[34] Accusation of treason: *var.* 1.9, on which see Schmidt, *Ostgermanen*, 343. *Annonae*: *var.* 2.5.

[35] *Castellum*: Ennodius, *carm.* 2.110 (p. 201); cf. *dictio.* 2 (pp. 121f.) for the basilica.

gain what profit he could in a confused situation, thereby causing fear in Ravenna.

More important than Gundobad's activities was the attitude of Constantinople. Hitherto Anastasius is not known to have taken a great interest in the rise of Clovis,[36] but writing to Clovis shortly before his defeat of Alaric Theoderic had hinted darkly that some 'alien malignity' might be becoming involved,[37] and Constantinople was not long in making its attitude clear: at some time in the last four months of 507 or in 508 a Byzantine fleet of 100 armed ships and 100 *dromones*, carrying 8,000 armed soldiers, devastated the Italian coast as far as Tarentum, 'a dishonourable victory', it was felt, 'which Romans snatched from Romans with the daring of pirates'.[38] Reports of the burning of crops by enemies in Apulia and an attack on the population of Sipontum, also in Apulia, probably reflect the activities of the fleet; we may note as well that work was carried out on the *moenia* of Rome at about that time.[39] The Byzantine chronicle of Marcellinus fails to register the consulship of Theoderic's nominee Venantius in 507, noting merely in an unusual form of words the consulship of the emperor Anastasius in that year, an omission which may have been due to bad relations between Ravenna and Constantinople.[40] It is possible that the Byzantine fleet blockaded Italy endangering trade with the East, which could be a sign that finance was urgently needed before the Ostrogothic government could intervene in Gaul.[41] Doubtless it would have been within the power of the Vandals to have inter-

[36] A phrase in the letter of Avitus of Vienne to Clovis, 'Gaudeat equidem Graecia principem legisse nostrum' (*ep*. 46, p. 75.17) is hard to evaluate, and the text seems to be corrupt.

[37] Cassiodorus, *var*. 3.4.4, on which see Lippold, 'Chlodovechus', 162.

[38] Marcellinus, *chron. sa* 508; Jordanes, *rom*. 356. The date cannot be expressed more precisely, for the portion of the text of Marcellinus printed by Mommsen under the heading '508 I. Celeris et Venantii' could refer to events occurring in either the indiction or the consular year. I am grateful to Brian Croke for advice on this matter.

[39] Cassiodorus, *var*. 1.16 (burning of crops), 2.38 (attack on Sipontum), 1.25, 2.34 (work on *moenia*; the last letter can be dated to 509–10).

[40] 'Anastasii aug. III', an odd formula which omits the 'solius' Marcellinus customarily placed after the name of a sole consul, as for 497, 'Anastasii aug. II solius'.

[41] D. Claude, *Der Handel im Westlichen Mittelmeer wahrend des Frühmittelalters* (Göttingen, 1985), 110, on the evidence of Cassiodorus, *var*. 4.34. It is interesting to note that towards the end of his reign Alaric seems to have debased the currency: Avitus of Vienne, *ep*. 78 (p. 96.33 f.).

vened on behalf of the Ostrogoths, and it may be that the business of state on which the patrician Agnellus travelled to Africa within the period 507–11 involved a request for help,[42] but if this was so it met with no response. Theoderic's army was only summoned to meet on 24 June 508. Even if we accept that Clovis defeated Alaric towards the end of 507 the date is late, and Theoderic's slowness in acting may well be connected with pressure exercised by Constantinople.[43]

The army headed into Gaul. Its movements are obscure, but we are told that a great victory which involved the death of over 30,000 Franks was won by Count Ibba, who is also known to have exercised authority of some kind in Narbonne, while Duke Mammo is known to have plundered parts of Gaul in 509.[44] The Franks and Burgundians had penetrated as far as Arles, which city they subjected to a difficult siege until it was relieved by the Ostrogoths in operations noteworthy for the defence of the bridge across the Rhone by Tuluin; perhaps attention should be drawn to the fact that the thought of surrender does not seem to have crossed the minds of the Catholic inhabitants of the city, and the biographers of Caesarius of Arles commented laconically that the city passed from the rule of the Visigoths to that of the Ostrogoths.[45] According to one report Theoderic was himself in Gaul, raising the siege the Franks had mounted against Carcassonne.[46] It was some time before peace returned to Gaul, and the provincials were still eligible in the indictional year 510–11 for tax relief on account of losses caused by enemy action.[47] When the frontier stabilized the Ostrogoths had advanced as far as the river Durance[48] and were confronted with the task of governing their new territory in Gaul as well as picking up the pieces of the Visigothic kingdom. But before we examine this it will be as well to turn briefly to Clovis and, as far as is possible, Byzantine policy.

[42] Cassiodorus, *var.* 1.15.

[43] Cassiodorus, *chron. sa* 508 and *var.* 1.24, combined, yield the date for the meeting of the army. On Theoderic's tardiness I agree with Wolfram, *History*, 309; on the possibility of Clovis having defeated the Visigoths late in 507, Binding, *Burgundische-Romanische Königreich*, 202.

[44] Ibba: Jordanes, *get.* 302; Cassiodorus, *var.* 4.17. Mammo: Marius of Aventicum, *chron. sa* 509.

[45] *V. Caes. Arel.* 1.24; Caesarius of Arles, *serm.* 298. See on Tuluin, Cassiodorus, *var.* 8.10.6f.

[46] Procopius, *BG* 1.12.44.

[47] Cassiodorus, *var.* 3.40.

[48] Ibid. 3.41.

THEODERIC AND ANASTASIUS

It is difficult to evaluate the role played by Constantinople in these events. The difficulty is part of the larger problem of establishing the thrust of Byzantine diplomacy in the West between the coming to power of Odovacer and the wars of Justinian, which by its nature must often have been covert. Nevertheless, the state which launched Theoderic against Odovacer wielded influence in the West, and the correspondence of Gundobad's son Sigismund with Constantinople verges on the sycophantic.[49] More interesting for our purposes is a story told by 'Fredegarius' according to which Childeric, the father of Clovis, was at one time in Constantinople, and said to the emperor Maurice, presumably an error for Marcian, 'Order me to go to Gaul as your servant'; he was sent away with gifts.[50] Perhaps there was a tradition of imperial involvement with the family of Clovis. The Byzantines lost little time in making public their attitude to the conflict which broke out in 507, but it may be that Theoderic's successes in 504 and 505 weighed more heavily than the attractiveness of the Franks in the forming of that attitude. As we have seen a fleet was sent which ravaged part of the Italian coast, quite possibly with the intention of pinning Theoderic down, and after Clovis defeated Alaric he received from Anastasius consular codicils. In the Church of St Martin of Tours he was clothed in a purple tunic and a chlamys, and a diadem was placed on his head. He then mounted a horse and rode to the *ecclesia* (cathedral), scattering gold and silver, and from that day he was called consul or augustus.[51]

The description of these events provided by Gregory of Tours is extraordinarily difficult, and has evoked a vast amount of discussion.[52] Briefly, it would seem that Clovis was appointed to an honorary consulship. He cannot have been *consul ordinarius*, for his name does not occur in the consular *fasti*, but it is possible, although it cannot be proved, that the title was meant to suggest

[49] 'vester quidem est populus meus; plus me servire vobis delectat quam illi praeesse': Avitus of Vienne, *ep*. 83.

[50] Fredegarius, *chron*. 3.11 (*MGH SRM* 2.96), well discussed by Wallace-Hadrill, *Long-Haired Kings*, 85.

[51] Gregory of Tours, *hist. franc*. 2.38.

[52] See esp. the summaries in Courcelle, *Histoire littéraire*, 242 ff., and Chrysos, 'Amaler-Herrschaft', 56 nn. 170 f.; most recently, McCormick, *Eternal Victory*, 335–7.

some sort of equality with Theoderic, who had held a consulship in 484. But the clothing described by Gregory is suggestive of an emperor rather than a consul, from which it could be argued that Clovis saw himself as more than a consul, that he was misinformed as to the nature of consular ceremonial, or that between Clovis and Gregory there was some inaccuracy in transmitting the details of what occurred at Tours. Gregory's words 'from that day he was called consul or augustus' are particularly confusing, for Gregory was certainly aware that an *augustus* was an emperor, and various emendations of the text have been proposed,[53] but it seems safest to let these words stand, with their implication that Clovis was not averse to being seen as an emperor. That Clovis may have felt some obscure identity with emperors is suggested by his beginning to build in Paris a church dedicated to the holy apostles (viz. SS Peter and Paul) or to St Peter, where he was buried in 511/12: Constantine had built in Constantinople a church dedicated to the holy apostles which became the burial place of emperors, on which the church of St Nazaro, originally called the *basilica apostolorum*, in Milan was modelled, and Clovis may have been locating himself in this imperial tradition.[54] It may also be worth recalling in this context that Gregory of Tours described Clovis as a 'novos [*sic*] Constantinus', and that his narrative twice seems to assimilate that king into the figure of the first Christian emperor. The invocation of Christ by Clovis when fighting the Alamanni recalls Constantine's faith at the battle of the Milvian Bridge, while the reference to Bishop Remigius, who baptized Clovis, as being equal to St Silvester and washing away the leprosy of the one being baptized suggests that Gregory was in this case using material similar to that later

[53] 'ab ea die tamquam consul aut augustus est vocitatus', emended by W. Ensslin to 'ut augustus' ('Nachmals zu der Ehrung Clodowechs durch Kaiser Anastasius', *Historisches Jahrbuch*, 56 (1936), 499–507 followed by Weiss, *Clodwigs Tauf*, 114), and by Reydellet to 'ab augustis' (*Royauté*, 408). It is of interest that the version of this story provided by the *Vita Remigii* suppresses the 'tamquam' (*V. Rem.* 20, *MGH SRM* 3.311; 'vocitatus' becomes 'appellatus'). The interpretation of Gregory is made no easier by the title given this chapter in the manuscripts: 'De patriciato Chlodovechi regis' (p. 35).

[54] For Clovis, see e.g. Gregory of Tours, *hist. franc.* 2.43 with M. Vieillard-Troiekouroff, *Les Monuments religieux de la Gaule d'après les œuvres de Grégoire de Tours* (Paris, 1976), 206f.; for Constantine, Eusebius, *V. Cons.* 1.16 with P. Grierson, 'The Tombs and Obits of the Byzantine Emperors (337–1042)', *Dumbarton Oaks Papers*, 16 (1962), 1–63; for Milan, R. Krautheimer, *Three Christian Capitals* (Berkeley, 1983), 80. See too K. H. Krüger, *Königsgrabkirchen der Franken, Angelsachsen und Langobarden bis zur Mitte des 8. Jahrhunderts* (Munich, 1971), 50, 470.

found in the Donation of Constantine to compare Clovis and Constantine.[55] If Clovis were coming to see himself as something of an emperor after his acknowledgement by Anastasius (which presumably occurred in 508, given that he had spent the winter of 507–8 at Bordeaux), it could hardly have passed unnoticed in Italy, in which case it may be significant that the inscription at Terracina describing Theoderic as 'semper Augustus' probably dates from the period 507–11.[56] But this is hypothetical, and it may be safest merely to note that Anastasius offered some kind of recognition to Clovis shortly after his defeat of Alaric which may have continued a relationship his father had enjoyed with Constantinople and which was to find expression in a letter in which his grandson Theudebert claimed to be the son of Justinian.[57] Procopius was later to observe that 'the Franks never considered that their possession of Gaul was secure except when the emperor had put the seal of his approval upon their title',[58] and perhaps 'seal of approval' is as close as we can get to the events in Tours. But any perception that Clovis had been rewarded for defeating the Visigoths would have worried Theoderic.

The events of 507–8 obviously drove a wedge between Theoderic and Anastasius, and as far as we can tell it was the former who took the initiative in restoring good relations. It was probably in 508, or a little later, that Theoderic sent the letter which Cassiodorus later selected to stand at the head of the *Variae*. The letter has frequently been cited as providing evidence for the way in which the relationship between Theoderic's kingdom and the empire were perceived, but the chief purpose of the stately periods of Cassiodorus was to seek a restoration of good relations:

> It is proper for us, most clement emperor, to seek peace, it being acknowledged that we have no grounds for resentment . . . Indeed, tranquillity is something to be desired by every state in which the peoples make progress and care is taken for what is useful for the nations . . . And so, most pious of princes, it is in accordance with your power and an honour for us, who have hitherto enjoyed your love, to seek concord with you.[59]

[55] Gregory of Tours, *hist. franc.* 2.30f.

[56] Fiebiger/Schmidt, *Inschriftensammlung*, no. 193 (*CIL* 10.6850–2), assuming that the draining referred to is that described in Cassiodorus, *var.* 2.32.

[57] On Theudebert and Justinian, Hauck, 'Randkultur', 43.

[58] Procopius, *BG* 3.33.4.

[59] Cassiodorus, *var.* 1.1.1f., misdated by Stein, *Histoire*, 146. Needless to say, the content of letters to the emperor need not have coincided with the instructions given to the ambassadors who bore them; cf. Procopius, *BG* 1–3.19–28.

It is quite possible that the mission was entrusted to the patrician Agapitus, for at about that time Theoderic sent him on a legation to Constantinople, remarking in the words of Cassiodorus that the task required a most prudent man, who could dispute with the most subtle and conduct himself in gatherings of experts in such a way that numerous skills acquired by learning did not overcome the cause he undertook. It needed great skill to speak against artificers and to conduct business with people who think that they can anticipate your every move.[60] The outcome of Theoderic's initiative is not reported, but there is no record of conflict in the following years, and Western consuls from 508 onwards were published in the East. We may take it, then, that the events surrounding the defeat of Alaric had no lasting impact on relations between Ravenna and Constantinople, but they can have done nothing to make Theoderic any less wary of the emperor.

Indeed, it may have been at about this time that Theoderic acted in a way which would have been thoroughly appropriate to an emperor by issuing the beautiful gold medallion, three solidi in weight, called the Senigalla medallion after the town near which it was discovered in 1894, and now in the Museo Nazionale in Rome.[61] The medallion has generally been dated to 500 and associated with Theoderic's visit to Rome on the occasion of his *tricennalia*, but a strong case has recently been made, arguing from the absence, on the medallion and issues which may be presumed to have been related to it, of formulae characteristic of *tricennalia* celebrations and the presence of some characteristics of victory, for dating it to 509 and seeing it in the context of Theoderic's military successes. The obverse presents a frontal picture of Theoderic, moustached and with long hair covering his ears, which it is tempting to regard as a real likeness.[62] He is shown with his right hand half raised, while in

[60] Cassiodorus, *var*. 2.6.2.

[61] M. R. Alfoldi, 'Il medaglione d'oro di Teodorico', *Rivista italiana di numismatica*, 80 (1978), 133–42, has effectively disposed of the thesis of E. Bernareggi, 'Il medaglione d'oro di Teoderico', *Rivista italiana di numismatica*, 71 (1969), 89–106, according to whom the medallion was issued towards the end of Theoderic's life, without, however, being able to prove conclusively that it was issued in 500. The date of 509 is powerfully argued for by P. Grierson, 'The Date of Theoderic's Gold Medallion', *Hikuin*, 11 (1985), 19–26. Illustrations of the medallion are common, e.g. as a frontispiece to Wroth, *Catalogue*; Hahn, *Moneta*, no. 1; S. Sande, 'Zur Porträtplastik des sechsten nachchristlichen Jahrhunderts', *Institutum romanum Norwegiae acta*, 6 (1975), 65–106, pl. 221.

[62] So Sande, 'Porträtplastik', 82f., citing Cedrenus, *ep*. 1.2.2 ('aurium lingulae, sicut mos gentis est, crinium superiacentum flangellis operiuntur').

his left hand he holds a globe on which stands a Victoria, facing him and bearing a wreath and a palm; the legend reads REX THEODERICUS PIUS PRINC[EPS] I[NVICTUS] S[EMPER].[63] On the reverse is Victoria, facing right. She holds a wreath in her right hand and in her left a palm branch; the legend on this side reads REX THEODERICUS VICTOR GENTIUM. The title 'rex' without qualification is precisely that applied to Theoderic in narrative sources and inscriptions, but the issuing of medallions was a familiar imperial activity,[64] the iconography of this medallion is imperial in the traditions it draws upon, and the titles applied to Theoderic are certainly imperial. Whatever his constitutional position was, the Romans could have been forgiven for seeing in Theoderic an emperor.

THE NEW SITUATION IN THE WEST

To the west, however, the situation had greatly changed. The powerful Visigothic kingdom of Toulouse had been, as a Spanish chronicler put it, 'destroyed',[65] and Theoderic found it necessary to see to the government of the territories extending to the Durance which the Ostrogoths had occupied during the war, as well as of the land still occupied by the Visigoths. Before long, perhaps in 508, Gemellus was sent to Gaul as *vicarius praefectorum*, and Theoderic was able to write to the provincials that after a long period the Roman way of doing things had been restored to them: 'and so, having been recalled to your old freedom by the gift of God, clothe yourselves with manners befitting the toga, eschew barbarism and put aside the cruelty of your minds, because it is not fitting for you to live according to strange customs in the time of our just rule'.[66] It was probably in 510 that Liberius, a former praetorian prefect of Italy, was appointed praetorian prefect of Gaul, so restoring, in a fashion one would have thought appropriate to an imperial appointee, an office which seems to have lapsed when Odovacer ceded territory in the south of Gaul to Euric. He was still holding

[63] Hazardous expansions and mis-transcriptions have been frequent. The legend as given here is fairly secure, although the second-last word may have been intended as 'invictissimus' or 'inclytus': Schramm, *Herrschaftszeichen*, 227.

[64] J. M. C. Toynbee, 'Roman Medallions: Their Scope and Purpose', *Numismatic Chronicle*, 6th ser. 4 (1944), 27–44.

[65] 'regnum Tolosanum destructum est': *chron. caesaraug. s.a.* 507.

[66] Cassiodorus, *var.* 3.16–18; the quotation is from 3.17.1.

the office in 533 or 534, implying a tenure of remarkable length by the standards of the later empire.[67] As we have seen the Western consul for 511, Felix, was a man of Gallic background.

But the advent of Ostrogothic power in Gaul was not without problems. In 513 Bishop Caesarius of Arles was accused of treachery and summoned to Ravenna. It was scarcely a novel experience for him: some years earlier he had been accused by Alaric of wishing Arles to become subject to the Burgundians and consequently exiled to Bordeaux, and later, during the siege of Arles, he had been accused of supporting the Franks and Burgundians.[68] The accusation against Caesarius puts one in mind of the charge of betraying the country (*proditio patriae*) made a few years earlier against the bishop of Aosta, on which occasion Theoderic felt that it fell within his own competence, rather than that of the bishop's ecclesiastical superiors, to enquire into the charge.[69] So it was that Caesarius was brought to Ravenna under guard. But when Theoderic saw him coming he rose from his seat and laid aside his head-dress; when Caesarius greeted him he returned his greeting, asking Caesarius whether the journey had been a hard one and enquiring after his Goths and the people of Arles. When the bishop departed Theoderic is said to have commented that God would not spare those who had accused this man with the face of an angel and whose presence had been enough to make him shake, although we may note that the ability to cause trembling in kings was part of the stock-in-trade of a successful holy man. Theoderic, describing himself as 'the king who is your son', presented the bishop with a silver dish weighing some 60 pounds and 300 solidi; much to the king's gratification Caesarius sold the dish to ransom prisoners.[70] Caesarius is reported to have performed two miracles while in Ravenna. He restored to life the son of a widow and, more interest-

[67] Ennodius, *ep*. 8.22, 9.23, 9.29; Cassiodorus, *var*. 8.6. On the date, O'Donnell, 'Liberius', 44f.; *PLRE* 2.678; Sundwall, *Abhandlungen*, 134. M. Rouche, *L'Aquitaine des Wisigoths aux Arabes 418–781* (Paris, 1979), 377 somewhat oddly sees Liberius as a specialist in contacts with the Germans, appointed to this office to thwart the Romanizing propaganda of Anastasius and Clovis. For Liberius' relinquishing the office, *PLRE* 2.679; *fasti* are provided on pp. 1246f.

[68] *V. Caes. Arel*. 1.16–18, 21f.

[69] Cassiodorus, *var*. 1.9.

[70] *V. Caes. Arel*. 1.26f. The narrative has the feel of authenticity; Caesarius' good reception at Theoderic's court is confirmed by Ennodius, *ep*. 9.33. In general it may be compared with the encounter Theoderic is reported to have had with the anchorite Hilary (AASS 3 May, 173); the narratives are similar in more than one respect, but for trembling cf. as well Eugippius, *V. Sev*. 19.2.

ingly for our purposes, rendered assistance to the deacon Helpidius, described by the biographers of Caesarius as one of Theoderic's intimates. The deacon was troubled by an infestation of demons, who frequently caused stones to fall on his house in the manner of rain. Caesarius dealt with the problem by sprinkling holy water in the house, and Helpidius was freed from such experiences.[71] The bishop then departed for Rome, where he was greeted by Pope Symmachus and senators and their wives, all of them thanking God and Theoderic that they had deserved to cast eyes upon him. Symmachus was quick to legislate in favour of the see of Arles, which had been involved in a long-running dispute with the see of Vienne, and Caesarius returned home a happy man.[72]

There remained the territories which the Visigoths continued to occupy. Alaric was survived by two sons, the illegitimate Gesalic and Amalaric, his son by Theoderic's daughter Theodegotha. Amalaric was still very young in 507, so the Visigoths raised Gesalic as their king at Narbonne,[73] but after Gundobad occupied this city he was forced to retreat to Spain.[74] In the palace at Barcelona Gesalic was responsible for the murder of Count Goiaric, who a few years previously had been responsible for the compilation of the *Brevarium Alaricianum*; this may be a sign of tensions among the Visigoths. He was certainly opposed by Theoderic, whose general Ibba defeated him in 511.[75] In search of help he fled to Africa, where he received some financial support from Thrasamund, another sign that the policy of the Vandal king towards Theoderic during this period was less than entirely benign. He returned to Spain by way of Aquitaine, but 12 miles outside Barcelona he was defeated a second time by Ibba. Again he fled, this time eastward, but he was captured and put to death, probably in 513, having crossed the river Durance, presumably seeking help from the Burgundians. The race of his murderers is unknown.[76] However, as early as 511 Amalaric

[71] *V. Caes. Arel.* 1.28.

[72] *V. Caes. Arel.* 1.30, *chron. caesaraug.* 1.42; *ep. arel. gen.* 25 (*MGH Ep.* 3.35f.).

[73] *Chron. caes. s.a.* 508 (where the statement that Gesalic reigned for 7 years is incorrect); Isidore of Seville, *hist. goth.* 37 (correctly crediting him with 4 years); Procopius, *BG* 1.12.43.

[74] *Chron. gall.* 690f.; Isidore of Seville, *hist. goth.* 37.

[75] Murder of Goiaric: *chron. caesaraug.* 510; Goiaric's work on the *Brevarium* is known from *Cod. theod.* ed. Mommsen, p. xxxii, l. 6. Defeat by Ibba: *chron. caesaraug. sa* 510; cf. Isidore of Seville, *hist. goth.* 37.

[76] *Chron. caesaraug. sa* 513; Isidore of Seville, *hist. goth.* 38; Procopius, *BG* 1.12.46; Cassiodorus, *var.* 5.43f.

had been made king, with Theoderic acting as the guardian of his grandson.[77] Perhaps we are to see in Theoderic's intervention a comradely attempt to prop up a threatened Visigothic state, but the persistence with which he opposed Gesalic, whose claim to rule was not contemptible, and who would certainly have provided stronger leadership than Amalaric, 'a very young child' at the time of his father's death in 507,[78] together with his failure to impose Amalaric on the throne at any time prior to his own death in 526, may incline us to believe that he was merely taking advantage of an awkward moment in the history of the Visigoths to expand his own power. Theoderic gives every appearance of having exercised real authority: the royal treasure was removed from Carcassonne; commanders and armies were sent into Spain and tribute received annually from Gaul and Spain was remitted to the army of the Goths (i.e. Ostrogoths) and Visigoths, while the two peoples intermarried.[79] The manner of dating Church councils in accordance with 'the year of King Theoderic', while implying that the Visigothic kingdom enjoyed a certain independence, in that the consular dates used in Italy were not employed, nevertheless clearly showed where power officially lay.[80] But much power in the Visigothic territories accrued to the Ostrogoth Theudis, sent to act as Amalaric's *tutor*.[81] We have already (above, Ch. 3) touched on the power of this man, who married a wealthy Hispano-Roman woman, from whose estate he raised some 2,000 soldiers as a personal following; such was his power, Procopius states, using terms strikingly similar to his evaluation of Theoderic himself, that 'while in name (*λόγῳ*) he was a ruler of the Goths by the gift of Theoderic, in fact (*ἔϱγῳ*) he was an out-and-out tyrant'. Theudis, while carrying out Theoderic's orders and remitting tribute annually, refused to go to Ravenna, but Theoderic feared to act against him for fear that the Franks would intervene or the Visigoths would revolt.[82]

In these ways Theoderic acted to stabilize and benefit from the situation created by the victory of Clovis in 507. But damage had

[77] *Chron. caesaraug. s.a.* 513, referring to Theoderic's 'tutela', but the date is contradicted by the statement that Theoderic ruled in Spain for 15 years.

[78] Procopius, *BG* 1.12.43.

[79] Procopius, *BG* 1.12.47–9.

[80] Schmidt, *Ostgermanen*, 346; D. Claude, 'Gentile und territoriale Staatsideen im Westgotenreich', *Frühmittelalterliche Studien*, 6 (1972), 1–38 at 9. In general, Wolfram, *Intitulatio*, 79 on Theoderic's wish to represent the entire Gothic people.

[81] Jordanes, *get.* 302.

[82] Procopius, *BG* 1.12.50–4; see 1.1.29 for his evaluation of Theoderic.

been done, and a particularly worrying element must have been the conduct of the Vandals. They had failed to intervene when a Byzantine fleet ravaged part of the Italian coast, and a few years later Thrasamund had assisted Theoderic's enemy Gesalic. There is another possible indication of bad relations at this time. Writing in the ninth century Agnellus of Ravenna stated that Theoderic, having defeated all his enemies, sent the army of Ravenna to Sicily, depopulated it, and subjected it to his authority; these events occurred, he believed, in the thirtieth year of Theoderic's reign.[83] This expedition is referred to in no other source, and of course need not have occurred, but Agnellus' report may be worth brief consideration. As it stands it could be taken in conjunction with the coming to power of the Vandal king Hilderic in 523, who, as we shall see, rapidly moved away from good relations with Theoderic; it would not have been beyond the power of a Vandal king to stir up trouble in Sicily. But as Agnellus considered Theoderic to have reigned for thirty-four years,[84] the thirtieth year would have run its course before Hilderic became king. Perhaps 'thirtieth' could be replaced by 'third', in which case Agnellus' story could be related to a report of unrest in Sicily at the beginning of Theoderic's reign, but our only source for this implies that Theoderic did not intervene.[85] After mentioning the expedition to Sicily, Agnellus immediately continues: 'And in those times the sky was seen by everyone to burn from the north', which concords with an entry in the chronicle of Marcellinus under the year 512: 'The sky was often seen to burn from the northern area.' Perhaps, then, we should read 'twentieth' for 'thirtieth', in which case it may be possible to see in the permission Theoderic gave the people of Catania within the years 507–11 to use stones from a ruined amphitheatre to strengthen the walls of their city a step taken in the anticipation of trouble from the Vandals.[86]

Obviously Theoderic was now confronted with a situation which required his old system of alliances to be overhauled: the defeat of the Visigoths, the rise of the Franks, the antagonism of the Byzantines, and what we may presume to have been the hostility of the Vandals necessitated a search for new allies. At some time in

[83] Agnellus, *Codex* 39 (ed. Testi-Rasponi, 110).
[84] Ibid. 39 (ed. Testi-Rasponi, 112.205f.).
[85] Cassiodorus, *var.* 1.3.3.
[86] Cassiodorus, *var.* 3.49, but the activities of the Byzantine fleet may have furnished the motive.

the period 507–11 the learned Amalaberga, the daughter of Theoderic's sister Amalafrida, was sent to marry king Herminifrid of the Thoringians, old enemies of the Franks.[87] During the same period Theoderic adopted as son in arms (*filius per arma*) an unnamed king of the Heruls. The relationship envisaged between Theoderic and the king was a close one, for the letter of Cassiodorus to the latter describes Theoderic as adopting him, making him, and even procreating him as his son. He forwarded horses, swords, shields, and other instruments of war, and sent legates who were to hold talks with the king in his native tongue.[88] It is tempting to identify this man with the king Rodulf who was killed when the Lombards, possibly influenced by the Byzantines, launched an unprovoked attack on his people, perhaps in 508, when Theoderic was otherwise engaged.[89] Recorded as it is in no Western source, the significance of this battle can only be perceived dimly, but it may have been an important stage in the rise of the Lombards, and may perhaps have had an importance not much less than that of Clovis' victory over the Visigoths in 507. As did the Alamanni following their second defeat at the hands of Clovis, the Heruls fled, some to Byzantine territory and others to Italy, while still others, if Procopius can be believed, made their way as far as Thule.[90] Theoderic's contacts across the Alps were nothing if not extensive: towards the end of his life an embassy arrived from the Hesti, who

[87] Cassiodorus, *var.* 4.1 (Krautschick, *Cassiodor*, 54 suggests a date of 507, on the basis of little evidence); Anon. Vales. 70 (mistakenly describing the bride as Theoderic's sister and implying too early a date); Jordanes, *get.* 299; Procopius, *BG* 1.12.21f. (dated too early, but correct in seeing the alliance so concluded as anti-Frankish); Gregory of Tours, *hist. franc.* 3.4. On the Thoringians see the rich note of Claude, 'Universale und partikulare Züge', 34 n. 108; of interest is the connection suggested by B. Schmidt, 'Theoderich der Grosse und die damaszierten Schwerter der Thuringer', *Ausgrabungen und Funde*, 14 (1969), 38–40.

[88] Cassiodorus, *var.* 4.2; on the institution, Procopius, *BP* 1.11.22. Theoderic's father Theodemer had adopted Hunimund king of the Suevi after defeating him (Jordanes, *get.* 274); but Hunimund was 'inmemor paternae gratiae' (ibid. 275). Note as well the strong language used in Gregory of Tours of the relationship between Guntram and his adopted son Childebert: *hist. franc.* 7.33, 8.4, 8.13, 9.11.

[89] Procopius, *BG* 2.14.11–22. That the man created *filius per arma* and the two Rodulfs were identical: Lotter, in Wolfram and Daim, *Fölker*, 56 (whose suggestion as to the date of Rodulf's defeat is to be preferred to that of e.g. Claude, 'Universale und partikulare Züge', 36); Hauptfeld in ibid. 124; J. Werner, *Die Langobarden in Pannonien* (Munich, 1962), 133. Unreliable are N. K. Lukman, *Skjoldunge und Skilfinge* (Copenhagen, 1943), 134, and Beyerle, 'Süddeutschland', 59; Byzantine influence in the Lombard attack is detected by Stein, *Histoire*, 151.

[90] Cassiodorus, *var.* 4.45; Marcellinus, *chron. s.a.* 512; Procopius, *BG* 2.15.

were to give their name to Estonia,[91] while grave goods indicative of contacts with Ostrogoths during the late fifth and early sixth centuries have been discovered in Lithuania.[92]

Data of this kind are such as to impose caution. The correspondence of Cassiodorus provides an invaluable guide to Theoderic's foreign relations, but it only covers approximately a quarter of his reign, and it may be that the impression we receive from it of heightened diplomatic activity after the victory of Clovis merely reflects the fact that Cassiodorus' earliest correspondence dates from this time; perhaps correspondence from other periods would have made them seem equally full of activity. Evidence drawn from archaeology, on the other hand, tends to be difficult to date and interpret. Relations between the kingdom of the Ostrogoths and the other peoples of the *Völkerwanderung* must have been much more complex and were doubtless subject to more change than we can now detect; further, alliances contracted between various kings and Theoderic may have answered the needs of the former at least as much as those of the latter, but it is difficult to determine what their needs may have been. But such evidence as we have suggests a dislocation of alliances around the period of the victory of Clovis over the Visigoths. The next major change occurred with the accession of the emperor Justin in 518.

THE END OF THE ACACIAN SCHISM

In July 514 Pope Symmachus died, and as we have seen the emperor Anastasius professed satisfaction at the elevation of the new pope, Hormisdas, even though he was not likely to have disowned the positions taken by Symmachus, who had probably chosen him as his successor.[93] But Anastasius may have found himself needing

[91] Cassiodorus, *var*. 5.2; but note that Krautschick, *Cassiodor*, 59 seeks to relate this letter, with the preceding one, to the period of Cassiodorus' quaestorship. N. Wagner, *Getica* (Berlin, 1967), 91 suggests that this was an attempt by the Hesti to renew old relations. Tacitus, *Germ*. 44f. deals with Goths and Aesti in adjacent chapters; the work was known to Cassiodorus (*var*. 5.2.2).

[92] J. Werner, 'Der Grabfund von Taurapilis, Rayon Utna (Litauen) und die Verbindung der Balten zum Reich Theoderichs', in G. Kossack and J. Reichstein, ed., *Archäologische Beiträge zur Chronologie der Völkerwanderungzeit* (Bonn, 1977), 87–92.

[93] Hormisdas is seen as *papabile* in letters written by Ennodius in 510–11 (*ep*. 8.33, 8.39).

peace with Rome at that time, for Vitalian, the *magister militum per Thracias* and an adherent of the Council of Chalcedon, was in revolt. Vitalian may have been a Goth, and it has been suggested that he was in league with Theoderic, another Gothic *magister militum*, the two of them conspiring so that Vitalian would come to enjoy in the East the power Theoderic held in the West, although this does not seem very plausible;[94] it may, however, be significant that the Eastern consul for 515, Procopius Anthemius, was the son of Anthemius, who had been emperor in the West 467–72.[95] But as part of the terms needed to secure peace with the rebel, Anastasius seems to have agreed to the holding of a council in July 515 to discuss union with the Roman Church;[96] certainly he wrote to Hormisdas concerning a council to be held in Heraclea near Constantinople in letters of 28 December 514 and 12 January 515, but for whatever reason his letters were only received in Rome on 14 May and 28 March 515 respectively.[97] Hormisdas took counsel with Theoderic. Given that relations between Constantinople and Ravenna may still have been cool in 515, such a course of action would have been prudent even if Theoderic played no part in the rebellion of Vitalian, but in any case Hormisdas seems to have enjoyed good relations with the king, whose only recorded gift to the Roman Church was made during his pontificate.[98] In August 515 the pope sent to Constantinople a mission comprising his old friend Ennodius, who had become bishop of Pavia in 513, Bishop Fortunatus of Catania, the priest Venantius of the Roman Church, the deacon Vitalis, and the notary Hilary.[99]

Needless to say the men sent on such missions needed to be sound: the bishops Vitalis and Misenus, sent to Constantinople by Pope Felix III, had shared Communion with Acacius. But the composition of this legation gave no grounds for supposing that the

[94] For Vitalian, *PLRE* 2.1171–6; the evidence for his being 'possibly of Gothic descent' is given p. 1171. His plotting with Theoderic: Schwartz, ACO 4.2, p. vi, with P. Charanis, *Church and State in the Later Roman Empire* (Madison, 1939), 63; Ensslin, *Theoderich*, 289.

[95] *PLRE* 2.99.

[96] Sources differ on the peace terms: Victor of Tunnuna believed Anastasius to have promised to unite all the churches of the East with the Roman Church (*chron. s.a.* 514); Theophanes believed there would be a synod attended by the bishop of Rome and all the bishops (*chron.* AM 6006, ed. de Boor, p. 160, with Theodore Lector, *epit.* 146.10ff.). See too John of Antioch, *frag.* 214a.

[97] *Coll. avel.* 109, 107.

[98] *Lib. pont.* 269.7–9 (Hormisdas takes counsel), 271.15 (Theoderic's gift).

[99] *Lib. pont.* 269.7.9; *coll. avel.* 115.1, 12.

Roman position would be compromised. It was led by a man who did not know Greek and believed that Acacius had acted under the control of the Devil, while Venantius was one of the Roman priests who had remained loyal to Pope Symmachus in 502.[100] Whether because of those to whom it had been entrusted, or because Anastasius came to feel that he had nothing to fear from Vitalian, the embassy failed, the emperor writing to the pope that the living should not be driven from the Church on account of the dead and concluding his letter with a quotation of some of Christ's words to the disciples, 'My peace I give to you, my peace I send upon you.'[101] But in July 516 Anastasius wrote to the senate of Rome, asking it to treat on his behalf with 'the most glorious king' and the pope, apparently to no avail, for the senate replied a few months later to the effect that Theoderic had urged it to adhere to papal instructions.[102] In April 517 a new mission, comprising Ennodius, Bishop Peregrinus of Miseno, and the subdeacon Pullio, sallied forth from Italy.[103] It too failed. In July 517 Anastasius wrote his last letter to Hormisdas, noteworthy for its quiet dignity, its stress on mercy, and its conclusion: 'We are able to put up with being hurt and contradicted; we cannot be commanded.'[104] A story was told in Rome of how the emperor, having failed in an attempt to bribe the delegates, sent them home on a dangerous boat under an armed guard, and of how he wrote to Hormisdas, 'we wish to command, not be commanded.'[105] The tale may be disregarded, although the dangerous boat is oddly anticipatory of a possible interpretation of a reference to the boat which carried Pope John to Constantinople some years later; it is chiefly of interest for the highly coloured version of the concluding sentence of Anastasius' letter it presents, another sign of the distortions so often found in the *Liber pontificalis*. It must have seemed an unpropitious moment for ecumenical relations. But on 9 July 518 the old emperor died, 'struck down', the author of this text continued, 'by the will of the

[100] Ennodius' lack of Greek: above, Ch. 5 n. 121. It is tempting to connect this deficiency with the appointment of the bishop of Catania to the mission; cf. Caspar, *Geschichte*, 133 n. 6. Diabolical control over Acacius: *dictio*. 6.8 (p. 323.21). Venantius: *MGH* AA 12.443 no. 17, cf. 402 no. 73, 414 no. 65.

[101] *Coll. avel.* 125; Christ's words are quoted from John 14: 27.

[102] *Coll. avel.* 133, 114, the latter possibly a product of the papal chancery: Caspar, *Geschichte*, 142.

[103] *Coll. avel.* 126f.; *lib. pont.* 100.

[104] *Coll. avel.* 138.

[105] *Lib. pont.* 269.12 to 270.3.

divinity',[106] and on the following day the *excubitor* Justin succeeded him.

Thereafter events moved quickly. Justin was a staunch supporter of the Council of Chalcedon, as were his nephew Justinian, clearly the power behind the throne long before his own accession in 527, and Vitalian, who was loaded with honours in the few years of life which remained to him prior to his assassination in 520.[107] On 1 August 518 Justin wrote to Hormisdas announcing his accession,[108] and on 7 September the legate Gratus was dispatched to the pope bearing a letter from Justin asking him to send as envoys bishops who embraced peace, perhaps a hint that the missions led by Ennodius had created a negative impression even in Chalcedonian circles, and a letter from Justinian suggesting that the pope might care to come to Constantinople.[109] We know from Justinian's letter that Gratus was expected to visit 'the unconquered king', and as the letters to Hormisdas only arrived in Rome on 20 December it may plausibly be conjectured that Gratus had held discussions with Theoderic, presumably in Ravenna, *en route* to Rome.[110] Hormisdas took counsel with the king, and this having been done it was probably in January 519 that Bishop Germanus of Capua, the priests John and Blandus, the deacons Felix and Dioscorus of the apostolic see, and the notary Peter were sent from Rome to Constantinople.[111] Most members of this legation are not otherwise recorded, but it is at least possible that Felix is to be identified with the man of that name who became pope in 526, while Dioscorus was the deacon through whom Theoderic had ordered the return of the churches of Rome to Symmachus. He played a role different to that

[106] Ibid. 270.4; almost exactly the same words were applied to the death of Pope Anastasius, ibid. 258.

[107] Justin: Theodore Lector, *epit.* 151; Victor of Tunnuna, *chron. s.a.* 518; Zachariah, *hist. eccl.* 7.14, 8.1. Justinian's power: Cyril of Scythopolis, *V. Sabae*, 68; Procopius, *anec.* 6.19 and *build.* 1.3.3, 1.4.29, with Averil Cameron, *Procopius*, 9 n. 35 and Stein, *Histoire*, 222f. Vitalian: *PLRE* 2.1174f.

[108] *Coll. avel.* 141. For what follows the best treatment remains Caspar, *Geschichte*, 150ff.; see too A. A. Vasiliev, *Justin the First* (Cambridge, Mass., 1950); L. Magi, *La sede romana nella corrispondenza degli imperatori e patriarchi bizantini (VI–VII sec.)* (Rome, 1972), 35ff.

[109] *Coll. avel.* 143, 147; also *lib. pont.* 270.5f.

[110] *Coll. avel.* 147.5 for a visit to Theoderic, with which may be compared the statement in the 1st edn. of the *lib. pont.* that Gratus went to the apostolic see 'cum consilio regis Theodorici' (101). Theoderic's consent for Hormisdas' dealings with Constantinople is indicated at p. 270.7, 17.

[111] *Lib. pont.* 270.6–8.

of the other legates, reporting to Hormisdas in separate letters.[112] Such was the respect of Hormisdas for Dioscorus that he felt he would be a desirable patriarch of Alexandria, his home city, but this was not to come about; however, following the death of Pope Felix in 530 he was elected pope at the same time as Boniface, causing a brief schism in the Roman Church.[113] We may well see in Pope Felix and his would-be successor Dioscorus the two Roman deacons who voyaged to Constantinople in 519. The papal emissaries carried various letters, including a fulsome one addressed to the emperor.[114] On 25 March, the Monday in Holy Week, they were met 10 miles outside the city by Vitalian, Pompeius, who was a nephew of the late emperor, and Justinian, and these notables accompanied them for the remainder of their journey.[115] They were not long in accomplishing their aim. On Maundy Thursday they had the satisfaction of watching the patriarch John sign the *libellus* they proffered in the palace, and then proceeding to his church, where he erased from the diptychs the names of Acacius and of Fravitta, Euphemius, Macedonius, and Timothy, the four subsequent patriarchs of Constantinople; similarly the names of Zeno and Anastasius were no longer to be recited at the altar.[116] John's capitulation was total, but it was popular, the clergy of Constantinople being reported to have commented immediately afterwards that they could not remember so many people ever taking holy Communion before.[117] After thirty-five years the Acacian schism had been healed on terms dictated by Rome.

It has been widely felt that the restoration of unity between the Churches of Rome and Constantinople, the latter so clearly under the imperial thumb, was 'an event of evil augury for Ostrogothic power'.[118] To be sure, for the first time in Theoderic's reign Italian Catholics were now in communion with the man to whom, as

[112] See Caspar, *Geschichte*, 151 n. 4, but note Haller, *Papsttum*, 536f.

[113] Patriarch of Alexandria: *coll. avel.* 175.2. Double election in Rome: *lib. pont.* 281.2–4; Dioscorus died a few weeks later

[114] *Coll. avel.* 149; Caspar's suggestion that Ennodius was called in to write this letter (*Geschichte*, 154 n. 5) is based on little evidence.

[115] *Coll. avel.* 167.5, 223.1. Typically, *lib. pont.* goes beyond this, claiming that the legates were met by Justin and Vitalian (270.10f.).

[116] *Coll. avel.* 223.4–6, 167.10f.

[117] Ibid. 167.

[118] Hodgkin, *Italy*, 437; cf. Llewellyn, *Rome*, 45, L. Obertello, *Severino Boezio* (Genoa, 1974), 1.82f.; H. Chadwick goes so far as to see the settlement of the schism as having been undertaken 'so as to make possible the ultimate overthrow of the Gothic kingdom' (in Gibson, ed., *Boethius*, 8). But contrast Demougeot, *Formation*, 822, and Haller, *Papsttum*, 254f.

Hormisdas put it, the Trinity had entrusted the governance of the worldly empire,[119] but we may nevertheless ask precisely what impact this had on the king. Hormisdas' epitaph, written by his son Silverius, who was himself to become pope in 536, expressed the conviction that Greece had yielded to the pope, but the patriarch John felt that the emperor was the one responsible for the peace, and the perspective on its establishment revealed by his letters is such as to minimize the victory of the papacy, while his successor Epiphanius, although becoming patriarch on 25 February 520, only troubled himself to inform Hormisdas of this in a letter of 9 July, which elicited from the pope a reply noteworthy for its coolness.[120] Hormisdas, unable to secure the appointment of Dioscorus as patriarch of Alexandria, received in 521 the unwelcome news that Paul, an old enemy of Severus who had become patriarch of Antioch in succession to him in 519, had been forced by lack of popular support to resign his see.[121] A Byzantine government which insisted on dealing with the ecclesiastically recalcitrant 'more mildly and clemently',[122] but which, as we shall shortly see, was already showing in its dealings with the apostolic see signs of imperiousness prompted by a desire not to alienate all elements which found it difficult to accept the Council of Chalcedon, a desire which was to culminate in the humiliation endured by Pope Vigilius at the time of the later council of Constantinople (553), was not likely to tempt a pope into political disloyalty towards the non-interventionist Theoderic, particularly given that Theoderic's one important intervention in papal affairs to this date, his decision of 506 or 507 in favour of Symmachus, had established a pope of whom Hormisdas was the nominated successor. In any case, the last letters of Hormisdas to Constantinople preserved in the *Collectio avellana* are dated 26 March 521, and it is possible that after that date the pope was not in contact with the court.[123] It could be suggested that the position of Theoderic's government was weakened by the settlement of 519, in that dissatisfied Italian aristocrats may have become more likely to have sought dealings with Constantinople, and indeed we know that

[119] *Coll. avel.* 201.1.

[120] Epitaph of Hormisdas: *lib. pont.* 274 n. 25, ll. 7f. John's perspective: *coll. avel.* 161. Epiphanius and Hormisdas: *coll. avel.* 195, 205. Note as well the context in which Matt. 16: 18 is cited by Epiphanius: Magi, *Sede romana*, 46f., 71.

[121] *Coll. avel.* 216.4, 217.2–4, 241.

[122] Ibid. 232.4.

[123] Ibid. 236–40. I accept the estimate of Hormisdas' position *vis-à-vis* the East offered by Caspar, *Geschichte*, 181–3.

Albinus, Symmachus, and Faustus *niger* followed the negotiations between the Churches and subsequent developments with a lively interest,[124] but we have suggested that ties between them and Constantinople, not necessarily of a kind which would have been to Theoderic's taste, long antedated 519. Finally, one cannot avoid being struck by the role of Theoderic himself in the moves towards ending the schism. It is possible that he was interested in this from the beginning of his reign, for the legations of Faustus and Festus to Constantinople seem to have been concerned with ecclesiastical as well as political issues, and the same could be said of that of Andromachus, whose mission took place during the reign of Odovacer.[125] The reference to Theoderic in correspondence between the emperor Anastasius and the senate, the repeated references in the *Liber pontificalis* to Theoderic's counsel being sought by Rome as negotiations proceeded during the reigns of both Anastasius and Justin, the dealings of Justin's legate Gratus with Theoderic, and the circumstance that Eulogius, who conveyed letters between Constantinople and Rome in 519–20, was entrusted by Justin with 'certain affairs' with Theoderic, suggest that if anything the king supported the restoration of unity between the Churches.[126] In any case, relations between Ravenna and Constantinople may have been improved with the accession of Justin, as the Western consular appointment for 519 suggests.

EUTHARIC: NEW APPOINTEES

As early as 507 Ennodius had concluded his panegyric by expressing the hope that an heir to the kingdom would play in Theoderic's lap.[127] But the years passed and as far as we know his marriage to Audefleda produced no sons and only one daughter, Amal-

[124] Albinus raised the point that some had condemned the Council of Chalcedon orally, others in writing: *coll. avel.* 173.1. Faustus was interested in the expression 'unus de Trinitate crucifixus': Schwartz, *Publizistiche Sammlungen*, 115–17. Symmachus: *coll. avel.* 219.1, 221.2.

[125] Gelasius, *ep.* 10 (ed. Thiel).

[126] Reference in correspondence: *coll. avel.* 113.3, 114.1. Theoderic's counsel: above, nn. 98, 111, although the repetition of the point is enough to raise the faint suspicion that a phrase has been misplaced. Gratus: *coll. avel.* 147.5. Eulogius: *coll. avel.* 199.2.

[127] *Pan.* 93; see too *ep.* 9.30.10.

asuintha.[128] It became important for Amalasuintha to be married well, and, according to Jordanes, to this end Theoderic located in Spain a man descended from the race of the Amals, who could be described as youthful and a man of prudence, *virtus*, and bodily wholeness.[129] Unfortunately this description of Eutharic Cilliga, apparently so straightforward, seems on a closer reading most perplexing: the form of his descent from the Amals, which no other text refers to, is not clear;[130] he cannot have been youthful if we accept the assertion of another source that he was almost the same age as the emperor Justin;[131] and the references to his *prudentia* and *virtus* seem almost designed to suggest he was possessed of both Roman and barbarian characteristics, and may have been phrased to conceal a more unpleasant reality.[132] His very name is awkward, for its second component, Cilliga, is not Germanic, and is only applied to him in inscriptions and the narrative of Anonymous Valesianus.[133] In short, Eutharic is a mysterious character. However, in 515 he and Amalasuintha were married,[134] and before long they were the parents of Athalaric.[135] There is no indication that Anastasius ever reacted to the rise of Eutharic, but Justin shared the office of consul with him in 519 and adopted him as son at arms. As so often it is difficult to interpret these events, for two passages of Cassiodorus

[128] This is the implication of Jordanes, *get.* 297f. [129] Ibid. 298.

[130] H. Wolfram, 'Theogenie, Ethnogenese und ein kompromittierer Grossvater im Stammbaum Theoderichs das Grossen', in Jaschke and Wenskus, eds., *Festschrift Beumann*, 80–97. A good recent discussion: P. Heather, 'Cassiodorus and the Rise of the Amals: Genealogy and the Goths under Hun Domination', *Journal of Roman Studies*, 79 (1989), 103–28.

[131] Cassiodorus describes him as 'paene aequaevus' with Justin (*var.* 8.1.3), who was born *c.*450–2 (*PLRE* 2.648–50). But Cassiodorus can describe Clovis and Alaric as 'regii iuvenes' (*var.* 3.2.2) when the former was about 40 (cf. Gregory of Tours, *hist. franc.* 2.43) and the latter had been king since the death of his father Euric in 484. Amalasuintha would have been in her early twenties at the oldest; see above, p. 51–2 for the date of her parents' marriage.

[132] Cf. *var.* 3.23.3 for 'Romanorum prudentia . . . et virtus gentium'; such vocabulary suggests that Jordanes may have been following Cassiodorus' lost history closely at this point. He is elsewhere described as being harsh and an enemy of the Catholic faith (Anon. Vales. 80), although this may reflect a later perception.

[133] References in *PLRE* 2.438. On the name Cilliga, Schönfeld, *Wörterbuch*, 62, with Wrede, *Sprache*, 67f.

[134] Cassiodorus, *chron. sa* 515; Jordanes, *get.* 298.

[135] Jordanes, *get.* 304 states that when Theoderic died in 526 Athalaric was just 10 years old (*adhuc vix decennem*) and has been widely followed: Stein, *Histoire*, 249; Ensslin, *Theoderich*, 301. Yet Jordanes, *rom.* 367 and Procopius, *BG* 1.2.1 provide an age of 8 in 526, and I doubt whether the younger age, supported by concordant and precisely expressed sources, should be laid aside.

suggest quite different estimates of the role of the emperor in Eutharic's becoming consul: a letter in his *Variae* written after the death of Theoderic states that Eutharic 'was adorned by [Justin] with the palm enwoven robe of the Consul', but the contemporary account provided in his *Chronicle*, while it mentions the senate and *plebs*, excludes the emperor, and the only reference to Constantinople occurs when Cassiodorus mentions the stupefaction of Symmachus, a legate of the East, at Eutharic's lavish consular games in Rome.[136] But the recognition accorded Eutharic by his being the first Western consul appointed by Justin, all the stronger for the emperor himself being the Eastern consul of the year, was impressive, however Cassiodorus chose to interpret it, and his adoption by the emperor implied imperial approval of the man who must have seemed likely to succeed Theoderic. Cassiodorus took advantage of the situation by producing a chronicle, nominally at Eutharic's request, which covered the history of the world from Adam to the consular games celebrated by Eutharic in 519, and an oration in his praise.[137] The year in which peace was concluded between the Churches must have seemed one of bright promise for the Ostrogothic kingdom.

Another development which occurred a few years later must also have seemed to strengthen Theoderic's position. We have seen that in the second decade of the century his nominations to the consulship and, if his appointees to the urban prefecture are any guide, his choices for other offices of state were generally men of no high family background. But in 522 Boethius' two sons, Symmachus and Boethius, were the two consuls. We cannot be sure whether Ravenna or Constantinople took the initiative, but Theoderic and Justin must have collaborated in this double appointment of consuls from the West, and we may be correct in seeing it as a response to the attainment of Church unity, just as the appointment of two

[136] Cassiodorus, *chron. s.a.* 519; *var.* 8.1.3, quoted from Hodgkin's translation, Anon. Vales. 80 (Rolfe's 'Theoderic made Eutharicus consul' does not accurately translate 'Theodericus dato consulatu Eutharico'). Note as well the assertion of Cassiodorus in an oration: 'et nos gloriamur de sententia boni principis; laetamur de consensu senatus' (*MGH AA* 12.468.19–21; unfortunately the text breaks off immediately). If Eutharic was made Justin's 'per arma filius' (*var.* 8.1.3) when he became consul it is perhaps odd that Cassiodorus describes Theoderic as his 'glorious father' (*chron. s.a.* 519).

[137] Cassiodorus, *chron. praef.* for the origin of the chronicle. Fragments of the oration are printed in *MGH AA* 12.465–72; for the occasion of its delivery, p. 463.

Western consuls in 494 may have been connected with Theoderic's defeat of Odovacer on behalf of Constantinople—although perhaps 522 was a little late for the commemoration of an event which occurred in 519. One notes with a sense of irony that the consulship of the boys' grandfather in 487 had not been recognized in the East.[138] Boethius pronounced an oration in honour of Theoderic in the senate,[139] and must have been overjoyed at this extraordinary honour, which may have reflected in some way his great design of making Greek thought understood in the West. Looking back after he had been persuaded that Boethius was plotting against him, Theoderic may have found such a role as Justin may have played in initiating or accepting the consulship of his sons compromising,[140] but this must have been the last thing on his mind when Boethius was appointed by him to the office of *magister officiorum* during the year when his sons were consuls, with tenure from 1 September.[141] The office marked an unexpected development in the career of Boethius, who had earlier complained that the fairly nominal burdens imposed by the office of consul which he held in 510 kept him from his real work.[142] Now, twelve years later, Boethius was probably in his early forties and must have been aware that, however impressive the volume of work he had done, most of the task of translating and commentating he had set himself remained ahead of him, yet he was prepared to move to Ravenna and assume the heavy workload of the *magister officiorum*, among whose functions it was to calm the nerves of senators when they were introduced to the king. It is obvious from the description of the office provided by Cassiodorus that it would have involved working closely with Theoderic,[143] and we may see in the appointment of Boethius and his accepting the office a sign of a *rapprochement* with the king.

It is hard to establish whether the king had begun to look with more favour on the nobles, or the nobles had begun to look with more favour on the king. But Theoderic's new policy continued. His

[138] *PLRE* 2.233 s.v. Boethius 4.

[139] *Ordo generis Cassiodororum*, ed. Mommsen, p. vi; Boethius, *phil. cons*. 2.3.8, 'tu regiae laudis orator'.

[140] Chadwick, *Boethius*, 45f.; Chrysos, 'Amaler-Herrschaft', 458.

[141] *PLRE* 2.235 and Sundwall, *Abhandlungen*, 238 discuss the date; the office is known from Anon. Vales. 85.

[142] PL 64.201B.

[143] Cassiodorus, *var*. 6.6; cf. Jones, *Later Roman Empire*, 368f. When King Witigis sent an embassy to Justinian, he thought it worth his while to try to enlist the support of the *magister officiorum* in Constantinople: Cassiodorus, *var*. 10.33.

consular nominee for 523, Maximus, was a descendant of Petronius Maximus, who had been emperor for a short while in 455.[144] Maximus was a member of the *gens Anicia*, but perhaps more important than this was the circumstance that his ancestor had become emperor by the expedient of murdering Valentinian III, who had himself been responsible for the murder of Boethius' grandfather in the preceding year. Maximus was to go on to marry a member of the Ostrogothic royal family in 535, but the year was not a propitious one for such a marriage; in 537 he was among a group of senators suspected of treason and expelled from Rome following its conquest by Belisarius, and subsequent to his murder at the hands of Goths years later, Justinian was to order that half a property he had been given by Theodahad was to pass to Liberius.[145] But such things could not have been foreseen in 523, and on 13 August of that year, one week after the death of Pope Hormisdas, a Tuscan, John, ascended the papal throne.[146] While Theoderic could not have appointed John pope, his elevation seems to have been of a piece with that of Boethius, his sons, and Maximus, and will require more extended discussion.

JOHN, DEACON AND POPE, AND BOETHIUS

On 18 September 506 the Roman deacon John, hitherto an adherent of Laurentius, made his submission to Pope Symmachus. We have noted already that John and Boethius seem to have been allies in more than one theological controversy (above, Ch. 4), and it will be as well to examine the evidence more closely now. In about 512 an Eastern bishop wrote to Pope Symmachus arguing that Christ exists 'both of and in two natures' (*et ex et in duabus naturis*), a formula which may be seen as an attempt to find common ground between Chalcedonian and what we may term Monophysite understandings of Christ.[147] From a Roman point of view it could well have been construed as being quite innocuous, for Pope Gelasius

[144] Procopius, *BG* 1.25.15.
[145] *PLRE* 2.748. 'Fl. Maximus'.
[146] *Lib. pont.* 272.5–7, 275.1.
[147] Symmachus, *ep.* 12 (ed. Thiel). For what follows, Schurr, *Trinitätslehre*, 108–35; Obertello, *Severino Boezio*, 1.47–9; C. Leonardi, 'La controversia trinitaria nell'epoca e nell'opera di Boezio', in Obertello, ed., *Congresso internazionale*, 109–22. I am not persuaded by the comments of Haller, *Papsttum*, 535 on the date.

himself had affirmed almost exactly the same thing.[148] But Symmachus was unhappy, as shown by a letter he sent to the Eastern bishops;[149] so too Ennodius, who early in 513 produced his *In nomine Christi*, a work which contains passages very similar to the letter of Symmachus and is noteworthy for its hostility to Acacius.[150] Boethius felt differently. In what is now known as his fifth tractate, although it was certainly not the fifth in order of composition, Boethius described the reading at a council which he attended of a letter arguing that Christ consisted of two and in two natures. He had kept quiet during the subsequent discussion, but in the tractate, addressed to John the deacon, he proceeded at length to defend the position taken in the letter.[151] Just as we have suggested that Boethius and John were supporters of Laurentius during the schism in Rome, both adhering to the party which sought an end to the Acacian schism, we have reason to associate them in about 513, with Boethius at any rate more favourably disposed to the Eastern initiative than Symmachus, Ennodius, and, by implication, those who had attended the council in Rome at which the letter had been discussed.

Fresh controversy broke out when, in July or August 519, a group of Scythian monks arrived in Rome from Constantinople. They proposed a theological system which they seem to have felt would have rendered the Council of Chalcedon more palatable in the East, affirming in particular that 'one of the Trinity was crucified' (*unus de trinitate crucifixus est*).[152] One of their leaders, Leontius, said he was related to Vitalian, and they enjoyed the support of Justinian,[153] so papal disagreement with their proposals would have been awkward. The monks spent over a year in Rome, during which

[148] In his *De duabus naturis*, ed. Schwartz, *Publizistische Sammlungen*, 93.15ff.; note esp. 'ex quibus et in quibus unus et idem'.

[149] *Coll. avel.* 104.

[150] Ennodius, *dictio*. 6.322f. On the date, Sundwall, *Abhandlungen*, 70.

[151] Ed. Stewart and Rand, 72–128, arguing for Christ being 'et in utrisque naturis . . . et ex utrisque' (116.27f.). The arguments of M. Nedoncelle, 'Les Variations de Boèce sur la personne', *Revue des sciences religieuses*, 29 (1955), 201–38 at esp. 236f., for dating this treatise to 523–4 are not strong.

[152] For the Scythian monks, Caspar, *Geschichte*, 161–4; C. Moeller, 'Le Chalcédonisme et le néo-chalcédonisme en Orient de 451 à la fin du VIe siècle', in A. Grillmeier and H. Bacht, eds., *Das Konzil von Chalkedon*, I (Würzburg, 1954), 637–720.

[153] Leontius: *coll. avel.* 216.6. Justinian's increasing sympathy with the position of the monks can be traced e.g. in *coll. avel.* 187, 188, 191.

period they apparently participated in tumultuous discussions in which senators were involved; Pope Hormisdas finally compelled them with some violence to leave.[154]

Reactions to the monks' proposal were amazingly along the positions taken during the Laurentian schism. Faustus *niger*, the old patron of Pope Symmachus, asked the priest Trifolius whether the formula 'one of the Trinity was crucified' could be discovered in the teachings of the Fathers, but Trifolius was not able to give a favourable report: 'This teaching proceeds from the fount of Arius and is in agreement with all his heirs.'[155] The deacon Dioscorus, formerly an emissary of Pope Symmachus to Theoderic and now one of the legates of Hormisdas in Constantinople, advised the pope to have nothing to do with their formula, as did his other legates.[156] After delaying what must have seemed an unconscionable period, Hormisdas finally gave his verdict in letters written to Justin and Bishop Epiphanius on 26 March 521, in which he came down against the teaching of the monks in a considered fashion,[157] but his true feelings are more evident in a letter he had already sent to one of his contacts in the city, Bishop Possessor, in which he expressed himself with a good deal more vigour.[158] Faustus, Dioscorus, and Hormisdas had all supported Pope Symmachus, and just as that pope was hostile to the theology of the East, so now did they turn against the Scythian monks.

Quite different was the response to the monks' teaching in circles which may be thought to have supported Laurentius during the schism. Boethius took up his pen again, writing two tractates, the first and second in the sequence in which they have been preserved, addressed respectively to his father-in-law Symmachus and John the deacon.[159] The most detailed examination of these works which has been undertaken suggests a date of 523 for their composition.[160] Another scholar who wrote in support of the Theopaschites was himself a Scythian monk, Dionysius Exiguus, who translated two

[154] John Maxentius, in ACO 4.46–62; this refutation of Hormisdas' letter to Bishop Possessor contains incidental descriptions of their time in Rome.

[155] Schwartz, *Publizistiche Sammlungen*, 115–17; the quoted words are at 115.5–7. That the 'beatus Faustus senator' was probably Faustus *niger*: Sundwall, *Abhandlungen*, 119f.; *PLRE* 2.456, against Momigliano, 'Anicii', 234.

[156] *Coll. avel.* 216.5ff., 217.5ff.

[157] Ibid. 236f.

[158] Ibid. 231.

[159] Ed. Stewart and Rand, 2–36.

[160] Schurr, *Trinitätslehre*, 136–225, esp. 224f.

letters of Cyril of Alexandria on behalf of their leaders John and Leontius,[161] and various pieces of evidence suggest that Dionysius, translator *extraordinaire*, may have stood towards the centre of a group of intellectuals who had supported Laurentius. His translation of a Life of St Pachomius may well have been undertaken on behalf of Proba, the daughter of the senator Symmachus and hence the sister-in-law of Boethius;[162] his translation of Gregory of Nyssa's *De conditione seu opificio hominis* was carried out for Eugippius, a friend of Paschasius, one of the adherents of Laurentius, himself the compiler of a collection of excerpts from the works of Augustine, using the library of Proba;[163] he translated a letter written by representatives of the Church of Alexandria to Festus and Pope Anastasius' legates to the emperor Anastasius in 497;[164] a collection of papal decrees he compiled concludes with the eirenic letter of Pope Anastasius to the emperor of this name, a document which must have been in line with the thinking of the supporters of Laurentius;[165] and it is possible that the 'dearest brother Laurentius' for whom Dionysius prepared a codex of canons was none other than the anti-pope. It must be admitted that in this case the form of address may have been unexpected, but the work could have been undertaken prior to 498, when Laurentius was a priest of the Roman Church, or following his banishment to the estates of Festus.[166] We may note as well that two works of Dionysius concerning the calculation of the date of Easter adhere to

[161] ACO 1.5.2.295. In general, Schurr, *Trinitätslehre*, 168–85.

[162] See the comments of F. Mahler, in H. von Craneburgh, ed., *La Vie latine de saint Pachome* (Brussels, 1969), 37–42.

[163] Translation for Eugippius: PL 67.345f. Eugippius and Paschasius: above, Ch. 4 with n. 50. Excerpts and Proba: CSEL 9.1–4. See on the circle of Eugippius, Lotter, *Severinus*, 34–7.

[164] *Coll. avel.* 102 (pp. 473.22f.).

[165] PL 67.311–16. Note, however, that the priest Julianus of the church of St Anastasia, for whom Dionysius compiled the decrees (PL 67.231A) was probably an adherent of Symmachus during the schism, attending as he did the synod of 6 Nov. 502 (*MGH AA* 12.443 no. 25); he will have been the 'Julianus presbyter tituli Anastasiae' who attended the synod of 499 (ibid. 414 no. 61).

[166] F. Maassen, *Geschichte der Quellen und der Literatur des canonischen Rechts in Abendlands bis zum Ausgang des Mittelalters* (Leipzig, 1870), 960; but cf. Dionysius' 'carissimus frater noster Laurentius' with Jordanes' 'nobilissime frater Vigili' (*rom.* 1) to denote a man who may have been pope. The identification of Laurentius with the anti-pope is made by Richards, *Popes and Papacy*, 86, but denied by Steinacker, 'Römische Kirche', 54 n. 119; 'peut-être' is the verdict of Mahler in von Craneburgh. ed., *La Vie de saint Pachome*, 33.

the Eastern method of calculation, as the supporters of Laurentius had done during the schism.[167]

There is some overlap between the circle of Dionysius and that of the African Fulgentius, bishop of Ruspe. During his second period of exile, that is, approximately 518–23, Fulgentius corresponded with Galla, a daughter of the senator Symmachus, and another woman whom he described to Galla as 'your sister Proba, the holy virgin of Christ'; if the reference to Proba as the sister of Galla is to be taken literally, as it probably is, Fulgentius corresponded with the two sisters-in-law of Boethius.[168] His correspondents included a consul of the golden first decade of the century, Theodorus, as well as Eugippius, abbot of the monastery which grew up around the relics of St Severin at Lucullanum and, as we have seen, a friend of the deacon Paschasius, an adherent of Laurentius, and Dionysius Exiguus.[169] Fulgentius was a supporter of the Scythian monks who came to Rome canvassing support for a variant of the formula 'one of the Trinity suffered in the flesh', writing at some length on the question in response to a query from a group in Rome conveyed, significantly, by John the deacon;[170] the fact that Boethius wrote in support of them suggests that we are dealing with a fairly clearly defined group.

There has been a good deal of discussion on how Boethius' theological activity should be interpreted, it having been suggested, on the one hand, that his interest was purely academic, and on the other that he was working in conjunction with Justinian as part of a

[167] *Liber de paschate,* PL 67.483–514, and the letter to Boniface and Bonus, PL 67.513–20; see C. W. Jones, 'The Victorian and Dionysian Paschal Tables in the West', *Speculum*, 9 (1934), 408–21.

[168] Fulgentius, *ep*. 2 to Galla, *ep*. 3f. to Proba; Proba is referred to as Galla's sister at *ep*. 2.31. See further on Galla, Gregory the Great, *dial*. 4.14. Büdinger, 'Eugipius', 805, felt that Proba was the daughter of Probinus (consul 489) or Probus (consul 502); G.-G. Lapeyre, *Saint Fulgence de Ruspe* (Paris, 1929), 234 suggests that the relationship between Galla and Proba was only spiritual; so too S. T. Stevens, 'The Circle of Bishop Fulgentius', *Traditio*, 38 (1982), 327–41 at 334; C. P. Hammond Brammel, 'Products of Fifth-Century Scriptoria Preserving Conventions Used by Rufinus of Aquileia', *Journal of Theological Studies*, n.s. 30 (1979), 449 believes that if Proba was not the daughter of Symmachus she may have been descended from Anicia Faltonia Proba, a correspondent of Rufinus. But Cassiodorus (*inst*. 2.3.1) and Eugippius (CSEL 9.1.2.4) unambiguously refer to Proba the daughter of Symmachus as a virgin.

[169] *Ep*. 5 to Eugippius, 6 to Theodorus (consul 505). Is it too much to see in the Venantia to whom *ep*. 7 was addressed a connection of one of the Venantii who were consuls in 507 and 508?

[170] *Ep*. 16f.; see for John *ep*. 17.1.

programme aimed at restoring the unity of the empire. The former viewpoint perhaps owes something to a romanticized view of Boethius as a pure intellectual who would have disdained to concern himself with issues exciting contemporary interest, while the latter cannot be sustained on the evidence available to us.[171] The evidence we have been considering here points in a different direction. Boethius, Dionysius, and Fulgentius were all writing works of a similar tendency, and can be shown to have had friends in common. It does not strain credulity to see in them and their friends the survivors of groups who had supported Laurentius in the schism; nor to see the deacon John, who became pope in 523, as having been among their number. We may therefore take it that the activity of these men, while arising from contemporary issues, was Italian in its inspiration and owed nothing to exterior political influence. But perhaps our discussion can be extended a little in a speculative direction. When Pope Symmachus' rival Laurentius was forced to quit Rome he went to live on the estates of his patron Festus, and as Festus was a neighbour of the patrician Agnellus who had an estate at the castrum Lucullanum, it is possible that it was there that Laurentius ended his days (above, Ch. 4). During the pontificate of Pope Gelasius the relics of the holy man Severin had been taken to the castrum Lucullanum near Naples, and Eugippius became the abbot of a religious community there.[172] One wonders whether Laurentius, if he were nearby, entered into contact with Eugippius, or whether contact had already been made. Some twenty years before the relics of Severin arrived at Lucullanum, the deposed emperor Romulus Augustulus had been sent by Odovacer to live there.[173] At that time Romulus was still very young, for one of our sources mentions his *infantia*,[174] and we do not know for how long he survived. A letter of Cassiodorus written in the period 507–11 is addressed to one Romulus, and is concerned with the affairs of himself and his mother; it is quite possible, although it cannot be proved, that this Romulus is to be identified with the last

[171] Interest as academic: Schurr, *Trinitätslehre*, 217–22. Co-operation with Justinian: W. Bark, 'Theoderic vs. Boethius: Vindication and Apology', *American Historical Review*, 49 (1944), 410–26, with which cf. Obertello, *Severino Boezio*, 54–6; Daley, 'Boethius' theological tracts', 189f.

[172] Relics of Severin: Eugippius, *V. Sev*. 46.2. Eugippius as abbot: Isidore of Seville, *de vir. ill*. 26.

[173] Anon. Vales. 38; Jordanes, *get*. 242; Marcellinus, *chron. s.a*. 476.

[174] Anon. Vales. 38.

emperor.[175] One possible sign of a connection between Romulus and the circle connected in various ways with Lucullanum is a reference in a letter of Fulgentius to a 'holy brother Romulus',[176] but it would be unwise to place much weight on this.

Our enquiries into Boethius and the circles in which he moved may seem to have taken us down some curious byways. But we may conclude that those medieval authors who identified the deacon John referred to in some of the theological tractates of Boethius with the man who became pope in 523 were correct.[177] One further circumstance which has never, to my knowledge, been acknowledged makes the identification even more likely: one of the most famous phrases in the works of Boethius, the definition of man as 'a rational and mortal animal', is also to be found in a letter written by John the deacon to Senarius.[178] Further, we have seen that it is highly likely that the same John had supported Laurentius for at least part of the Roman schism; and for what the *argumentum e silentio* is worth, Pope Hormisdas does not seem to have seen fit to avail himself of John's talents in his negotiations with the East; further, in the years preceding 523 Boethius, together with Dionysius Exiguus and Fulgentius, had been members of a group continuing to argue for the principles which, in another form, we have suggested animated the move to install Laurentius as pope. Here, indeed, is a context in which Boethius may be set at about the time of his appointment as *magister officiorum*.

But the context is not merely suggestive of intellectual and theological enthusiasm; for we are dealing with a group which suddenly found itself powerful. The stage was set by the appointment of the sons of Boethius as consuls in 522; then followed the appointment of Boethius as *magister officiorum* on 1 September of that year and the installation of his friend John as pope on 13

175 Cassiodorus, *var*. 3.35. See e.g. Burns, *Ostrogoths*, 74, ed. Krautschick, 'Zwei Aspekte', 358; Ruggini, in Obertello, ed., *Congresso internazionale*, 75 n. 9 is sceptical.

176 Fulgentius, *ep*. 6.1.

177 John Scotus Erigena, ed. E. K. Rand (1906), 47.19 (at the beginning of his commentary on Boethius' 2nd tractate); Peter Abelard, *theol. christ*. 1.134 (CCCM 12.130). Among modern writers, I agree with Pfeilschifter, *Theoderich*, 171 and J. Matthews, in Gibson, ed., *Boethius*, 24, against Caspar, *Geschichte*, 185 n. 3 and Daley, 'Boethius' theological tracts', 162 (= 163) n. 17.

178 'rationale animal atque mortale', *phil. cons*. 1.6.15 (see Gruber, *Kommentar zu Boethius*, 155f., for other examples in the works of Boethius); cf. 'animal rationale mortale', John to Senarius (ed. A. Wilmart, *Studi e testi*, 59 (1933), 170–9 at 179.13).

August 523. We do not know whether Theoderic played any role in the elevation of John. He had given judgement concerning Pope Symmachus and was later to intervene in the election of Pope Felix IV, but in both cases he became involved in times of controversy. But an aristocracy which seems to have been substantially excluded from high office for at least a decade, and a linked current within the Roman Church which had been oppositional during two pontificates, suddenly came to power at almost exactly the same time. And if we were to choose to see at least some elements of this group as harbouring sentiments which led them to look with nostalgia on the old empire in the West, we could perhaps see some significance in the appointment of Maximus, who claimed the emperor Petronius Maximus as a forebear, as Western consul in 523.

However, the rapid rise of this group was to be parallelled by its fall. Within a few years the consuls of 522, together with their mother Rusticiana and their aunts Proba and Galla, were to have their properties confiscated;[179] their father Boethius and grandfather Symmachus were to be executed; and their father's friend, John, was to experience Theoderic's anger and die while still under the king's displeasure. To the tragedies of Theoderic's last years we must now turn.

[179] Procopius, *BG* 1.2.5; I presume the children of Symmachus to have included Rusticiana, Proba, and Galla.

7

The Last Years

According to Jordanes, subsequent to his defeat of Odovacer in 493 Theoderic ruled Italy 'prudently and peacefully for 30 years',[1] while Anonymous Valesianus asserts that during Theoderic's reign of thirty-three years, 'Italy enjoyed felicity for 30 years, so that even those who were travelling enjoyed peace'.[2] It thus becomes a task for the historian to establish what happened in approximately 523 to change the character of Theoderic's rule. Needless to say, it is by no means impossible that we have to deal with circumstances of a very general kind, among which weight should be given to the fact that Theoderic was becoming an old man. The significance of this is difficult to assess, particularly as Theoderic lived in a society in which those who survived the dangerous years of early childhood often remained active in a vigorous old age; among them one thinks of Cassiodorus, who was still writing at 93, and Liberius, who seems to have held military commands not long before his death at the age of 89.[3] Nevertheless, Theoderic would have been about 70 in 523, and by the end of the year he had seen the deaths of many of the most important figures in his public life: even during the portion of his life spent in Italy he had, by the end of 523, outlived the emperors Zeno and Anastasius, kings Alaric, Clovis, Sigismund, and Thrasamund, and no less than five popes. Perhaps advancing years made him more vulnerable to feelings of disquiet if events unfolded contrary to his wishes, and the manner of his death in 526, which was occasioned by an attack of diarrhoea, may imply that his last years were overshadowed by illness. Beyond this, it seems best to enquire into

[1] *Rom.* 349. The suggestion that the 30 years began in 489 (Wes, *Das Ende*, 173) seems implausible given the context in which Jordanes' remark occurs.

[2] Anon. Vales. 59. Demougeot suggests that the 33 years began in 493 and the 30 in 496–7 (*Formation*, 804 n. 43); see, rather, Momigliano, 'Anicii', 235 and J. N. Adams, *The Text and Language of a Vulgar Latin Chronicle (Anonymous Valesianus II)* (London, 1976), 5, although overlooking the significance of the defeat of Odovacer leads the latter to make too much of imperial recognition. For 'pergentibus' as 'travellers', ibid. 111f.

[3] Cassiodorus: *PLRE* 2.268. Liberius: ibid. 680f.

specific circumstances which may have given him cause for unhappiness.

One's attention is immediately caught by the death of his son-in-law Eutharic. He was certainly alive in 519, when he celebrated his consulship, and his role in the disturbances which broke out between Jews and Catholics in Ravenna in 519 or 520 implies that at that time he held some authority in the town while Theoderic was absent in Verona. Thereafter nothing is known of him, save that by the time Theoderic died in 526 Eutharic was already dead, which means his death cannot be dated at all closely.[4] His death would not, however, have been a total surprise if Eutharic was indeed of about the same age as the emperor Justin, for Justin had been born *c.*450–2 and was himself to die in 527.[5] The death of a son in law who had shared the office of consul with the emperor and been adopted by him as son in arms, and so was presumably not only Theoderic's presumptive successor but a successor accepted in Constantinople, must have been a blow, but that his death is announced in none of our sources may be held to suggest that it was not in itself of the greatest moment.

Before his death Eutharic had become the father of a son, Athalaric, and a daughter, Matasuentha.[6] Two other grandsons of Theoderic are known, Amalaric, the son of Theodegotha and Alaric the Visigoth who, despite the power Theudis wielded in the Visigothic territories, was to become king in reality as well as name following the death of Theoderic,[7] and Sigiric, the son of his daughter Ostrogotho Areagni by her husband Sigismund the Burgundian; the couple also had a daughter, probably Suavegotho.[8] But Ostrogotho died and Sigismund, who, having become a Catholic, succeeded his father Gundobad as king in 516, took a second wife, whose name is unknown to us; by this marriage he became the father of Gisclahad and Gundobad.[9] The jealous stepmother is a motif well known in folklore, and, according to a circumstantial story told by Gregory of Tours, Sigismund's second wife prevailed

[4] Jordanes, *get.* 304 and Procopius, *BG* 1.2.2, *BV* 1.14.6; attempts such as that of Schmidt, *Ostgermanen*, 353 to argue for 'um 522' seem ill-conceived.

[5] On the age of Eutharic and that of Justin, above, Ch. 6 n. 31.

[6] Jordanes, *get.* 80f.

[7] *Chron. caesaraug. sa* 525.

[8] On Suavegotho, above, Ch. 2 n. 82 ('Suavegotta').

[9] Gregory of Tours, *hist. franc.* 3.6; Marius of Aventicum, *chron. s.a.* 523; *Passio Sigismundi regis*, 9.

upon her husband to put Sigiric to death, alleging that he planned to kill his father, bring Italy under his sway, and take over the kingdom Theoderic had held there. In any case, Sigismund had Sigiric executed in 522, shortly afterwards repenting and entering a monastery to perform penance.[10] In 523 the Franks invaded Burgundy. Gregory of Tours, ever with an eye for the influence of women on their menfolk, attributed the invasion to the influence of Clovis' widow Chlotilde, who sought revenge by means of her sons for the murder of her parents by Gundobad, the father of Sigismund. An unlikely story, one would have thought, especially since Gundobad is accused of the murder of Chlotilde's parents in no other source; be this as it may the Burgundians handed Sigismund and his second wife and children over to the Franks, who murdered them and were believed to have disposed of the bodies by throwing them in a well.[11]

It is hard to imagine Theoderic rejoicing at these developments, but we may doubt whether they caused him enormous concern. Sigismund had been showing disquieting tendencies to ally himself with Constantinople, having been created *patricius* and, possibly, *magister militum per Gallias*, and while it is true that his predecessors Gundioc and Gundobad had held this office, they had received their appointments before 476, and not from Constantinople.[12] We know that Theoderic found it necessary to intercept the correspondence of Sigismund with Constantinople.[13] Further, while Theoderic cannot have been alarmed when Sigismund became a Catholic, no more than he had been when Clovis was converted, there may have been tension between the monarchs arising from or reflected in the close relationship Sigismund enjoyed with Bishop Avitus of Vienne and that which Theoderic seems to have had with his rival Caesarius, the bishop of Arles.[14] Moreover, there is no hint that he had seen Sigiric as a potential heir to his kingdom, for while Gregory of Tours asserts that the second wife of Sigismund claimed

[10] Gregory of Tours, *hist. franc.* 3.5; Marius of Aventicum, *chron. s.a.* 522; *Passio Sig.*, 8.

[11] Gregory of Tours, *hist. franc.* 3.6, *in glor. mart.* 74; Marius of Aventicum, *chron. s.a.* 523; *Passio Sig.*, 9.

[12] *PLRE* 2.1009.

[13] Avitus of Vienne, *ep.* 84 (p. 101.27f.).

[14] See on Caesarius A. Malorny, *Saint Césaire évêque d'Arles* (Paris, 1894), and W. Klingshirn, 'Charity and Power: Caesarius of Arles and the Ransoming of Captives in Sub-Roman Gaul', *Journal of Roman Studies*, 75 (1985), 183–203, but it would be good to have more available on both Caesarius and Avitus. Note that Sigismund is known to have made a pilgrimage to Rome: Ensslin, *Theoderich*, 284f.; Caspar, *Geschichte*, 127.

that Sigiric sought to take over the kingdom Theoderic had held in Italy, the story, itself made implausible by the belief attributed to the wife that Theoderic had already died,[15] does not imply that Theoderic wished to be succeeded by Sigiric. If Sigiric were murdered before the death of Eutharic the question would not have arisen, and even if Eutharic predeceased Sigiric, Theoderic's grandson Amalaric the son of Theodegotha was alive and probably of an age to govern, to say nothing of his daughter Amalasuintha, who, as it turned out, proved a moderately competent ruler of the Ostrogoths in the name of her son Athalaric after Theoderic died.[16]

Indeed, the confusion which overcame the Burgundian state following the death of Sigismund provided its powerful neighbours with the opportunity to make useful gains. In 523 Theoderic sent Duke Tuluin, who had already distinguished himself fighting in Gaul, to see to Ostrogothic interests, and while the Franks and Burgundians were fighting he was able, possibly with the help of some Gepids, to occupy a block of territory peacefully.[17] As a result of this the Ostrogothic border advanced considerably towards the north, probably from the Durance, which it had reached after the earlier war in Gaul, to the Isère.[18] The annexation of fresh territory is surely to be seen as a simple piece of expansionism, in no sense a *wergeld* for the murder of Sigiric,[19] and it is hard to see the fall of Sigismund, with whom Theoderic's relations seem to have been poor, and the consequent advance as being likely to have brought him anything other than satisfaction. Yet again, the borders of Theoderic's kingdom had been pushed back, and it may be significant that Gregory of Tours, an author not prone to exaggerate in this direction, asserts that between the death of Clovis and 533 the Goths occupied much of the territory they had lost in Gaul.[20] Such

[15] 'regnum quod avus eius Theudoricus Italiae tenuit': *MGH SRM* 1.112.9f.

[16] The possibility of Sigiric uniting the Ostrogothic and Burgundian kingdoms was suggested by Binding, *Burgundische-Romanische Königreich*, 245f.

[17] Cassiodorus, *var*. 8.10.8: 'triumphus sine pugna, sine labore palma, sine caede victoria'; Procopius, *BG* 1.12.24–32. Cassiodorus, *var*. 5.10f., written within the period 523–6, describes Gepid soldiers passing through Venetia and Liguria on their way to Gaul.

[18] Malorny, *Saint Césaire*, 129–32; see the map in Hodgkin, *Italy*, 3, facing p. 319. The chronicle attributed to Fredegarius presumably refers to the period after the occupation of these lands when it describes Theoderic as ruling 'a finibus Pannoniae usque ad Rodano fluvio, a Terreno mare usque Alpis Poenucas et Isera fluvio' (*chron*. 2.57, *MGH SRM* 2.82.2f.).

[19] To the contrary, Rosenfeld, 'Ost- und Westgoten', 250.

[20] *Hist. franc*. 3.21.

gains, we may presume, were more likely to have occurred when Theoderic was alive than after his death.

Events in Africa gave more cause for concern. As we have suggested the most important of the alliances Theoderic contracted in the West was that sealed in 500 by the marriage of his daughter Amalafrida to Thrasamund the Vandal. The failure of the Vandals to intervene when the Byzantine fleet ravaged part of the Italian coast in 507 or 508 and the subsequent aid given by Thrasamund to the Visigoth Gesalic must have been disagreeable; still more, if it occurred, the attack on Sicily which may have taken place in 513. We also know that shortly after his accession in 519 the emperor Justin was in touch with Thrasamund,[21] and relations between his people and the Vandals must often have been of concern to Theoderic. Thrasamund died in 523[22] and the succession passed to his cousin Hilderic. The new king, a man in his mid-sixties, was the son of King Huneric (477–84), a fierce persecutor of Catholics,[23] and of Eudocia, the daughter of Valentinian III, who had been captured by Huneric's father Geiseric in 455. The blood which flowed in the new king's veins was thus an odd mixture, but it was not long before he revealed which direction he would take. Despite having given an oath to Thrasamund that he would not open the churches for the Catholics in his kingdom or restore their privileges, he was quick to recall the exiled bishops, open the churches, restore full liberties to Catholics, have one Boniface consecrated bishop of Carthage at the request of the people of the city, and see to the election of bishops in other places;[24] on 5 February of 'the second year of the most glorious king Hilderic' a council met at Carthage.[25] There was nothing here to disturb Theoderic, for Hilderic was merely restoring to the Catholics of Africa rights those of Italy had never lost. But Hilderic is described as having been a 'guest friend' (ξένος) of Justinian, which suggests that he had

[21] *Coll. avel.* 212.1. [22] On 6 May, according to *PLRE* 2.1117.

[23] Victor of Vita, *hist. pers.* 3.19. Gregory of Tours believed that Hilderic succeeded Huneric: *hist. franc.* 2.3 (p. 45).

[24] Victor of Tunnuna, *chron. sa* 523.2; *Laterculus regum VVandalorum et Alanorum* 16 (*MGH AA* 13.459); *consularia italica* (*MGH AA* 9.269 no. 26); Ferrandus, *V. Fulg.* 54; *lib. pont.* 271.1 f. If the attribution of the recall of the bishops to the pontificate of Hormisdas (ob. 6 Aug. 523) made in the last source is correct, the observation of the *Laterculus* that Hilderic acted in the *exordium* of his reign is confirmed.

[25] Courtois, *Vandales*, 305–9; its proceedings in *Concilia Africae*, CCSL 259.255–82.

spent time in Constantinople and enjoyed good relations with Justin's nephew,[26] and given Theoderic's concern for the dealings of other Western states with Constantinople this must have worried him. In time the Vandals put Amalafrida in prison and destroyed 'all the Goths', presumably the remnants of those Theoderic had sent to accompany her to Africa in 500, which meant, as Procopius saw it, that the Vandals became enemies instead of allies and friends to Theoderic and the Goths in Italy.[27] These events cannot be precisely dated, but they ushered in a phase of his dealings with the Vandals in which Theoderic's position was severely weakened. Among the exiled bishops recalled by Hilderic was Fulgentius of Ruspe,[28] and one wonders whether his recall at a time when Vandal statecraft was turning away from the Goths had any impact on the way Theoderic regarded those Italians who had been associated with him. But it is time we turned our attention to Italy.

Knowing as we do the circumstances which attended the end of Theoderic's reign there is a natural inclination for students to project his difficulties backwards. But care is called for. It is true that our most detailed source for the last years of Theoderic dates his turn for the worse to shortly after the fracas with the Jews of Ravenna which occurred as early as 519–20, but this seems to have been a purely local disturbance. The words of Anonymous Valesianus are striking: 'Shortly after that the Devil found an opportunity to steal for his own a man who was ruling the state well and without complaint' (Anon. Vales. 83, trans. Rolfe). Indeed, so great is the alteration in the tone of his narrative that some scholars have unwisely held that the part of the text commencing with chapter 79, the account of Theoderic's alleged illiteracy, was the work of a second author.[29] The reference to diabolical intervention is followed by brief accounts of three matters. Firstly, the Anonymous notes that Theoderic ordered that a chapel of St Stephen ('that is, an altar', he or an interpolator confusingly adds) beside the springs in a suburb of Verona was to be overturned (Anon. Vales. 83). It is difficult to know what to make of this, but given the military importance of Verona at the time and the fact that the word used for suburb, *proastium*, almost certainly indicates a place outside the city walls,

[26] Procopius, *BV* 1.9.5. [27] Procopius, *BV* 1.9.3 f.

[28] *V. Fulg.* 55, noting the 'wonderful goodness' of Hilderic.

[29] So Cessi, ed. cit. p. clxvi ff., followed by C. H. Coster, *Late Roman Studies* (Cambridge, Mass., 1968), 77. But see Chadwick, *Boethius*, 291 n. 67; Barnish, '*Anonymous Valesianus II*', 572 ff.

we may be justified in interpreting Theoderic's measure as having been undertaken in the interests of defence but remembered in a hostile way at a later time.[30] Secondly, we are told that Theoderic forbade the Romans the use of arms, even down to a small knife (Anon. Vales. 83). No indication is given as to why Theoderic took this step, but there is no reason to doubt that he did take it, for similar orders had been issued in earlier times, the use of arms having being forbidden in 364.[31] Thirdly, various omens are reported. A poor Gothic woman, we are told, gave birth to four dragons (*dracones*), two of which were borne in the clouds from west to east, being cast into the sea in view of the people, while the other two, which had one head, were carried off. There appeared a star with a torch called 'comet' which was seen for fifteen days, and there were frequent earthquakes (Anon. Vales. 84). Here, it may seem, we are dealing with something sinister, for that comets portend the deaths of kings is a fact known to every reader of Suetonius, and Anonymous Valesianus may well have been such a reader, while an earthquake had occurred in Italy in 492, the year before the final overthrow of Odovacer.[32] But the comet was known to Byzantine authors, who date it to 519.[33] It would be possible to interpret the narrative of the Anonymous in such a way as to suggest that reasons for tension between Goths and Romans accumulated from 519/520;[34] on the other hand, given that after the Anonymous' lengthy treatment of events which occurred in 519 or, perhaps, 520 (Anon. Vales. 80–4) he immediately proceeds to describe the accusation made against Albinus, which probably occurred in 523, the narrative need not imply a sustained crescendo of hostility, and as some of the events were of local import and others unexceptionable it

[30] For walls built at Verona, Anon. Vales. 71. *Proastium* must mean a settlement outside the city walls (cf. Procopius, *BV* 1.17.11 on Decimum as a προάστειον of Carthage, *BG* 2.29.31 on Classis as one of Ravenna). I therefore accept the interpretations offered by Pfeilschifter, *Theoderich*, 157 and Mor, in *Verona in età gotica e longobarda*, 71, against Stein, *Histoire*, 249.

[31] *Cod. theod.* 15.15.1; the right of using arms had been restored in 440 when attack by Geiseric was feared (*Nov. Val.* 9) and during the troubled reign of Majorian (*Nov. Maj.* 8). See further, Barnish, '*Anonymous Valesianus II*', 386.

[32] *Fasti Vind. priores, sa* 492 (*MGH AA* 9.318); see too Agnellus, *Codex* 39 (ed. Testi-Rasponi, 108). For Suetonian echoes in the Anonymous, Barnish, '*Anonymous Valesianus II*', 575f. Another testimony from the 6th cent. to the significance of comets: Gregory of Tours, *curs. stell.* 34.

[33] *Chron. pasc.* records it under the consulship of Justin and Eutharic (ed. Dindorf, 612); John Malalas, *chron.* 411.

[34] So Obertello, *Congresso internazionale*, 64.

may be preferable to see them as having been recalled after Theoderic's death and given their negative interpretation by the author in the light of the events which were to follow.

ALBINUS AND BOETHIUS ACCUSED

'After these things', our source continues, 'the king began to grumble against the Romans whenever he found an opportunity' (Anon. Vales. 85). Cyprian the *referendarius*, whose task it was to present matters before the king, charged (*insinuans*) the patrician Albinus with having sent the emperor Justin a letter which was hostile to Theoderic's kingdom; we know from another source that the charge was made at Verona.[35] The accused person was a man of some distinction, for Albinus, a member of the Decii, was a pillar of the senate. He had almost certainly been Theoderic's first consular nominee, just as his father Basilius had been the first nominee of Odovacer (480). He had been connected with the negotiations concerning the end of the Acacian schism,[36] and there is no reason to doubt that he had been engaged in correspondence with Constantinople, whatever its topic.[37] A passage in the Byzantine author Suidas has occasionally been construed to involve in the process an African, Severus, who forwarded to Ravenna a compromising letter he had obtained; while this interpretation should be rejected,[38] Cyprian is known to have undertaken a mission to Constantinople prior to his appointment as *comes sacrarum largitionum* on 1 September 524 which, if it occurred in about 522, would have put him in

[35] Anon. Vales. 86. The functions of *referendarius*: Cassiodorus, *var.* 6.17; Procopius, *BP* 2.23.6, *anec.* 14.11, with Mommsen, *Gesammelte Schriften*, 6.421. On *insinuans*: G. B. Picotti, 'I senato romano e il processo di Boezio', *Archivio storico italiano*, 7th ser. 7 (1931), 205–28 at 209. Charge at Verona: Boethius, *phil. cons.* 1.4.32.

[36] One would like to know the extent of the *distantia* from which he wrote to Pope Hormisdas: *coll. avel.* 173.1.

[37] Stein, *Histoire*, 255 and Obertello, ed., *Congresso internazionale*, 64, see it as connected with the accession of Pope John in Aug. 523; Coster, *Later Roman Studies*, 76f., suggests Albinus was corresponding with Justin concerning the emperor's anti-Arian measures; Wolfram, *History*, 330 and Chadwick, *Boethius*, 52, surely with greater plausibility, connect it with the Ostrogothic succession following the death of Eutharic.

[38] Picotti, 'Senato romano', 209–11 and 'Osservazioni', 205–9, against e.g. Sundwall, *Abhandlungen*, 157, 243; Bertolini, *Roma*, 88f.; Demougeot, in Obertello, ed., *Congresso internazionale*, 107.

a good position to obtain damaging material or at least plausibly claim he had encountered such material.[39] A dramatic scene occurred in which Albinus, summoned before the king, denied the charge, whereupon Boethius commented in the king's consistory 'Cyprian's charge is false, but, if Albinus did it, both I and the entire senate have done it acting together. The business is false, lord king.' Cyprian hesitated, and then produced what are described as false witnesses against both Albinus and his defender, Boethius. Theoderic, angered at the 'Romans', was later believed to have been trying to find a way of killing them, and hence was prepared to trust false witnesses rather than senators.[40]

When Boethius came to write the *Consolation of Philosophy* he did not intend to provide a full narrative of his fall.[41] Nevertheless, the account provided there allows three elements of the case against him to be disentangled. In the first place, it was said that he had wished that the senate be safe. This, of course, is Boethius' statement of the charge, but as he goes on to explain that he was accused of hindering the activities of an informer who sought to produce documents by which he would show the senate guilty of treason it is clear that he was summarizing events something like those described in Anon. Vales. 85.[42]

Secondly, forged letters were produced by which people attempted to show that Boethius had hoped for Roman liberty. He had no doubt that the letters were forged, his evidence being the confession of the informers.[43] Presumably this was connected with the correspondence Albinus was accused of carrying on with the emperor, and the charge could have been a very serious one indeed: Theoderic had been made nervous by Sigismund's attempting to communicate with Constantinople, the Vandal king Gunthamund had imprisoned the poet Dracontius for merely praising another lord, generally identified by scholars as the emperor Anastasius, while we know that an earlier Vandal king, Huneric, had been concerned lest bishops send 'letters across the seas', apparently in

[39] Cassiodorus, *var.* 5.40.5 on Cyprian's mission to the East.

[40] Anon. Vales. 85 f.; on 'in conspectu regis', here translated 'in the king's consistory', Picotti, 'Senato romano', 214.

[41] Gruber argues on the basis of *phil. cons.* 1.4.25 that Boethius composed an extensive memorandum on this subject, of which *phil. cons.* 1.4 contains only the chief ideas (*Kommentar*, 123).

[42] Boethius, *phil. cons.* 1.4.20f., with 1.4.23, 1.4.31, 1.4.36.

[43] Ibid. 1.4.26. It may be worth noting a claim in Procopius that Justinian was a victim of sophistries practised by his *referendarii*: *anec.* 14.11.

connection with succession to the throne, and during the Gothic war Totila was to accuse senators of having brought in the Greeks to attack their fatherland.[44] There is no reason to believe that the Goth would have been slow to act on a charge of this kind. Doubtless this was the charge reflected in the comment of Procopius that Theoderic put Boethius, as well as his father-in-law Symmachus, to death 'on the ground that they were setting about a revolution'.[45] Procopius' wording is general, and it is of interest that he describes Hilderic as moving against Amalafrida and the Goths in Africa at much the same time because of their alleged 'revolutionary designs',[46] but the language of Boethius implies a much more precise charge, for the 'libertas Romana' which he was accused of having desired was probably a coded term for Byzantine intervention in Italy. Belisarius is said to have claimed that his army came to Africa in the interests of the freedom of the people,[47] and as early as 534 a law of Justinian spoke of Africa as having regained her *libertas* after 105 years of captivity under the Vandals.[48] A letter Cassiodorus wrote to Justinian in the name of the senate in 535 represents the figure of Roma as saying to the emperor, 'If Libya deserved to receive freedom through you, it would be cruel for me to lose what I have ever been seen to possess',[49] but this was not the only interpretation which could be placed on the situation of Italy under the Ostrogoths: Belisarius, whom an Italian source describes as having been sent 'to free all Italy from the captivity of the Goths', is represented as having told the people of Naples that the army of Justinian had come to secure their freedom, the poet Arator was to read his *De actibus apostolorum*, referring to Rome as always free, at a time when the city was held by the Byzantines but threatened by the Goths, and Narses was to place on the Pons Salarius at Rome an inscription mentioning 'freedom restored to the city of Rome and

[44] Sigismund: above, n. 13. Gunthamund and Dracontius: Dracontius, *Satisfactio*, 93ff. (*MGH AA* 14.118), on which Courtois, *Vandales*, 265 no. 10. J. Fontaine suggests that Dracontius may have praised Theoderic, *Naissance de la poésie dans l'occident chrétien* (Paris, 1981), 252, see too F. Chatillon, 'Dracontiana', *Revue du moyen âge Latin*, 8 (1952), 177–212 at 195f. 'Letters across the seas': Victor of Vita, *hist. pers.* 3.19. Totila's charge: Procopius, *BG* 3.21.12.

[45] Procopius, *BG* 1.1.34.

[46] Procopius, *BV* 1.9.4.

[47] Ibid. 1.16.9, 1.16.14, 1.20.20.

[48] *Cod. iust.* 1.27.1.1. See further on 'Roman freedom' Ioannes Lydus, *Powers*, 1.6, 2.8; and in general, J. Moorhead, '*Libertas* and *Nomen Romanum* in Ostrogothic Italy', *Latomus*, 46 (1987), 161–8.

[49] Cassiodorus, *var.* 11.13.5; see too Ennodius, *pan.* 42, for a description of Theoderic's sword as 'vindex libertatis'.

all Italy'.[50] Such usage, then, implies that Boethius was accused of desiring the restoration of Italy to the empire.

The third accusation mentioned by Boethius sounds the most strange to modern ears. In addition to the charges mentioned above it was alleged that his ambition for office had led him to pollute his conscience with sacrilege and seek the assistance of the most vile spirits, a charge which Boethius felt a misunderstanding of his interest in philosophy must have made more plausible.[51] It is clear from the Theodosian code that the practice of magic and sorcery had been of great concern to the government in the early fifth century,[52] and Theoderic had already found himself constrained to deal with this problem among senators, for at some time in the years 510–11 Basilius and Praetextatus had been accused of dabbling in the magic arts.[53] Despite the reality of magic for the people of late antiquity, and the fact that it seems to have been practised in situations of competitiveness, such as we know Theoderic's court to have been, we may nevertheless doubt whether this charge, unknown outside the account of Boethius, was what turned Theoderic against him. Surely the second charge, that which Boethius expressed as his hoping for Roman liberty, would have been decisive.

THE EXECUTION OF BOETHIUS AND SYMMACHUS

Boethius later wrote that the king wished to destroy the entire senate, despite the innocence of its members,[54] but for the time being he and Albinus, presumably still at Verona, were held in the baptistery of a Church, while Theoderic summoned Eusebius, who almost certainly held the office of *praefectus urbis Romae*, to Pavia.[55] At this point Albinus drops out of the story, never to be heard of

[50] Belisarius: *lib. pont.* 290.10; Procopius, *BG* 1.8.13. Arator: *De actibus apostolorum*, 1.1070–3 (ed. CSEL 72 p. 75). Inscription of Narses: Fiebiger/Schmidt, *Inschriftensammlung*, no. 217 (*CIL* 6.1199).

[51] Boethius, *phil. cons.* 1.4.37, 39, 41.

[52] *Cod. theod.* 9.16; 9.38.3f., 7f.; 9.40.1; 9.42.2, 4; 11.36.1, 7.

[53] Cassiodorus, *var.* 4.22f., with Gregory of Rome, *dial.* 1.4; see too *var.* 9.18.9 for the edict of Athalaric on *malefici*.

[54] Boethius, *phil. cons.* 1.4.32; cf. the words attributed to him by Anon. Vales. 85.

[55] Anon. Vales. 87, reading Ticinum for Ticini: see H. Tränkle, 'Philologische Bemerkungen zum Boethiusprozess', in *Romanitas et Christianitas: Studia I. H. Waszink oblata* (Amsterdam, 1973), 329–39.

again.[56] Sentence was passed on Boethius without him being heard. It is not clear what body conducted the trial, for while it was customary for senators accused of serious crimes to be tried by a commission of five of their number, the *judicium quinquevirale*, presided over by the *praefectus urbis Romae*, of which the group which had heard the case against Basilius and Praetextatus forms an example, Boethius wrote that the senate had made the desire of the safety of its order a crime, at least by its decrees concerning himself, and it is certainly possible that the senate, meeting in Rome, condemned him,[57] but whichever body it was doubtless had an eye for what Theoderic wanted. Boethius, who describes himself as terrified at the prospect of spear or sword, cannot have enjoyed his time in prison, although it may be going too far to see the letter *theta* of the *pi* and *theta* which Boethius observed on the dress of Philosophy, as described in the first prose of the *Consolation*, as having been suggested by the *theta*, an abbreviation for *thanatos*, Greek for 'death', which was sometimes stamped on the clothing of condemned prisoners.[58] However long he spent there was enough to allow the completion of the *Consolation*[59] before he was put to death 'in agro Calventiano', almost certainly a part of Pavia.[60] A graphic account describes him being throttled with a cord so tightly that his eyes rattled, after which he was dispatched by the blow of a club, but elsewhere we read that both he and

[56] The Albinus mentioned in *lib. pont.* 298.16 as being in Constantinople during the Gothic war was not him: Cameron and Schauer, 'Last Consul', 130.

[57] On the *judicium quinquevirale*: C. H. Coster, *The Iudicium quinquevirale* (Cambridge, Mass., 1935), who sees it as having been employed in the present instance, as does Ensslin, *Theoderich*, 318, against which Matthews, in Gibson, ed., *Boethius*, 23 with n. 23. Note that Arvandus, accused of treason in 469, was judged in the senate house by a panel of ten: Sidonius Apollinaris, *ep.* 1.7.9. The senate's decrees: Boethius, *phil. cons.* 1.4.23.

[58] Boethius, *phil. cons.* 2.5.34. *Theta* and *thanatos*: H. Chadwick, 'Theta on Philosophy's Dress in Boethius', *Medium Aevum*, 49 (1980), 175–9; id., *Boethius*, 225. The proceedings of the conference edited by Obertello in 1981 contain various suggestions for the meanings to be attached to the *Pi* and *Theta*: *πίστις* and *θεωρία* (C. S. Starnes, p. 37 with n. 20), *πρακτική* and *θεωρετική* (A. Quacquarelli, p. 241), *πίστις* and *θεωρία* or *θεολογία* (L. C. Ruggini, p. 98 n. 83).

[59] I accept that this work was finished, with E. Reiss, *Boethius* (Boston, 1982), 129ff., 133–7, and by implication Gruber, *Kommentar*, whose chart following p. 16 strongly suggests this, against e.g. H. Tränkle, 'Ist die *Philosophiae Consolatio* des Boethius zum vorgesehenen Abschluss gelangt?', *Vigiliae Christianae*, 31 (1977), 148–56.

[60] Anon. Vales. 87, to be preferred to Marius of Aventicum, *chron. s.a.* 524 'in territorio Mediolanense': see F. Gianini, in Obertello, ed., *Congresso internazionale*, 41–7.

Symmachus were killed by the sword, and we may hope that this was the case.[61]

Had Boethius known that Symmachus was in danger he would doubtless have been distraught, for while in prison he took pleasure in the thought that his father-in-law, 'a most valuable adornment of the human race . . . a man entirely made up of wisdom and virtues' was flourishing, but after Boethius had been accused Theoderic, fearing that Symmachus was going to act in a way hostile to his rule, ordered that he be put to death.[62] One has the impression that the execution of Symmachus made a greater impact on public opinion than that of Boethius, for some of our sources refer to Theoderic's acting against 'Symmachus and Boethius', contrary to the order in which he dealt with them, and by the end of the century a legend was circulating according to which Theoderic was led to hell by Pope John and Symmachus, without Boethius being mentioned.[63] Theoderic had the estates of the children of Symmachus, of whom Rusticiana, Galla, and Proba are known, and those of the children of Boethius, of whom Symmachus and Boethius are known, confiscated. They were returned after his death by Queen Amalasuintha, and when Rusticiana fell on hard times and was forced to beg for bread in 546 the Gothic king Totila shielded her from charges made by his compatriots that she had destroyed statues of Theoderic out of revenge.[64]

The timetable of these events is not as clear as one would like, especially as there are problems in the chronology of Anonymous Valesianus which will be best considered separately (see Appendix 2). But for the time being we may note that Cyprian's initial accusation was made when he was *referendarius*, and hence prior to his appointment as *comes sacrarum largitionum*, which may be dated, thanks to a reference to the 'third indiction' in his letter of appointment, to 1 September 524, and while Boethius was *magister officiorum*, hence subsequent to his appointment to that office, which probably occurred on 1 September 522, and prior to his being re-

[61] Anon. Vales. 87 (cord and club), *lib. pont.* 276.3 (sword).

[62] Boethius, *phil. cons.* 2.4.5 (Symmachus flourishing); and Anon. Vales. 92.

[63] 'Symmachus and Boethius': *BG* 1.1.32, 38; *lib. pont.* 276.3. Legend: Gregory of Rome, *dial.* 4.31.3f. The location of hell is discussed by A. Gurevich, *Medieval Popular Culture: Problems of Belief and Perception* (Cambridge, 1988), 130f.

[64] Procopius, *BG* 1.1.34 (estates confiscated), 1.2.5 (estates restored), 3.20, 27–31 (Rusticiana and Totila).

placed by Cassiodorus, which probably occurred in 523.[65] Given the argument advanced above that the accession of Pope John on 13 August 523 was of a piece with the rise of Boethius in the esteem of Theoderic, and given further that it would have been unlikely that a man so close to Boethius would have been elected pope after his friend had been accused of treason, we may conclude that the fall of Boethius from office occurred towards the end of 523. The time of his execution is more difficult to determine. The chronicle of Marius of Aventicum locates it firmly in the year 524, which date has been widely followed, often by the most learned scholars.[66] But this text is an unsound guide to Italian affairs.[67] On the other hand, Procopius testifies that a few days after the executions of Symmachus and Boethius the head of a large fish was placed before Theoderic at dinner. The king thought it was the head of Symmachus and, overcome by a bad conscience, rushed to his bedroom. He told his doctor Helpidius what had happened and died 'not long afterward'.[68] Similarly, the *Liber pontificalis* dates the executions of Symmachus and Boethius to the period when Pope John was absent from Italy on an embassy to Constantinople, which suggests that they occurred late in 525 or, perhaps more plausibly, in 526, a conclusion which is made more likely by a confused entry in the *Fasti Vindobonenses priores*.[69] Hence, although with some lack of confidence, we may nominate 526 as the year of the execution of Boethius and his father-in-law.[70] It must immediately be said that there are difficulties with this date beyond the evidence of Marius, for Anonymous Valesianus locates the execution of Boethius to before the sending of Pope John to Constantinople and that of Symmachus during the time he was away, difficulties which can be partly but not entirely overcome by dating John's expedition late

[65] Appointment of Cyprian: Cassiodorus, *var.* 5.40. Appointment of Boethius: *PLRE* 2.235; Sundwall, *Abhandlungen*, 238. Replacement by Cassiodorus: Mommsen's introduction to his edn. of the *Variae*, p. xxix; *PLRE* 2.267.

[66] As by Sundwall, *Abhandlungen*, 103; *PLRE* 2.236; Gruber, *Kommentar*, 11 f.; Wolfram, *History*, 331.

[67] Morton, 'Marius of Avenches', 111–15.

[68] *BG* 1.1.35–9.

[69] *Lib. pont.* 276.1–4. The *Fasti Vindobonenses priores* locate the deaths of Symmachus and Boethius 18 days before the death of Theoderic (*MGH AA* 9.332). It must immediately be acknowledged that these events are located in the consulship of Maximus, viz. 523, but accuracy is not the strong point of this text, which gives 504 as the year in which the emperor Anastasius died. Nevertheless, the executions of Symmachus and Boethius are placed very shortly prior to the death of Theoderic.

[70] So too Richards, *Popes and Papacy*, 118; Reiss, *Boethius*, 83.

rather than early.[71] But the later we date the execution of Boethius the easier it is to account for the perfection of the work he wrote in prison, the *Consolation of Philosophy*.

BOETHIUS' ENEMIES

Our attempt to uncover the reasons for Theoderic's turning against Boethius may begin by establishing what is known of those responsible for it. The *referendarius* Cyprian was the son of Opilio, who had held the office of *comes sacrarum largitionum* under Odovacer, but he displayed some unusual characteristics: he had served in Theoderic's army, apparently one of a small number of Romans to have done so, both he and his sons knew the Gothic language, and he made a practice of going on horse rides with Theoderic; as Cassiodorus expressed it, he was a man of action rather than reading.[72] Turning to the informers who testified against Boethius, we find three named.[73] Boethius states that Opilio, known from elsewhere to have been the brother of Cyprian, and Gaudentius had been accused of 'countless and manifold frauds'. Theoderic had decreed that they go into exile, but not wishing to do this they took refuge in a church, whereupon the king ordered that if they failed to leave Ravenna by a specified day they were to be branded on the forehead; on that very day they submitted their information against Boethius.[74] But Boethius' evaluation need not be taken at face value. Opilio was a keen supporter of the Gothic state, for when King Theodahad sent him with Liberius and some other senators to Constantinople in 534 to explain his treatment of Amalasuintha, he alone maintained that Theodahad had not behaved badly.[75]

[71] Anon. Vales. 87f., 92; the date of John's journey is discussed below.

[72] Cyprian's father: Cassiodorus, *var.* 8.16.2. This letter is addressed to Cyprian's brother, another Opilio. Cyprian: Cassiodorus, *var.* 8.21.3 (military service); 5.40.5, 8.21.7 (knowledge of Gothic), 5.41.3 (horse-riding with Theoderic), 5.40.4 (action rather than reading). Wolfram, *History*, 331, oddly describes Cyprian as ex-consul (yet he was never consul) and patrician (yet he only acceded to this office following the death of Theoderic).

[73] It will be clear that I take a view different to that expressed by Reiss, *Boethius*, 87, 89, although I do not deny that artifice is to be found in the *Consolation*. Similarly, W. Goffart suggests that the names 'Vigilius', 'Jordanes', and 'Castalius' which occur in the *Getica* may be 'speaking names' (*Narrators*, 103–5; cf. 106, 'The anonymous Constantinopolitan whom we know of as Jordanes'). I doubt whether this is more than possible.

[74] *Phil. cons.* 1.4.17f.

[75] Procopius, *BG* 1.4.15, 21, 23–5.

Opilio's wife is known to have been associated with a 'Basiliana familia', and it is tempting to see this family as that of Basilius, another informer against Boethius, a man who Boethius claimed had been driven out of the king's service and was forced by debts to lay information against him.[76] It is possible that we are dealing here with people connected with the Decii, for the Western consul of 524, Venantius Opilio, perhaps to be distinguished from Opilio the informer against Boethius,[77] combined in his name suggestions of kinship with both the informer and the Decii (cf. Basilius Venantius, consul 508, and Decius Marius Venantius Basilius, consul 484), which raises the possibility that the 'Basiliana familia' was linked with the Decii, and that the attack on Boethius may have owed something to people connected with this *gens*, perhaps disquieted by the rise of an Anicius such as Boethius. But if this were so, nothing suggests that those involved were influential members of it. It is, of course, possible that those named as the enemies of Boethius were merely acting on behalf of other people who were clever enough to choose them to do the dirty work, but none of our sources breathes the slightest hint that this was so. We may doubt whether Boethius' enemies were socially out of the top drawer: it is highly unlikely that the informer Basilius is to be identified with the senator of that name who had been accused of magic in 510–11, for Anonymous Valesianus contrasts false witnesses and senators,[78] while Gaudentius may have been the *consularis Flaminiae* mentioned in the inventory of an archive which was drawn up at Ravenna, perhaps in 502–3.[79]

But the narrative of his fall provided by Boethius yields a good deal of information about enemies he made at court in addition to those who moved against him in 523, and it will be worth our while investigating the information given there. One Cunigast is accused by him of attacking the fortunes of the weak, behaviour which Boethius opposed, yet any evil reputation he may have possessed

[76] 'Basiliana familia': Cassiodorus, *var.* 8.17.5. Basilius: Boethius, *phil. cons.* 1.4.16.

[77] J. Moorhead, 'Boethius and Romans in Ostrogothic Service', *Historia*, 27 (1978), 604–12, is not necessarily correct on Opilio. Prior to his consulship Opilio had constructed a basilica in honour of St Iustina at Padua: *CIL* 3.3100, repr. in G. Cuscito, 'Testimonianze archeologico monumentali sul più antico culto dei santi nella "Venetia e Histria"', *Aquileia nostra*, 45/6 (1974/5), 631–68 at 655.

[78] Anon. Vales. 86.

[79] Tjäder, *Papyri*, pap. 47f., B. 3, 6 (vol. 2, p. 192 with p. 297 nn. 22f., proposing 502/3 for the indictional year referred to in the text).

did not stop him being appointed by the government to judge a complaint two Romans brought against a Goth some time after the death of Theoderic in a letter which addressed him as *vir inlustris*.[80] More importantly, Boethius describes the *praepositus* Triwila as beginning wrongful deeds and carrying them to their completion; indeed, he observes, barbarian avarice was vexing wretched folk.[81] As it happens, Triwila is known from a variety of other sources, which allow him to be seen in a more positive light. Acting in another case as a *saio*, together with the *apparitor* Ferrocinctus he was ordered by Theoderic to take action against Faustus *niger*, who had been accused of taking property belonging to Castorius.[82] He helped the Jews of Ravenna make their charges against the Catholics of that city before Theoderic, being described in connection with this as 'a heretic who supported the Jews',[83] and assisted Ennodius in the purchase of a house at Milan after the deacon's repeated approaches to Boethius had failed to gain their end.[84] More information concerning Triwila is preserved in an odd story told by Gregory of Tours, according to which Theoderic was survived by his wife Audofleda and his daughter Amalasuintha. When Amalasuintha grew up she took as her lover the slave Traguila. The couple fled, whereupon the disapproving Audofleda sent an army after them, Traguila being killed and Amalasuintha brought home. The angry daughter poisoned her mother, whereupon the Italians called in one Theodad, the king of Tuscany, to rule over them; subsequently Theodad had Amalasuintha murdered in a hot bath.[85] In reality Theoderic was survived by his daughter Amalasuintha and her son Athalaric, but the narrative of the succeeding events by Procopius has something in common with that of Gregory. Procopius states that mother and son fell out, and when Athalaric, at the insistence of a group of Goths he persistently calls 'the barbarians', was removed from the Roman teachers to whose care Amalasuintha had entrusted him, he joined a gang of young Gothic drunkards and subsequently died. Amalasuintha sought the sup-

[80] Boethius, *phil. cons.* 1.4.10; Cassiodorus, *var.* 8.28.
[81] *Phil. cons.* 1.4.10.
[82] Cassiodorus, *var.* 3.20.
[83] Anon. Vales. 82, where his name appears as Trivanis.
[84] Ennodius, *ep.* 9.21, where his name appears as Triggua; see further above, Ch. 5.
[85] Gregory of Tours, *hist. franc.* 3.31. For this and what follows, J. Moorhead, 'Culture and Power among the Ostrogoths', *Klio*, 68 (1986), 112–22 at 117–19. The form of the name 'Traguila' is that printed by Krusch, *MGH SRM* 1.126.15, 19.

port of Theodahad, who is known to have had estates in Tuscany, and made him king, but Theodahad later murdered her, according to one report by having her strangled in the bath.[86] If we assume that Gregory of Tours is telling a garbled tale in which Theoderic's wife Audofleda stands for his daughter Amalasuintha, and Theoderic's daughter Amalasuintha represents her own real-life son Athalaric—and such confusions should surprise no careful reader of Gregory of Tours—his data can be brought tolerably close to what we know occurred. We have no reason to suppose that Athalaric and Triwila were lovers, or that they ran away together, but it would have made perfect sense for a boy in rebellion against the Romanizing ways of his cultivated mother to find himself in the company of Triwila. Indeed, as Amalasuintha is known to have restored their estates to the families of Symmachus and Boethius, it would have been entirely reasonable for Athalaric to have sought out as his companions those who had come into conflict with Boethius.

Nevertheless, Theoderic's court was not populated entirely by Goths such as Cunigast and Triwila, and such evidence as we have suggests that relations between Boethius and some of the Romans who served there cannot have been good. At a time of severe famine when a praetorian prefect imposed on the people of Campania a compulsory purchase of grain by the state at a fixed price, known as *coemptio*, Boethius did battle with the prefect and, with Theoderic's knowledge, was able to stop the *coemptio* being imposed.[87] It has been widely suggested that the praetorian prefect concerned was Faustus *niger*, whose tenure of office, certainly 509–12 and perhaps longer, included the year of Boethius' consulate, 510.[88] As it happens, Faustus' tenure of the office was certainly not above reproach. Described on one occasion by Cassiodorus as 'that well-known contriver', he may well have been forced to leave Rome for four months and surrender his office because of his intriguing (see above, Ch. 5), and on another occasion he was

[86] Procopius, *BG* 1.2.1 to 1.4.27, with Jordanes, *get.* 306 for the strangulation of Amalasuintha.

[87] Boethius, *phil. cons.* 1.4.12; for other unpopular levyings of *coemptio* by praetorian prefects, Ennodius, *V. Epiph.* 107 and Procopius, *anec.* 22.14–21. On the retention of *coemptio* in Ostrogothic times, consult P. Herz, *Studien zur römischen Wirtschaftgesetzgebung* (*Historia Einzelschriften*, 55; Stuttgart, 1988), 348–51.

[88] Dates of Faustus' office: *PLRE* 2.455 f.; with Krautschick, 'Bemerkungen', 122. Faustus as the prefect mentioned by Boethius: Obertello, *Severino Boezio*, 37; Chadwick, *Boethius*, 24; E. Demougeot, in Obertello, ed., *Congresso internazionale*, 101.

accused of causing terror to the people of Campania, using his office to pursue private hatreds against them.[89] It would be attractive to be able to suggest that Faustus, the leading supporter of Pope Symmachus during the Roman schism, and Boethius, who probably found himself on the other side, remained on bad terms, using the incident alluded to by Boethius in the *Consolation* as evidence for this. But scepticism seems called for. The most straightforward way of reading the evidence of the *Consolation* is to see the 'public administration' which Boethius sought to conduct in accordance with philosophical principles as referring to his holding of the office of *magister officiorum*, to which he was appointed perhaps a decade after Faustus relinquished the prefecture, it being difficult to accept that he would have interrupted the intellectual pursuits in which he was engaged at the time of his consulate to fight on behalf of the inhabitants of Campania.[90]

But even if we deny the identification of Faustus with the praetorian prefect of Boethius' story, we know that Boethius stood on the toes of one of his protégés. His opinion of one Decoratus was curt: 'You saw there the mind of a nasty buffoon and an informer.'[91] Decoratus seems to have died in 524 and been succeeded in the office of *quaestor* by his brother Honoratus, on which occasion Cassiodorus recalled the dead man toiling over the tasks of an advocate, 'qua se unicuique bonorum probitate coniunxerit';[92] it comes as something of a shock to realize that Boethius and Cassiodorus were writing such conflicting assessments of Decoratus at much the same time. Decoratus received three letters from another member of Faustus' circle, the deacon Ennodius, two of the letters also being addressed to another advocate at Ravenna, Florus. In one of his letters Ennodius complained to Florus and Decoratus at not receiving more letters from 'your friend my lord', and it seems reasonable to see here an allusion to Faustus, for Florus was among those whom Faustus had educated, just as 'the lords who love you' to whom Ennodius asked Florus and Decoratus to pass on his respects presumably included Faustus.[93]

[89] Cassiodorus, *var*. 3.27.1; cf. perhaps ibid. 4.50, and, more generally, 2.38, although this applies to a town in Apulia.

[90] 'Public administration': Boethius, *phil. cons*. 1.4.7. My doubts about Faustus are shared by L. C. Ruggini, in Obertello, ed., *Congresso internazionale*, 94 n. 87.

[91] *Phil. cons*. 3.4.4.

[92] Cassiodorus, *var*. 5.4.4.

[93] Letters to Florus and Decoratus: Ennodius, *ep*. 7.6 ('your friend, my lord', 7.6.3), 7.10 ('the lords who love you', 7.10.2; I agree with Vogel, *MGH AA* 7.352 that Faustus is meant here). Ennodius, *ep*. 4.17 is addressed to Decoratus alone; that Florus was educated by Faustus is known from *ep*. 1.11.2.

Decoratus, quite possibly a provincial from Spoleto,[94] may have been typical of the parvenus at Theoderic's court with whom Boethius enjoyed bad relations. In the corpus of Ennodius' writings one of his letters to Florus and Decoratus is followed immediately by another letter to Ravenna, addressed to the deacon Helpidius, in which Ennodius, finding himself in the midst of unidentified troubles, sought the aid of his friend and asked to be informed frequently of the welfare of Helpidius and of 'those who love us'.[95] As we have seen, when Ennodius despaired of obtaining a house at Milan through the generosity of Boethius it was Helpidius and the Triwila attacked by Boethius to whom he turned. When Boethius arrived in Ravenna in 522, over a decade after these incidents, he must have found in place networks of power and influence which had been built up over the years and which, in the case of such a man as Cyprian, whose father, another Opilio, had held the office of *comes sacrarum largitionum* in the days of Odovacer,[96] must have extended back to before the advent of Theoderic.

Another whose contacts at Ravenna extended back to the reign of Odovacer was Cassiodorus, whose father had served this king as *comes rei privatae* and *comes sacrarum largitionum*; that Cassiodorus was later prepared to step into the shoes of the deposed Boethius as *magister officiorum*, to write letters praising Cyprian and Decoratus, and still later to include them in the edited version of his correspondence published as the *Variae*, suggests that, while he had great respect for Boethius as an intellectual, he was less impressed by him in other ways.[97] Yet another was Petrus Marcellinus Felix Liberius, whom we have encountered more than once in the story of Theoderic. A former loyal servant of Odovacer, he could be described as making himself worthy of Theoderic's love by only entering Theoderic's service after Odovacer had been defeated. Appointed praetorian prefect, he had seen to the settling of the Goths in Italy, and when he was succeeded in this office by Theodorus, one of the sons of Basilius, in 500, he received the dignity of patrician. In 507 he celebrated the consulship of his son Venantius, and a few years

[94] *PLRE* 2.350, to which may be added the circumstance that his brother Honoratus had served as an advocate in Spoleto: Cassiodorus, *var*. 5.4.6.

[95] Ennodius, *ep*. 7.7.

[96] Cassiodorus, *var*. 8.16.2, 8.17.2.

[97] Ibid. 1.4 (father of Cassiodorus), 5.4.4 (praise of Decoratus), 5.41 (praise of Cyprian). I am not persuaded by F. Troncarelli's argument that Cassiodorus was an agent in the early diffusion of Boethius' *Consolatio*: *Tradizioni perdute: La 'Consolatio philosophiae' nell'alto medioevo* (Padua, 1981).

later he became praetorian prefect of Gaul. A friend of Ennodius and supporter of Pope Symmachus, he enjoyed some kinship with Avienus, son of Faustus *niger*.[98] This is not, of course, to say that Boethius had no antecedents at Ravenna, for the man almost certainly to be identified as his father had served Odovacer in high offices, as indeed had his father-in-law Symmachus,[99] and as we have seen Symmachus seems to have delivered a speech supporting the rights of those who had been adlected under Odovacer. But the fate of Boethius was that of a senator of Rome who had entered the very different environment of Ravenna, where he inevitably displaced existing holders of power. Given the lofty ideals with which he approached his office, and the lack of charity suggested by his negative evaluations of his colleagues in the *Consolation*, he may not have been the easiest colleague to get on with, and in the dog-eat-dog environment at the court to which the letters of Ennodius testify, it may not be surprising that those who had previously held favour did not delay in moving against him.

WHY HIS ENEMIES ATTACKED BOETHIUS

There is an immense body of literature on the issues which prompted the fall of Boethius. We may set aside any notion that he was a martyr: the religious tolerance which animated Theoderic's reign is not likely to have been set aside in the case of a man whose theological writings scarcely mention Arianism, and no source with any claim to being contemporary sees him as a martyr, whereas Pope John, who was to die in disgrace in 526, was quickly viewed in this light.[100] Relations with the East present a more plausible reason, especially given that Boethius seems to have been part of a group around Laurentius, a claimant to the papal throne who had

[98] Liberius is discussed by Cantarelli, 'Patrizio Liberio', and O'Donnell, 'Liberius'. For Liberius and Avienus, Ennodius, *ep*. 9.7.

[99] *PLRE* 2.232 f. (Boethius' father), 1045 (Symmachus).

[100] I disagree with the assessments of Boethius made by E. K. Rand, *The Founders of the Middle Ages* (Cambridge, Mass., 1929), 179, 322f., and M. Cappuyns, 'Boèce', *Dictionnaire d'histoire et de géographie ecclésiastiques*, 9 (1937), 348–80 at 358; see rather W. Bark, 'The Legend of Boethius' Martyrdom', *Speculum*, 21 (1946), 312–17, to which the following contributions of C. H. Coster and H. R. Patch, 'Procopius and Boethius', *Speculum*, 23 (1948), 285–7 add little. For John as a martyr, below, n. 137.

been prepared to support a Byzantine theological line, and that he remained in sympathy with Eastern positions, but it is one which needs to be expressed with care. A decade after the execution of Boethius the Byzantine conquest of Italy was under way, but there is no need to project expansionist designs on the part of the Byzantine government back to the time of Theoderic; indeed, it may be argued that Justinian's plan to invade Italy did not crystallize until the Vandal war had been won. Even if it were to be accepted that Boethius and other senators were involved in intrigues with Constantinople, it is hard to see what practical impact these could have had, for the fate of Italy depended on the Gothic army, a force which, given the position the Goths had come to enjoy in Italy, could be expected to offer stiff resistance to invasion.

Perhaps it makes better sense to locate the fall of Boethius against the tensions concerning the succession which must have mounted as Theoderic grew older, and the role of the senate in discussions which must have been taking place. As we have seen, the presumptive succession of Eutharic had been indicated by the joint consulship he held with the emperor Justin in 519 and his adoption by this emperor as son at arms. If Eutharic died in the early 520s, it would have been natural for leading members of the senate to have corresponded with the emperor concerning future developments. We know that senators were used as ambassadors by both Odovacer and Theoderic when they sought imperial recognition of their positions in Italy, just as there was to be a strong senatorial presence in the embassy Theoderic was shortly to send to Constantinople under the leadership of Pope John, and one cannot help notice the weight given the senate in the accounts of both the Anonymous and Boethius. It is also worth bearing in mind that senators may have had different ideas concerning the succession to those of Theoderic, and his cousin Theodahad comes to mind as a possible subject of senatorial correspondence.[101] It is easy to envisage a situation in which correspondence on the succession, particularly if it involved a despised member of the family, could have been drawn to Theoderic's attention in such a way as to arouse his antagonism.

But this does not necessarily answer the question as to why Cyprian struck. A military man, he need have had little in common

[101] S. J. B. Barnish, 'Maximian, Cassiodorus, Boethius, Theodahad: Literature, Philosophy and Politics in Ostrogothic Italy', *Nottingham Medieval Studies*, 34 (1990), 16–32 at 30.

with Theodahad or any other senatorial favourite, but there may have been other reasons for his acting in the way he did. When Cassiodorus wished to sum up the character of Boethius' father-in-law Symmachus he was able to do so in a few well-chosen words: 'A patrician and consul ordinarius, a man and a philosopher who was a very recent imitator of Cato of old, but who transcended the virtues of the people of old by his most holy religion.'[102] Cato was best remembered for the quality commemorated in a famous tag of Lucan, according to which the conquering side pleased the gods, but that which had been conquered pleased Cato. This line of Lucan is directly alluded to by Boethius in the *Consolation*, and when people in late antiquity thought of Cato it was with respect to his *aequitas* and his refraining from vices.[103] But Cassiodorus' juxtaposition of philosophy and imitation of Cato in his description of Symmachus (*philosophus quis antiqui Catonis fuit novellus imitator*) is of a piece with a passage in Procopius:

> Symmachus and his son in law Boetius [*sic*] were men of noble and ancient lineage ... But because they practised philosophy and were mindful of justice in a manner surpassed by no other men, relieving the destitution of both citizens and strangers by generous gifts of money, they attained great fame and thus led men of the basest sort to envy them. Now such persons slandered them to Theoderic and he, believing their slanders, put these two men to death, on the ground that they were setting about a revolution.[104]

Such was the generosity of the family that Boethius' widow Rusticiana 'was always lavishing her wealth upon the needy', and the charity practised by her sister Galla was still remembered in Rome at the end of the century.[105] Boethius' enemies at court, who included, if we are to believe his account in the *Consolation*, the debtor Basilius, were prone to attack the wealthy: people he described as 'palatine curs' had sought in their hope and ambition to devour the wealth of Paulinus, a former consul who is perhaps to be identified with the patrician Paulinus to whom Theoderic had given all the ruined granaries of Rome, and an attack was launched

[102] *Ordo generis Cassiodororum*, ed. Mommsen, p. v.

[103] Boethius, *phil. cons*. 4.6.33: 'Et victricem quidem causam dis, victam vero Catoni placuisse familiaris noster Lucanus ammonuit' (with reference to Lucan, 1. 128). See too Ennodius, *V. Epiph*. 135 (*aequitas*), Cassiodorus, *var*. 2.3.4 (refraining from vices; on the allusion to Martial in the line 'ad circum nesciunt convenire Catones', *var*. 1.27.5, see above, Ch. 5 n. 67).

[104] Procopius, *BG* 1.1.32–4 (trans. Dewing).

[105] Rusticiana: Procopius, *BG* 3.20.27. Galla: Gregory the Great, *dial*. 4.14.3.

against Albinus—another former consul, who was rich enough to have built a basilica near Rome—which one of our sources explicitly attributes to the greed of his accuser, Cyprian.[106] That Opilio and Cyprian may have been disloyal and guilty of greed is indirectly suggested by a passage in a letter written by Cassiodorus to the senate on the occasion of his appointment to the office of *comes sacrarum largitionum* in 527–8 which describes Opilio and his brother as faithful to friends, devoid of avarice, and far from cupidity, which could be held to be an official reply to criticisms which were then circulating.[107] Doubtless the attack on Boethius was motivated by more than a desire on the part of some courtiers to enrich themselves, and it is not impossible that men such as Opilio and Cyprian, who had been loyal servants of Theoderic for decades, genuinely feared that the rise of Boethius and those around him posed a threat to the principles to which they had devoted their lives; it may well be that modern scholars have too easily made their own the negative evaluations of Boethius' enemies contained in Procopius and Anonymous Valesianus as well as the *Consolation*. Nevertheless, the rise of Boethius must have threatened the positions which the provincial parvenus had been carving for themselves at Ravenna, particularly given that it came at the end of a period of more than a decade in which Theoderic seems to have given preferment to those of undistinguished background, and the accession of Boethius' friend Pope John could only have made them more insecure, for as we have seen the circles in which Faustus *niger*, Ennodius, and their friends moved were those of the popes Symmachus and Hormisdas, while the circle of Boethius and his father-in-law Symmachus included John. The attack on Albinus and its extension to Boethius may, in very large part, be attributed to the members of a threatened court group who struck to preserve their power.

POPE JOHN'S EMBASSY TO CONSTANTINOPLE

The charges made against the recently appointed Boethius left Theoderic in a bad mood, and the following events are best introduced by following the narrative of one of our sources, Anonymous

[106] Paulinus: Boethius, *phil. cons.* 1.4.13, with Cassiodorus, *var.* 3.29, 'probably identical' according to *PLRE* 2.847. Albinus: Boethius, *phil. cons.* 1.4.14: Anon. Vales. 85 ('Cyprianus actus cupiditate'), with *lib. pont.* 263.1f.

[107] Cassiodorus, *var.* 8.17.4.

Valesianus. No longer a friend of God but an enemy of his law, trusting in his own right arm and believing that the emperor Justin was greatly afraid of him, Theoderic summoned the pope to Ravenna and ordered him to go to Constantinople. The pope was to tell Justin, among other things which are not named, 'to restore those who had been reconciled to the Catholic religion'. John's reply began with words modelled on those Jesus addressed to Judas at the Last Supper: 'What you are going to do, king, do quickly [cf. John 13: 27]. Behold I stand in your sight [cf. perhaps Apoc. 3: 20]. I do not promise that I will do this, nor shall I say it to him. But in the other matters which you have enjoined upon me, I should be able to gain success, with the help of God.' Theoderic, angry, ordered the construction of a ship, on which John was placed with five other bishops, of whom Ecclesius of Ravenna, Eusebius of Fano, and Sabinus of Campania are named, and the senators Theodorus, Inportunus, Agapitus, and another Agapitus. But, we are told, God, who does not abandon his faithful worshippers, led them forward safely. John was met by Justin as if he were St Peter and given a promise that everything he sought would be carried out, except that those who had become Catholics could in no way be restored to the Arians.[108]

It must be said that this apparently straightforward account bristles with problems. No date is provided for the events it describes, but its position within the text implies that they occurred after the death of Boethius, and given that the pope celebrated the Easter liturgy in Constantinople according to the Western rite in 526[109] and that the expedition was conducted in haste—a delightful story concerning a horse John was given in Corinth, presumably to ride across the isthmus, implying that the legation put in there to avoid the *Umlandfahrt* around the Peloponnisos—we may conclude that it may well have left Ravenna early in 526, although a date towards the end of 525 is possible.[110] A departure at that time of the

[108] Anon. Vales. 88–91.

[109] Marcellinus, *chron. s.a.* 525, although the year is wrong.

[110] Horse in Corinth: Gregory the Great, *dial.* 3.2; cf. Chadwick, *Boethius*, 61. The papal mission of 519 followed a different route but it could afford to travel more slowly, and was travelling at a season more favourable to sea travel. The failure of W. Ensslin to accept that John travelled by Corinth weakens an otherwise impressive argument: 'Papste Johannes I. als Gesandter Theoderichs des Grossen bei Kaiser Justinos I.', *Byzantinische Zeitschrift*, 44 (1951), 127–34. John certainly left Rome after 1 Sept. 525: B. Krusch, 'Ein Bericht der päpstlichen Kanzlei an Papst Johannes I. von 526', in A. Brachmann, ed., *Papsttum und Kaisertum: Festschrift Paul Kehr*

year would have been unusual, as the period from 10 November to 10 March saw little naval activity, but the pope would have had plenty of time to reach Constantinople, celebrate Easter there on 19 April, and return home to die, exhausted, on 18 May.[111] The only part of their task to be mentioned, 'to restore those who had been reconciled to the Catholic religion', is ambiguous, for it is not clear whether the Catholic religion is that to which they had been converted or Theoderic's term for the Arianism he wished them to resume, and to make matters worse the text is corrupt at this point.[112] But conversions to Catholicism seem to have been at issue. The only recorded legislation against heretics at about this time was that issued by Justinian in 527, depriving them of earthly goods and dignities, but this does not exclude the possibility of Constantinople having taken measures while Theoderic was still alive; indeed, it could be argued that an exemption accorded the Gothic *foederati* in 527 owed something to John's intercession on their behalf in the preceding year.[113] But the *Liber pontificalis* asserts that John's mission was undertaken to secure the restoration of Arian churches which had been taken over for Catholic use.[114] While Arian churches are not known to have been resumed by the emperor Justin such conduct is not implausible: as early as 381 it was commanded that heretics hand their churches over to bishops of sound Trinitarian convictions, while in 484 the Vandal Huneric had, for his part, ordered that the Catholic churches of Africa were to be handed over to the Arians; certainly the wealth which Arian churches had accumulated by the sixth century would have made them a tempting target.[115]

It is therefore not clear whether John was sent to Constantinople

(Munich, 1926), 48–58. The evidence adduced by Stein (*Histoire*, 795) for John having been in Constantinople by Christmas 525 is not compelling. See further P. Goubert, 'Autour du voyage à Byzance du pape saint Jean I. (523–526)', *Orientalia christiana periodica*, 24 (1958), 339–52.

[111] Winter travel: Claude, *Handel*, 31 ff. On the time a voyage between Italy and Constantinople would have taken, Jones, *Later Roman Empire*, 1161; A. Guillou, *Régionalisme et indépendance dans l'empire byzantine au VIIe siècle* (Rome, 1969), 233; Magi, *Sede romana*, 20f.; T. S. Brown, *Gentlemen and Officers: Imperial Administration and Aristocratic Power in Byzantine Italy* A.D. *554–800* (Rome, 1984), 151.

[112] Mommsen, Cessi and Rolfe read 'reconciliatos', Moreau and Velkov 'reconciliatos [hereticos nequaquam]', but 'reconciliatus' is possible.

[113] *Cod. iust.* 1.5, esp. 1.5.12.

[114] *Lib. pont.* 275.5f.; see too Marcellinus, *chron. s.a.* 525.

[115] *Cod. theod.* 16.1.3; Victor of Vita, *hist. pers.* 3.3–6. For the wealth of Arian Churches, above, Ch. 3 n. 136.

to intercede for the return of Arian converts or Arian churches. Given that Anonymous Valesianus states that John was told to plead for the return of converts 'inter alia', a phrase such as to suggest the possibility that there was a hidden political agenda involved in the mission as to which our sources are silent, it may be that he had to concern himself with both. Whether the Byzantine government was acting against Arian believers or merely against Arian church buildings, its attentions were presumably confined to the East.[116] But given the arrest of Albinus and Boethius on charges of conspiring with Constantinople, and the concern the Vandals, at any rate, had traditionally displayed for their co-religionists in the East,[117] there is every reason to believe that the Byzantines were playing on Theoderic's nerves, for he would have taken any report that Arians were being persecuted by Justin very much amiss. Stories that he was planning some form of action against the Catholics of Italy in the event of John's mission failing, however exaggerated, may contain some kernel of truth.[118] Threats against the Catholics would have strengthened Theoderic's position against Justin, and it is quite possible that the Catholics most likely to have been threatened were Albinus and Boethius.

The composition of Theoderic's embassy is of particular interest in this regard. Its clerical members were distinguished and, as far as we can tell, eminently suitable for their task: Pope John, a friend of Boethius who was almost certainly still in prison, Bishop Ecclesius of Ravenna, who would have been used to dealing with the king, Bishop Eusebius of Fano, traditionally regarded as the greatest of the bishops and protectors of that see,[119] and Bishop Sabinus of Campania, perhaps to be identified with the Bishop Sabinus of the Apulian town of Canossa who, to judge by his attendance at councils in Jerusalem and Constantinople in 536, would have had a command of Greek which the embassy would have found useful.[120] But the lay members were at least as distinguished, comprising as they did three former consuls, Theodorus (consul 505), Inportunus

[116] Note the 1st edn. of the *lib. pont.*: 'ut redderentur ecclesias hereticis in partes Greciarum' (105).

[117] Victor of Vita, *hist. pers.* 2.3f., 2.24.

[118] *Lib. pont.* 275.10 (all Italy would be put to the sword; cf. 275.5f.); see too Theophanes, *chron.* AM 6016.

[119] AASS Apr. 2.541.

[120] Synods in 536: ACO 3.27.20, 113.19, 126.10, 154.21 etc.; see too Gregory the Great, *dial.* 2.15, 3.5, and R. Cessi, 'Un vescovo pugliese del sec. VI', *Atti del reale istituto Veneto di scienze, lettere ed arti*, 73 (1914), 1141–57.

(consul 509) and Agapitus (consul 517), as well as another Agapitus. Unfortunately it is very difgcult to assign pieces of data bearing upon an Agapitus to either of these Agapiti, but one of them had discharged an earlier embassy to Constantinople, and it is possible that he had taken with him the letter which Cassiodorus later chose to stand at the head of the *Variae*.[121] Here, however, Theoderic's choice may seem unexpected, for Theodorus and Inportunus were brothers of Albinus, Cyprian's accusation against whom had set in motion the fall of Boethius. It will be worth our while to examine these brothers in more detail.

Albinus, Theodorus, Inportunus, and their brother Avienus (consul 501) were, as we have seen, the sons of Caecina Decius Maximus Basilius (consul 480) and presumably the grandsons of Caecina Decius Basilius (consul 463), with whom Sidonius Apollinaris had thrown in his lot when visiting Rome. A politically alert man, Sidonius quickly realized that there were two great civilians in the city, Avienus, who devoted himself to the advancement of his relatives, and Caecina Decius Basilius, who was more useful to people other than members of his family,[122] and it may well be that the family of Basilius was not a cohesive group. Within the period 507–9 Theoderic told Albinus and Avienus that he had learned from the Green party in the Roman circus that certain thoroughly wicked men had been stirring up trouble for them; hence Albinus and Avienus were to take over the patronage of the faction, which their father Basilius had earlier exercised. But in 509 the Greens complained to Theoderic that Theodorus and Inportunus had been causing them such trouble that one of their number had been killed.[123] The violence which seems to have been afflicting the people of Rome in these years, causing problems for the prefect of the City, Agapitus, later one of the lay members of Theoderic's embassy, may well have been associated in some way with the 'civil wars' which the city had experienced during the turbulent years of the Laurentian schism and represent a continuation of them.[124]

[121] Embassy to Constantinople: Cassiodorus, *var.* 2.6. On the problems posed by the existence of two Agapiti, *PLRE* 2.32.

[122] Sidonius Apollinaris, *ep.* 1.9.3f.; see further above, Ch. 4.

[123] Cassiodorus, *var.* 1.20, cf. 1.33; 1.27. Sons did not always support the same faction as their fathers: 'Discordat multum contra suscepta voluntas;/Dilexit genitor prasinum, te russeus intrat' (*Anthologia Latina*, 191.4f. ed. A. Riese, 1.1.158, with reference to one Bumbulus).

[124] Violence in Rome: Cassiodorus, *var.* 1.20, 27, 30–3, 44.

Given that Albinus and Avienus seem to have taken the side opposite to that of Theodorus and Inportunus in the civil disturbances after the termination of the schism, it would be interesting to be able to demonstrate that they had supported different sides during the schism. This is not easy to do. However, even if Albinus had not supported Pope Symmachus from the beginning he certainly made his peace with him by the time Symmachus was called on to dedicate the church which he and his wife had built, perhaps significantly in honour of St Peter. Albinus received letters from Symmachus' great supporter Ennodius, as may have Avienus, in both cases during the years of the schism.[125] Theodorus and Inportunus, on the other hand, are conspicuously absent from the roll of Ennodius' correspondents both during and after the schism, and the former, whom early texts of the *Liber pontificalis* indicate was the most important of the lay ambassadors, was a correspondent of Fulgentius of Ruspe and so in contact with circles of a decidedly anti-Symmachan complexion. Further, as Fulgentius was a correspondent of Boethius' sisters-in-law Galla and Proba, Theodorus' circle may well have extended to the imprisoned philosopher.[126]

One other consideration points in the same direction. We have noted above (Ch. 4) that at the time of the Laurentian schism, Pope Symmachus enjoyed the support of the Roman plebs, and a passage in Cassiodorus makes it crystal clear that the Green faction was supported by the greater part of the populace of the city.[127] Therefore, that Theodorus and Inportunus were accused of stirring up trouble for the Greens suggests that they were of anti-popular, and hence anti-Symmachan orientation. Perhaps they had been adherents of Pope Symmachus' rival Laurentius, and now found themselves travelling to Constantinople in the company of a still more distinguished former Laurentian, Pope John. At first appearance it may seem extraordinarily audacious of Theoderic to have sent men so close to Boethius and Albinus to treat with Justin, even if the brothers of Albinus were not of one mind with him, for the members

[125] Albinus: Ennodius, *ep.* 2.21, 6.12; *PLRE* 2.193 identifies the Avienus of *ep.* 2.28 and 3.8 with the Decius of this name, and not the son of Faustus *niger*, one of Ennodius' most regular correspondents, contrary to Sundwall, *Abhandlungen*, 18f., 21. For the church built by Albinus and his wife, see above, Ch. 4 with n. 79.

[126] Theodorus as leader: *lib. pont.* 104f. Theodorus and Fulgentius: Fulgentius, *ep.* 6, *ep.* 2–4.

[127] 'Transit prasinus, pars populi maeret: praecedit venetus et potius turba civitatis affligitur' (*Var.* 3.51.11).

of embassies need not be loyal: when King Theodahad sent a group of senators which included Liberius and Opilio, the former enemy of Boethius, on an embassy to Justinian in 534, Opilio was the only one of the legates to act as Theodahad had instructed him, and Liberius seems to have decided to remain in the East, only returning to Italy during the Gothic war as the leader of a Byzantine task force.[128] But it is hard to imagine more weighty hostages than Albinus, assuming he was still alive, and Boethius, whose being in the power of Theoderic would have impelled their relations and friends among the king's envoys to exert themselves to the fullest on his behalf.

Such was the group Theoderic assembled to plead the cause of the Arians in the East. A ship seems to have been constructed to take them to Constantinople,[129] where they were well received. The *Liber pontificalis* gives a remarkably triumphalist account, according to which Justin 'abased himself as he bent down', adored the pope, was full of joy that he had deserved to see the vicar of Peter in his kingdom, and was crowned by his hands.[130] The language is hard to interpret, it not being clear whether the emperor greeted the pope with merely a kiss and a bow or full proskynesis, but the coronation went beyond what was necessary, for Justin had been crowned previously.[131] The embassy appears to have been largely successful, with Justin agreeing to Theoderic's demands on all points save the return of converts to Arianism.[132] Despite this, its members returned to Ravenna to face a frosty reception, which would have been the frostier had Theoderic been aware of reports that the pope had been splendidly received by Justin, and the lavish gifts from the emperor to the churches of Rome which John con-

[128] Procopius, *BG* 1.4.15, 23f., 3.36.6, 3.37.26f., 3.39.6–8, 3.40.12–14, 18, 4.24.1.

[129] Anon. Vales. 90, reading 'iubet ergo rex iratus navem fabricari' with Mommsen, Cessi, Moreau/Velliou, and Rolfe, although the last word is doubtful, other possibilities being 'praeparari' (so Morton, 'Marius of Avenches', 121f.) and 'forari'. It is hard to imagine a hole being bored into a ship, but the reading would suggest reminiscences of the story that Pope Hormisdas' second embassy to Constantinople returned in an unsafe vessel: *lib. pont.* 269.15.

[130] *Lib. pont.* 275.10–19; cf. Marcellinus, *chron. s.a.* 525 and Anon. Vales. 91, but compare 'occurrit ac si beato Petro' with 'occurrit beato Petro devotissimus ac si Catholicus' (65, of Theoderic's meeting with Pope Symmachus in 500).

[131] Ensslin argues for a kiss and bow ('Papste Johannes I.', 288), Stein for proskynesis (*Histoire*, 260). Justin's earlier coronation is known from Constantine Porphyrogenitus, *de caerim.* 1.93.

[132] Anon. Vales. 91. It is surely to be preferred to *lib. pont.* 276.4f. 'dum omnia obtinuissent', even though supported by the *abrégé félicien* and Duchesne's restored 1st edn. (104f.).

veyed back may have irritated the king.[133] John was placed under the king's displeasure and, together with the senators, was imprisoned in Ravenna.[134] One of the Agapiti had died on the return journey;[135] it is chilling to note that none of the other lay members is recorded in any subsequent source, no more than is Albinus. John may not have been strong physically, for he had complained of being unwell in the letter he wrote to Senarius when still a deacon, and is described as weak at the beginning of the embassy to Constantinople; on 18 May, which can have been only a few days after his return to Italy, he died.[136] While the funeral made its way to St Peter's a man possessed by a demon was healed, causing the people and senators to take pieces of John's garb as relics. He was remembered as a martyr.[137] There can have been little doubt as to who was responsible. It is not impossible that Boethius and Symmachus were executed at about the same time.

THEODERIC AND THE ROMANS

The traditions which our sources preserve concerning Theoderic's relations with the Romans during this period are hostile. The account of Pope John in the *Liber pontificalis* contains the only references in that source to Theoderic's being a heretic, and the persistence with which he was so described in it is enough to refute any suggestion that he was termed this simply by way of contrast to the orthodox emperor Justin.[138] Indeed, the king we find described in this source is a remarkably sinister figure: he is variously credited

[133] *Lib. pont.* 276.13–20, indicating that John brought the gifts in person.

[134] Anon. Vales. 93; *lib. pont.* 276.7. On the meaning of 'in offensa' in the former, H. Löwe, 'Theoderich der Grosse und Papst Johann I.', *Historisches Jahrbuch*, 72 (1953), 83–100; Chrysos, 'Amaler-Herrschaft', 451.

[135] *Lib. pont.* 276.2.

[136] 'Infirmus corpore' (to Senarius, ed. Wilmart 171.15); 'egrotus infirmitate', *lib. pont.* 275.7. The date of John's death: *lib. pont.* 276.8.

[137] Anon. Vales. 93; *lib. pont.* 276.8, 20f.; Gregory of Tours, *glor. mart.* 40; and Chadwick, *Boethius*, 63f. In 597 a Roman synod dealt with the problem presented by the faithful who snatched and tore into pieces dalmatics covering the bodies of dead popes; Gregory of Rome, *reg.* 5.57a, 4th can. (*falso* 6, *MGH Ep.* 1.364); cf. perhaps id., *dial.* 4.42.2.

[138] *Lib. pont.* 275.6, 17; 276.3, 5, 10; note, however, that early versions of the account of Pope Symmachus style Theoderic a heretic: pp. 97f. The evidence is wrongly interpreted by Pfeilschifter, *Theoderich*, 177 n. 4; cf. R. Cessi, 'La vita di papa Giovanni I', *Archivo muratoriano*, 19/20 (1917), 463–87 at 466.

with wishing to put all Italy to the sword, threatening to put all Italy to the sword if John's mission to Constantinople did not succeed, killing Symmachus and Boethius with the sword, wishing to kill John and the senators with the sword when they returned, and, in early versions, by implication with having made Justin agree to return the churches to the heretics 'because of the blood of the Romans'.[139] Reports that Theoderic threatened to cause the Catholics trouble were known to the Byzantine historian Theophanes,[140] and it is possible that a curious reference in Boethius' *Consolation* to setting temples on fire, butchering priests with an impious sword, and the murder of all good men owes something to such reports.[141] Anonymous Valesianus not only casts Theoderic prior to John's departure as Judas to the pope's Christ, but asserts that after his return Theoderic, acting no longer as a king but as a tyrant, gave orders on Wednesday, 26 August, in the fourth indiction, while Olybrius was consul, through the *scholasticus* Symmachus, a Jew, that on the following Sunday Arians were to take over the Catholic churches, a development only prevented by Theoderic's death on that very day, 30 August.[142] The precise dating lends verisimilitude to the account, and it may not seem implausible, for even if the matter of control over Arian churches were not one of those discussed by his legation to Constantinople there was precedent for the confiscation of Catholic churches by an Arian monarch in the order issued by the Vandal Huneric in 484, and the canons of the Council of Epaone in 517 envisage a situation in which heretics had taken over Catholic basilicas with violence.[143]

However, in the last year of Theoderic's life bricks were still being stamped with the claim that he governed 'for the good of Rome',[144] and quite beyond this various circumstances combine to suggest that we are dealing with a legend.[145] Contemporary Eastern

[139] *Lib. pont.* 275.7, 10, 276.4, 6, 104f. [140] Theophanes, *chron.* AM 6016.

[141] Boethius, *phil. cons.* 1.4.36. As Obertello points out, if this is an indirect reference to Theoderic's conduct the early parts of the *Consolation* may have been written last: 1979 edn., p. 65.

[142] Anon. Vales. 94.

[143] Victor of Vita, *hist. pers.* 3.3–14. Council of Epaone: can. 33 (ed. CC SL 148A).

[144] Reg[nante] d[omino] n[ostro] Theoderico bono Romae p[er] ind[ictionem] IIII, interpreted as being the indictional year 1 Sept. 525 to 31 Aug. 526 by Bloch, 'Ein datierter Ziegelstempel'.

[145] I agree with Picotti, 'Osservazioni', 224f., and Ensslin, *Theoderich*, 318. Gregory of Tours, *glor. mart.* 39 is wildly inaccurate.

sources betray no awareness that any persecution was planned, and indeed imply the continuation of good relations between Theoderic and the Romans. Procopius states that Theoderic died and 'left to his subjects a keen sense of bereavement at his loss', that his execution of Symmachus and Boethius was 'the first and last act of injustice which he committed towards his subjects', and that shortly before his death he was attended by his physician Helpidius, known from elsewhere to have been a Catholic deacon;[146] given that Procopius was in Italy a decade after the death of Theoderic we have every reason to accept what he says. There is no sign of disaffection among Catholic laity such as Cassiodorus, whose letters for the period breathe no hint of any plan to occupy churches or of relief felt by Romans when Theoderic died. Indeed, when Cassiodorus wrote to the senate of Rome on behalf of Theoderic's successor, the young Athalaric, he asserted that the dead king could be thought to live on in his offspring, while he told the Roman people that there would be no change in policy.[147] Needless to say such evidence is not conclusive: Cassiodorus may have revised his correspondence prior to publication, and it contains no direct allusions as to the fates of Boethius, Symmachus, John, and the pope's senatorial companions. But that the Gothic government sought to give an impression of continuity after the death of Theoderic is also indicated by coinage, for Amalasuintha issued quarter siliquae coins with the name of the emperor on the obverse and the monogram of Theoderic on the reverse.[148] Further, the *Variae* contain interesting information on Theoderic's role in the selection of a successor to Pope John.

It is easy to imagine the high feelings that must have run following the death of John on 18 May. A remarkably long period of *sede vacante* followed, it not being until 12 July that Felix IV was installed. This was the first time in Theoderic's reign that the hiatus between popes extended to more than a week, and it is clear that when Felix was finally ordained it was in accordance with Theoderic's command.[149] After Theoderic died on 30 August,

[146] Procopius, *BG* 1.1.31, 381f. *PLRE* 2.537 dates the discussion between Theoderic and Helpidius to 525/6; surely it must have been in the latter year.

[147] Cassiodorus, *var*. 8.2f. Note too the implication that Bishop Victorinus would feel sorrow at hearing of Theoderic's death: *var*. 8.8.2; Victorinus seems otherwise unknown.

[148] Grierson and Blackburn, *Coinage*, 37.

[149] *Lib. pont*. 106f., in preference to 279.5.

Athalaric wrote to the Roman senate, which we again find interesting itself in a papal appointment, advising it to accept the choice Theoderic had made, even though he had been of another religion. Theoderic had examined the man and found him worthy; there was no shame in the defeated party's having been overcome by the king.[150] Needless to say, that this letter was written after the death of Theoderic suggests that Felix experienced difficulties in imposing himself on the Roman Church which lasted for some time. Felix is an obscure figure, but one of Hormisdas' emissaries to Constantinople in 519 was a deacon of that name, and given both the small number of deacons in the Roman Church and the preponderance of former deacons among the popes of the period it seems likely that we are dealing with the same man. In 499 there had been no deacon named Felix in the Roman Church, from which it follows that Felix was presumably made deacon by Symmachus or Hormisdas.[151] This, of course, does not necessarily mean that Felix was committed to maintaining the policies of those popes, although Hormisdas' selection of him to participate in an important mission in 519 implies that he was considered sound by him. Perhaps, given this, we are confronted with a pope of the tendency represented by Symmachus and Hormisdas, contrary to that represented by Anastasius, Laurentius, and John, and, given the charges made against Boethius and the circumstances in which John's pontificate had ended, Theoderic's deciding in favour of such a man would not have caused surprise. But it would be straining credulity to believe that Theoderic was involving himself in the selection of a new pope at a time when he was planning to take over the Catholic churches of Italy. In Africa, confiscation had been accompanied by the exile of clergy. Needless to say it would have been just possible for Theoderic to have decided in favour of Felix and then to have determined the closure of the churches, leaving his successor to overthrow the latter policy and reassert the former. But it seems more likely that Theoderic never planned closure in the first place, and that, despite rumours, relations between Theoderic and his Roman subjects remained on the whole good.[152]

[150] Cassiodorus, *var.* 8.15.

[151] *MGH AA* 12.402. It is possible, however, that Peter of Altinum or Laurentius made him deacon.

[152] A final consideration may be of some weight: when Justinian launched his attack on the Vandals, religion was put forward as a reason for the invasion (*cod. iust.* 1.27, Procopius, *BV* 1.10.18–20), whereas it was not with respect to the Gothic war (e.g. Procopius, *BG* 1.5.8). My view is therefore more optimistic than that of Wolfram, *History*, 331 f.

THE CONSTRUCTION OF THE FLEET

At the very time of these events, Theoderic was striking out in an entirely new direction.[153] Traditionally the Goths had not been a naval people,[154] but suddenly, at some time during the period when Cassiodorus was serving Theoderic as *magister officiorum*, Theoderic ordered the rapid building of a fleet of 1,000 dromones, which was to be ready at Ravenna by the following 13 June.[155] It would have been a fleet of considerable size, for that which had devastated the coast of Italy in 507 or 508 and carried 8,000 men consisted of a mere 100 armed ships and 100 dromones, while a fleet of ninety-two dromones prepared by Justinian for deployment against the Vandals conveyed 2,000 men; and, whatever manpower had to be called upon to build Theoderic's fleet in such a hurry, the rowers which would have been required to man it must have been exceedingly numerous.[156] One wonders why Theoderic made the apparently sudden decision to commit resources on this scale to the construction of a fleet in such haste. In one of his letters Cassiodorus indicated that the fleet would have a double purpose: it would convey grain and, if necessary, act against enemy ships.[157] It being highly unlikely that Theoderic would have gone to the trouble of constructing so vast a fleet so quickly for the sole purpose of carrying grain, we must look for some military reason, and as it happens Cassiodorus supplies one: when Theoderic's kingdom had a fleet at its disposal, the Greek would not be able to reproach, nor would the African insult. It is highly likely that Cassiodorus used the word 'Graecus' in a pejorative sense.[158]

Fear of Byzantines and Vandals could have motivated Theoderic

[153] For what follows, Ruggini, *Economia e società*, 548–52 is most detailed. The sources are misread by Jones, *Later Roman Empire*, 829f., and Nagl, 'Theoderich der Grosse', 1761.

[154] Wolfram, *History*, 305.

[155] Cassiodorus, *var*. 5.16–20.

[156] Fleet of 507 or 508: Marcellinus, *chron. sa* 508. Justinian's fleet: Procopius, *BV* 1.11.15f. For manpower, cf. Nehlsen, *Sklavenrecht*, 126f.; H. Antoniadis-Bibicou, *Études d'histoire maritime de Byzance, à propos du 'thème des caravisiens'* (Paris, 1966), 157f.

[157] Cassiodorus, *var*. 5.16.2; cf. Claudian, *de cons. Stil*. 1.308, referring to vessels 'quae fruges aut bella ferant'.

[158] Cassiodorus, *var*. 5.17.3. On the sense of 'Graecus', F. Haenssler, *Byzanz und Byzantiner: Ihr Bild im Spiegel der Uberlieferung der Germanischen Reiche im früheren Mittelalter* (Berne, 1960), 24. Military purpose may also be implied by the date, 13 June, by which the fleet was to be finished; the army which went to Gaul in 508 had been summoned to move into action on 24 June (Cassiodorus, *var*. 1.24).

at almost any time from 523: the possibility that senators were conspiring with Justin to overthrow his rule which he became aware of at that time, and the accession of the pro-Byzantine Hilderic to the Vandal throne in the same year, would have given cause for concern in both directions. But I suspect that Theoderic became deeply worried only in the last year of his reign. Following the death of Thrasamund in 523 his widow Amalafrida, Theoderic's sister, fled to the Berbers. She was later captured and imprisoned, subsequently dying in prison, and it is tempting to connect the deteriorating relations between the Vandals and the Ostrogoths with Hilderic's possible perception of growing tensions between Byzantium and the Ostrogoths. It is not impossible that Hilderic was being encouraged to act against Amalafrida by his friend Justinian. The chronicle of Victor of Tunnuna describes the death of Amalafrida under the year 523, although the events described need not have occurred then and modern scholars generally agree that she died in 525 or 526.[159] Procopius explains Theoderic's failure to take revenge for the death of his sister by referring to his inability to gather a large fleet and to Hilderic's friendship with Justinian,[160] which suggests that the building of the fleet may have been in response to the death, perhaps not of natural causes, of Amalafrida, and gives added point to Cassiodorus' remark about the Greek and the African.

Another reference to Amalafrida's death is of assistance in establishing when it occurred. It is mentioned in a strongly worded letter of Athalaric to Hilderic which must have been written while Cassiodorus served the former sovereign as *magister officiorum*, hence in 526 or 527. Athalaric is credited with believing that Amalafrida had met with a violent loss of life and accused Hilderic of a kind of parricide, while observing that he seemed to despise the *virtus* of her Gothic relations. Three legates were sent to enquire into the matter.[161] No summary can do justice to the power with which the

[159] Courtois suggests 'sans doute peu de temps après la morte de Théoderic(?)' (*Vandales*, 401); Diesner, 526 ('Auswirkungen', 20); Bierbrauer, 525 ('Ostgotische Geschichte', 34); Krautschick, probably 525 ('Bemerkungen', 121 f.). See as well the discussion of F. X. Zimmermann, 'Der Grabstein der ostgotischen Königstochter Amalafrida Theodenanda in Genazzano bei Rom', in *Festschrift für Rudolf Egger*, 2 (Klagenfurt, 1953), 330–54.

[160] *BV* 1.9.5. The assertion that Theoderic found himself unable to gather a fleet cannot be held to deny that he attempted to build a fleet, contrary to Burns, *Ostrogoths*, 240 n. 162 (where Procopius is cited inaccurately).

[161] Cassiodorus, *var.* 9.1.

feelings of the Gothic government are expressed in this letter, which concludes with a reference to divine judgement and an apposite scriptural allusion (Gen. 4: 10), but it is noteworthy that Hilderic was reproached for the death of Amalafrida only after the death of Theoderic, which suggests that it occurred at about the time Theoderic himself died, and that the Goth at the end of his life may have been preparing a naval response to any insults from Africa. The fact that Athalaric was beset by 'coastal cares' at the beginning of his reign which involved the provisioning of ships and the fear of war provides further evidence for this,[162] and if we were to locate a reference of uncertain date in the *Variae* to repairing the walls of Syracuse to 526 or 527, we would have still more.[163] I therefore suggest 526 for the building of the fleet, which was perhaps unfinished when Theoderic died, on the second last day of summer.[164]

DEATH AND BURIAL

Hostile to the end, Anonymous Valesianus attributes Theoderic's death to God's not suffering those who worshipped him faithfully to be oppressed. Theoderic was stricken with diarrhoea, and after suffering for three days he lost his kingdom and his life, on the very day, according to the narrator, he had planned to take over the churches.[165] The Anonymous claims that his fate was the same as that of Arius, 'the founder of his religion', and one cannot help noticing that the Vandal Huneric was believed to have died a few months after he took over the Catholic churches, of the same complaint, 'like his father Arius': but the impeccably orthodox emperor Leo had met a similar end in 474, and there is no reason to doubt that Theoderic died naturally of diarrhoea.[166] We are here at the

[162] *Var*. 9.25.8–10.

[163] *Var*. 9.14; unfortunately all letters preceding this one in bk. 9 must have been written in 526 or 527 and all subsequent to it in 533 or later. Cassiodorus' editors Mommsen and Fridh declined to assign it a date, but Krautschick suggests it belongs to the earlier period, as the Gildilas to whom it was addressed is mentioned as having the same office in *var*. 9.11 (*Cassiodor*, 89).

[164] So already Gaudenzi, *Rapporti*, 73; Hartmann, *Geschichte*, 217.

[165] Anon. Vales. 95.

[166] Hilderic: Victor of Tunnuna, *chron. s.a.* 479 (misdated), but cf. Victor of Vita, *hist. pers*. 3.71; for the death of Arius, see e.g. Cassiodorus, *hist. trip*. 3.10.9f. (CSEL 71.151) and Gregory of Tours, *hist. franc*. 2.23 (early), 3 *praef*., 9.15, where a reference to Eusebius, *hist. eccl*. 10.4. Leo: John Malalas, *chron*. 376. Note, how-

beginning of the hostile traditions concerning Theoderic's death, of which a particularly graphic example is furnished by a bas relief of the twelfth century on the right of the main door into the basilica of St Zeno in, of all places, Verona. Procopius chose to situate the death of Theoderic in a very different context, for when he described the old king as 'having lamented and grieved exceedingly' at the execution of Symmachus and Boethius, he was borrowing from Herodotus a term used to describe Cambeses, just before his tragic end.[167] We are obviously dealing here with a very different perspective to that of the Anonymous, one suggestive of much more dignity. One would like to think it more accurate.

Before he died Theoderic, in the presence of the Gothic counts and the chief people of his race, appointed his grandson Athalaric to succeed him, and is reported to have instructed the Goths who were present to cherish the king, love the senate and Roman people, and always to ensure that, after God, the emperor was well disposed. These details are known from Jordanes, who had represented Zeno as having commended the senate and Roman people to Theoderic in 487.[168] Athalaric, a boy of perhaps only 8 years, was not to have an easy life, but in letters sent out after he had become king Cassiodorus stressed, time and time again, Theoderic's role in his accession.[169] He adopted a different strategy only when writing to the emperor Justin on behalf of the new king. This may have been a difficult task, for Theoderic had not obtained imperial assent to Athalaric's succession, as he had for that of Eutharic, and relations between king and emperor had recently been so poor. So it was that in this letter Cassiodorus expressed the hope that, Theoderic having died, relations between Ravenna and Constantinople would improve.[170]

Theoderic had seen to the preparation of his tomb, 'a work of astonishing size', outside the Porta Artemetoris in the north wall of Ravenna, and found the enormous piece of marble which was

ever, the tradition that Judas Iscariot died when his bowels gushed out (Acts 1: 18); given the Anonymous' earlier assimilation of Theoderic and Judas it is possible that the death of Theoderic was meant to recall that of the traitor.

[167] Procopius, *BG* 1.1.39, ἀποκλαύσας δὲ καὶ περιαλγήσας, cf. Herod. 3.64.2. I owe the reference to Herodotus to F. Bormann, 'Motivi tucididei in Procopio', *Atene e Roma*, 19 (1974), 138–50 at 147.

[168] Jordanes, *get.* 304, cf. 292 and *rom.* 348, with Anon. Vales. 96.

[169] *Var.* 8.2.4, 8.3.3, 8.4.2, 8.5.1, 8.6.2, 8.7.1, 8.8.2.

[170] *Var.* 8.1.

placed on top of it.[171] The tomb, which is unusual at Ravenna in that it was made of limestone rather than bricks, and was apparently unfinished at the time Theoderic died, is still to be seen just outside the town. It continues to fascinate observers by virtue of its brooding mass and elegant proportions, and to tax the ingenuity of the technically minded, driven to speculate how the dome, a single block 10.76 m in diameter and perhaps 230,000 kg in weight, was transported to the site and hoisted into position.[172] Its architectural antecedents have been much discussed, but it seems clear that, in very general terms, they are classical and early Christian rather than medieval or pagan Germanic, and the lack of any recent building of this kind in the West and various points of detail incline scholars to postulate an origin in Asia Minor for its stonemasons.[173] It is tempting to use such considerations as a basis for general deductions as to Theoderic's *Weltanschauung*, but as we have no idea of the extent to which he was personally involved in its preparation, and no means of assessing what its various features may have meant to him, it will be safer to content ourselves with one observation: despite its evident classical antecedents, the tomb was located outside the walls of Ravenna, in an area where Goths lived.[174]

The unexpected location of a building owing much to classical precedent in a barbarian setting may stand for the Roman and barbarian elements in Theoderic's Italy, an uneasy mixture which failed to achieve synthesis. But synthesis was never sought, and Italy remained a sub-Roman state in which ultimate power resided with non-Romans. In the long term, this situation could not have lasted. But Theoderic's Goths were only first-generation settlers, and despite the circumstances of his last years there was no reason why convergence could not have ultimately occurred. As for their king, who led them from wandering around the plains of Hungary to mastery of what was incomparably the most powerful state in

[171] Anon. Vales. 96; Agnellus, *Codex* 39 (ed. Testi-Rasponi, 112f.). See the discussions of R. Heidenreich and H. Johannes, *Das Grabmal Theoderichs zu Ravenna* (Wiesbaden, 1971), Deichmann, *Ravenna*, 1. *Geschichte* (1969), 213–19, 2. *Kommentar*, 1 (1974), 209–39, and Johnson, 'Theoderic's Building Program', 92–5. A recent brief summary: Effenberger, *Frühchristlicher Kunst*, 246.

[172] Heidenreich and Johannes, *Grabmal*, 63f.; G. Tabarroni, 'La cupola monolitica dal mausoleo di Teoderico', *Felix Ravenna*, 105/6 (1973), 119–42.

[173] Deichmann, *Ravenna, 2. Kommentar*, 1.230–3.

[174] Agnellus, *Codex* 39 (ed. Testi-Rasponi, 113 with n. 1); consult Testi-Rasponi's map, between pp. 116 and 117.

Western Europe, his statecraft was such as to attract praise from his Roman contemporaries while his military feats were to be reflected in the legends of the Germanic Middle Ages. There can be no doubt that his achievement was immense.

Conclusion

Writers close to Theoderic reveal a high degree of what might be called temporal self-consciousness. This is particularly to be seen in the brief notices in Cassiodorus' chronicle: by Theoderic's mighty works 'the wonders of old are surpassed' (s.a. 500); when concord returned to the Roman clergy and people in 514 it was 'to the praise of your times' (s.a. 514); and Eutharic's consular games were noteworthy for the display of different kinds of wild animals 'which the present age wondered at on account of their novelty' (s.a. 519). References to 'our time' and 'our times' in letters written on behalf of Theoderic by Cassiodorus occur very frequently.[1] It was in Theoderic's heart to 'change all things for the better' (*var*. 2.21.3), although elsewhere he would be credited with the more modest desire of merely equalling the ancients (*var*. 9.24.8). Columns and marbles were not to lie in ruins giving rise to sad thoughts of earlier periods, but rise up to be used again (*var*. 3.9.2); so too, in Spoleto, by the reuse of materials a mature appearance of newness could be given to things confused in antiquity (*var*. 4.24.1). One of the tasks of the *comes sacrarum largitionum* was to mint coins which would 'remind future periods of our times' (*var*. 6.7.3). Cassiodorus was obviously acutely aware of some claim to distinction possessed by Theoderic's reign, and it comes as no surprise to discover that he was one of the first Latin authors to make frequent use of the adjective *modernus*.[2] Ennodius, applying a traditional term, felt that Theoderic's was a golden age ('aureum saeculum': *pan*. 93, *ep*. 9.30.10; cf. perhaps *lib*. 136), and his political vocabulary is full of images of renewal ('reparatio', *pan*. 30, 'iuvenescere', *pan*. 56, 'resuscitare', *opusc*. 5.20, p. 303.5), as is that of Anonymous Val-

[1] *Var*. 1.17.3, 1.18.3, 2.3.4, 2.21.4, 2.23, 2.28.5, 2.35, 2.37, in the first two books alone. Note, however, that such terms were by no means new; cf. Majorian, *nov*. 1.

[2] *Var*. 3.5.3, 3.9.1, 3.31.4, 4.51.2, 8.14.2, 8.25.1; further references in *Thesaurus linguae Latinae*, 8.1211 f. Ennodius uses the word fairly often, but only once in a political context: Bishop Epiphanius is represented as having invited Gundobad, the 'antiquus dominus', to love Liguria which a 'modernus' (Theoderic) had embraced: *V. Epiph*. 161.

esianus ('recupere'/'recuperatio', 60, 67; 'restaurare'/ 'restauratio'/ 'restaurator', 67, 70, 71; 'renovare', 71).

Vocabulary of this kind points up what may appear something of a paradox in the image the Ostrogothic government promoted of itself: it liked to appear both conservative and innovatory. While it was Theoderic's purpose to construct new things, the conservation of the old was more important, because at least as much praise could be gained by safeguarding the old as by coming upon new things (*var.* 3.9.1, cf. 1.25.1). The government was obviously interested in praise (cf. *var.* 11.1.18–20), and in his panegyric, which was designed to supply this very thing ('nunc ecclesia dirigit laudatorem', *pan.* 77), Ennodius could comment 'your predecessors loved ignorance, for they did nothing worthy of being praised' (*pan.* 76). Some hostility to Odovacer is implied here, and perhaps, given the plural 'predecessors', to those who wore the purple prior to 476, for as we have seen Ennodius studiously sought to assimilate emperors and kings. What would distinguish Theoderic's times and make them worthy to be praised was the restoration of the old (*var.* 1.28.1), and we may conjecture that the 'old newness' of buildings restored by Theoderic, which would mean that the *antiqui principes* who built them owed their praise to Theoderic (*var.* 1.25.3), involved both innovation, in so far as it meant turning its back on Odovacer and the last period of the empire in the West, and conservation, in so far as it meant restoring the buildings and more generally the tenor of the empire in an earlier period.

Quite apart from the image Theoderic or his advisers sought to create, it is clear that in practical terms his government, despite its having a *rex* at the head, provided a high degree of continuity and remained essentially Roman. The structures of government, both those with power and those with authority, were maintained; the bureaucracy, which was frequently to be a powerful organ of historical continuity in periods of Byzantine history far darker than Ostrogothic Italy, flourished. Roman judicial procedures continued in operation and Roman law, although of a vulgar form, was the law of the land; the system of taxation remained in effect. Indeed, if we accept that the Goths were supported from taxation rather than thirds of land, their insertion into the fabric of Italian society must have been remarkably gentle. Obscure institutions of the late empire were retained: Theoderic levied the *siliquaticum*, a sales tax introduced by Valentinian III in 444 which could only have been collected in the small amount of territory in the West still under imperial control.[3] He even continued to pay *domestici* and *scholares*, surely a complete irrelevance in a state where political power was based on the Gothic army, going so far as to order, in late-Roman fashion, that the pay could be transmitted to their descendants.[4]

As with the state, so with the king. It would be simplistic to see his striking out at Boethius and his associates as a recrudescence of some barbarian *furor teutonicus* which had already been manifested in his murder of Odovacer, for there was nothing specifically barbarian in such conduct. Theoderic had already murdered Recitach at the instigation of Zeno and in the interests of his statecraft in Constantinople, and the imprisoned Boethius found impeccable classical antecedents for the treatment he received.[5] Even in its sombre end, there can be no doubt that the reign of Theoderic looks backwards into antiquity rather than forwards into the Middle Ages.

Nevertheless, the decline of the Ostrogothic state after the death of Theoderic was rapid, even more dramatic than the decline of the Visigoths following the defeat of Alaric by the Franks in 507. Theoderic was succeeded in 526 by his grandson Athalaric. The wheels of state continued to turn, oiled by the customary rhetoric; when Ambrosius, a former student of Ennodius, was appointed *quaestor*, it was expected that he would be a Pliny to Athalaric's Trajan (Cassiodorus, *var.* 8.13.4). Nevertheless, power resided in the hands of the boy's mother, Amalasuintha, who lost no time in having a letter sent in the name of her son to the emperor Justin which observed that 'Love and friendship should pass from parents to their offspring, while hatred should be buried in the tomb' (Cassiodorus, *var.* 8.1.2, trans. Hodgkin). It was a curious way of alluding to the tensions which had developed towards the end of Theoderic's reign, and some of Amalasuintha's actions, such as the restoration to the children of Symmachus and Boethius of the estates of their families (Procopius, *BG* 1.2.5) and the granting to the pope of the right of hearing charges made against members of the Roman Church (Cassiodorus, *var.* 8.24), may be seen as

[3] Cassiodorus, *var.* 2.4, 2.30.3, 3.25.1, 4.19, 5.31; some idea of Cassiodorus' sense of time may be gathered from his attributing the tax to 'antiquitas' (*var.* 4.19.2). On the *siliquaticum*, Ensslin, *Theoderich*, 201; Jones, *Later Roman Empire*, 826.

[4] Cassiodorus, *var.* 1.10 (cf. 6.6.1); Procopius, *anec.* 26.27f.

[5] See esp. *phil. cons.* 1.3.9.

attempts to resolve them. But the new ruler found it necessary to make gestures in other directions: Boethius' accuser Cyprian was made *patricius* (Cassiodorus, *var.* 8.21f.) while his ally Opilio received preferment (*var.* 8.16), and Theoderic's general Tuluin was made *patricius praesentalis* and, despite being a Goth, joined the senate (*var.* 8.9–11). Perhaps most ominously, Theodahad received pieces of land (*var.* 8.23).

Despite such gestures, not all the Goths were easy with the learned Amalasuintha and her pro-Roman ways. Around the young Athalaric there formed something of which there is scarcely a trace throughout the reign of Theoderic, an opposition party, in which Boethius' old enemy Triwila was involved. So threatened did Amalasuintha's position become that at one stage she planned to flee to imperial territory, only deciding to remain in Italy when she learned that three of her enemies had been put to death. The death of Athalaric, whose decline can only have been hastened by the company he was keeping, occurred in October 534, a time when Ostrogothic interests were being threatened by successes the Franks were enjoying: the Thoringians were defeated, causing Amalaberga to return to Italy, while the process of dismembering the Burgundian state continued without the Ostrogoths making further gains. More alarming was the annihilation of the Vandal kingdom by the army of Justinian in 533. Seeking to strengthen her position, Amalasuintha associated Theodahad with her on the throne. Given that the accession of Theodahad may have been an issue towards the end of Theoderic's reign, her act may have had a significance which we cannot appreciate, but the unpopularity of this scholarly land-grabber was almost certainly enough to make Amalasuintha's move politically foolish.

Theodahad quickly turned against his patron and had her put to death, so furnishing the emperor Justinian, who had been fishing in the troubled waters, with a pretext for invading Italy. In 535 a force led by Belisarius occupied Sicily, in the following year it entered mainland Italy, and by the end of 536 the Goths, infuriated by the apparent cowardice of Theodahad, had murdered him and chosen a man of undistinguished family, Witigis, as their new king. The war between the Ostrogoths and the Byzantines which was then in its opening stage was to continue for nearly 20 years, and ended in the total defeat of the former.[6] With this war, and the subsequent in-

[6] Wolfram, *History*, 334–62.

vasion of the Lombards, Italy entered the Middle Ages. The last Western consul held office under Theodahad in 534, and the last non-imperial Eastern consul in 541; the maintenance of this office by the Goths and its suppression by Justinian is itself enough to prompt reflections on their relative success in maintaining the classical past.

Procopius was aware of a story according to which Theodahad, alarmed by the progress of the army of Belisarius at the beginning of the Gothic war, asked a Jew known for his ability to predict the future what the outcome of the war would be. He was told to place in separate huts three groups of ten pigs, representing respectively Goths, Romans, and imperial soldiers. Some days later eight of the pigs representing Goths had died, while all but a few of those representing the soldiers were alive. The most interesting result occurred with the pigs representing Romans, all of which lost their hair and about half of which survived.[7] The failure of the Goths was dramatic, but of more long-term importance was the fate of the Romans and their civilization. In such a matter as the production of books, for example, and the oligarchy of letters which had flourished under Theoderic, Justinian's reconquest was disastrous,[8] and its impact was made much worse by the Lombard invasions which followed hot on its heels. In the days of Theoderic, the senators generally resided in Rome, but eighty-five reasonably certain references to landowners of senatorial rank in the correspondence of Pope Gregory I (590–604) show only four living there.[9] Hence, with respect to both Goths and Romans he had governed, Theoderic, then, may be seen as a man whose work failed to survive him for long[10] and who can be classed as 'late Roman' much more readily than 'early medieval'. His reign can more reasonably be seen as having looked backwards rather than forwards.

Yet his achievement was immense. Having gained control of a large kingdom Theoderic became one of the most powerful rulers of his period, influential beyond borders which, decade by decade,

[7] Procopius, *BG* 1.9.3–7.

[8] G. Cavallo, 'La circolazione libraria nell'età di Giustiniano', in *L'imperatore Giustiniano: Storia e mito* (Milan, 1978), 201–36, at 210f., 228.

[9] Brown, *Gentlemen and Officers*, 23.

[10] In general, Hartmann, *Geschichte*, 227f. (although the speculations as to Theoderic's feelings at the end of his days seem unwarranted), against e.g. A. S. Graf von Stauffenberg, 'Theoderich der Grosse und seine römische Sendung', in *Würzburger Festgabe Heinrich Bulle* (*Würzburger Studien zu Altertumswissenschaft*, 13; Stuttgart, 1938), 115–29, at 115f.

tended to expand throughout his reign. Over a time-scale of some centuries, Theoderic's reign was the only period of any length in which the political significance of Italy did not diminish. Within the borders of his kingdom Goths and Romans lived in reasonable peace, and even this, considering the career of the former prior to 493, must have been largely due to Theoderic and the respect in which he was held by both peoples. Some of his subjects saw him as something like an emperor, while the senate and the Church were, for the most part, content with his rule, despite his having been a barbarian and an Arian, and we would surely be justified in seeing this as a response to the same royal benevolence which was displayed towards the Jewish community. His reign ended in tragedy, but as we have suggested, the falls of Boethius, Symmachus, and John, which so coloured later perspectives, were brought about by highly contingent circumstances which do not reflect his reign at large, and most Romans during and immediately after these events viewed them as less tragic than have most subsequent commentators. Indeed, Theoderic is notable for the loyalty he attracted, and while it is true that some of our sources were written by those who had reason to feel grateful to him, it is worth insisting that our image of his reign is substantially based on the writings of Ennodius, a Catholic cleric, of Cassiodorus, who published the letters he wrote on behalf of Ostrogothic sovereigns at a time when the Byzantine army was already in Italy, an act which may be held to represent the triumph of conviction over prudence, and of Procopius, a representative of the power which overthrew the state Theoderic established. However much we may feel disinclined to accept at face value all the details of their evaluation, there is no reason to doubt that their professed admiration of 'that man who was so rarely and so nobly qualified by Nature for the cares of royalty' (Cassiodorus, *var.* 10.31.5, trans. Hodgkin) was sincere, that they were happy to live in his times, and that they felt praise of him to be justified.

Notwithstanding this, it is legitimate to wonder to what extent Theoderic could be blamed for the downward trajectory which the Ostrogoths and, indeed, Italy were to take following his death. The tragedies of his last years doubtless alienated some Romans, but their discontent was without immediate political import. A more powerful case against Theoderic could be mounted along the following lines: partly because of an existing situation, but also partly out of policy, he was content for the Goths and Romans to live as two separate peoples. This attitude would be at least symbolically expressed by a report in Procopius that after he died some Goths asserted that Theoderic 'would never allow any of the Goths to send their children to school' (*BG* 1.2.14). In the long term this policy would have been hard to sustain, for the later history of the Visigoths was to demonstrate a convergence with the Romans, and even in Theoderic's reign there are signs of prosperous Goths adopting Roman ways. Indeed, Theoderic's apparent reluctance to countenance Romanizing tendencies among the Goths was contradicted by what was occurring in his own family, with the result that, after his powerful leadership was withdrawn, Amalasuintha and Theodahad found themselves unpopular and were hard pressed to cope with the threats posed by such Goths as Triwila and Witigis, who may have constituted resentful members of a class of Gothic loyalists.[11] But the collapse of the Ostrogothic state can more simply be accounted for with reference to Theoderic's failure to provide himself with an adult male heir and Justinian's launching an invasion of Italy after his amazingly rapid defeat of the Vandals. There is no need to invoke any structural weakness in Theoderic's system of government, nor any defect of character in the king. He can hardly be blamed for not having been survived by a son, and the disastrous results of Justinian's wars in the West, not just for Italy but for the Byzantine empire as a whole, were soon to be all too clearly seen.

[11] Possible analogues are thoughtfully discussed by E. A. Thompson, 'The Visigoths from Fritigern to Euric', *Historia*, 12 (1963), 105–26.

APPENDIX 1

The Roman Synods of 502

The question as to whether Bishop Peter of Altinum was appointed visitor of the Roman Church to celebrate the Easter of 501 or that of 502 is connected with a problem in dating the two synods held in Rome shortly afterwards. The first of these was held on 23 October 'Rufio Magno Fausto Avieno v.c. cons.' (*MGH AA* 12.426.6) and the second on 6 November 'Fl. Avieno iun. v.c. consule' (ibid. 438.4). As it happens there were consular Avieni in both 501 and 502, and Mommsen, in his still standard edition of the Symmachan synods, assigned one of the synods to the consulship of each of the Avieni, a decision which has been followed by some of the most eminent scholars of ecclesiastical history.[1] Nevertheless many voices have been raised arguing for both the synods having been held in 502.[2] It is awkward for the latter group that their dating entails the papal chancery having used such different forms for the name of the same man in documents issued precisely a fortnight apart. But they seem to have the better of the argument, and two circumstances, of a completely non-technical nature, may be adduced in support of them.

Seventy-five bishops put their names to the proceedings of the synod of 23 October. The list of signatories runs to 76 names, but the absent-minded Hilary of Temesa signed twice (nos. 15, 43). The list of the bishops who attended the synod of 6 November extends to 79 names, while 65 bishops subscribed to its proceedings; both these lists include Symmachus of Rome, who for obvious reasons had not been present on 23 October. It is evident that in size alone the second group has more in common with the first than with the third, and while it is sometimes difficult to be certain of the identities of members of the second group, as the names are given without their sees, it is perfectly clear that there are at least a dozen bishops named in both the first and second groups but not the third.[3] On the other hand,

[1] e.g. Duchesne, *L'Église*, 122 n. 1; Caspar, *Geschichte*, 91 ff.; Haller, *Papsttum*, 236–40; C. Vogel, in Duchesne's edn. of the *lib. pont.* 3.88.

[2] e.g. Pfeilschifter, *Theoderich*, 71–4; Sundwall, *Abhandlungen*, 95 f.; Stein, *Histoire*, 136, 793 f. See more recently G. B. Picotti, 'I sinodi romani nello scisma laurenziano', in *Studi storici in onore di Gioacchino Volpe* (Florence, 1958), 743–86, esp. 763–6, and Alan Cameron, 'Junior Consuls', *Zeitschrift für Papyrologie und Epigraphik*, 56 (1984), 159–72 at 160 n. 7, 171.

[3] Eustasius (7 in 2nd list) = Eustasius of Cremona (32 in 1st list); Serenus (26) = Serenus of Nola (36); the two Mercurii (27, 77) = Mercurius of Sutri (13) and Mercurius of Gabii (48); Stephanus (33) = Stephanus of Naples (14); Felix (44) = Felix

only seven bishops occur in both the second and third lists but not the first.[4] Hence, more names are common to the first and second lists than to the second and third lists. It would be astonishing if bishops left Rome after a synod in October 501, returned for a synod in November 502, but left without subscribing; it is surely more natural to suppose that the synod of 23 October was held shortly before the following synod, but that some of the bishops who were present as it began did not remain until it ended. Given that they had been in Rome for several months this would be easy to understand. It is also noteworthy that not one see is represented by a different bishop in the first and third lists, surely remarkable if one was compiled over a year after the other.

A further matter deserves attention. The holding of a synod towards the end of autumn was itself unusual, for if Pope Hilary presided over one on 19 November 465 (*ep.* 15, ed. Thiel) we find Felix presiding on 13 March 487/15 March 488 (*ep.* 13, ed. Thiel), Gelasius on 13 May 495 (*coll. avel.* 103), Symmachus himself on 1 March 499 (*MGH AA* 12.399), and Gregory I on 5 July 595 (*reg.* 5.57a). Rather than believing that the long-suffering bishops, detained in Rome well into the October of 501, were summoned to meet again in November 502, it makes better sense to see the unseasonable month at which the November synod met, as well as the list of those who attended, as reflecting a date very shortly after the previous meeting.

We may therefore date both synods to 502.

of Attela (49); the two Innocents (48, 54) = Innocent of Ferentino (17) and Innocent of Bevagna (30); Gerontius (56) = Gerontius of Fidenae (9); Dulcitius (61) = Dulcitius of St Anthimius (76); Castus (64) = Castus of Porto (20); Servusdei (78) = Servusdei of 'Feraena' (72). I presume one of the Hilarys (36, 50) to have been Hilary of Temesa (15 and 43), and Adeodatus (67) to have been either Adeodatus of Formiae (52) or Adeodatus of Silva Candida (67).

[4] Eulalius of Syracuse (3, 4), Pacatianus of Forum Cornelii (18, 12), Basilius of Tolentino (51, 16), Florentius of Plestina (53, 42), Martianus of Ostra (?) (69, 17), Eusebius of Fano (71, 64), Proiectus of Forum Novum (73, 49).

APPENDIX 2

The Anonymous Valesianus

The chronicle written by this author, of such deceptive simplicity, would repay serious study. None of the critical editions which have been published (Cessi, Mommsen, Rolfe, Velkov/Moreau) is entirely satisfactory, and no study has succeeded in explaining its relationship to Eastern, perhaps Monophysite material. This is at its most puzzling when c. 57–9 is taken in conjunction with Zachariah, but the name given to Theoderic's father, the story of his use of a stencil, and the detail of the comet of 519, together with the volume of material bearing on emperors, suggests Eastern connections, although the author was familiar with the *Vita Severini* of Eugippius (45 f.). Further, the text is puzzlingly full of mistakes, such as the dating of Odovacer's death to the year in which Zeno died (55–7), and the description of Amalaberga as Theoderic's niece and the implication that she married Herminifred in 500 (70). Considerable discussion has been devoted to problems caused by the sequence of passages dealing with the events which darkened the last years of Theoderic's reign. An impressive case has been made for reordering them in the sequence 84, 88–91, 85–7, 92 f., so making the embassy of Pope John to Constantinople antedate the attacks on Albinus, Boethius, and Symmachus, and creating a much tighter sequence of events than that proposed in the preceding discussion.[1] It is admitted that there is no manuscript justification for this procedure, but in its favour it could be argued that the dating of the death of Boethius to 524 by Marius of Aventicum is of little weight, given the unreliability of this source, although against this must be set the possibility that Marius was interpreting data already contained in the Anonymous or a similar source, and hence supports the case for the ordering of the text found in the manuscripts as being the original one,[2] and that locating the moves against Boethius and Symmachus to the very end of Theoderic's reign makes better sense of the data of Procopius and the *Liber Pontificalis*. Our discussion above has gone some distance in this direction by suggesting 526 for the

[1] See most recently Morton, 'Marius of Avenches'. The traditional sequence is accepted by Obertello, *Severino Boezio*, 125–38, and *Congresso internazionale*, 61–9; see further Barnish, '*Anonymous Valesianus II*'.

[2] It cannot be too strongly emphasized that if Marius misdated the murder of Odovacer (Morton, 'Marius of Avenches', 111, assuming the omission of the consuls of 491), so did the Anonymous (54, although apparently to a different date), and that Marius and the Anonymous agree as to the date of the murder of Odoin (*sa* 507, 68 f. respectively), against *auct. haun.* (*s.a.* 504).

execution of Boethius and Symmachus, contrary to the Anonymous Valesianus, while nevertheless proposing late 523 for the attack on Boethius. The issues are complex, but three considerations impose themselves.

In the first place, the Anonymous describes Cyprian as *referendarius* at the time he accused Boethius ('qui tunc referendarius erat', 85). This official is known from a reference in Cassiodorus to have assumed the office of *comes sacrarum largitionum* for 'the third indiction' (*var*. 5.40.7). The possible dates for this are 509–10 and 524–5. But there is no reason to date any of the letters presented in book 5 of the *Variae* to outside the period 523–6, except for the last two letters, which were doubtless placed there to make the book conform to Cassiodorus' practice of ending each book with a letter or letters to distinguished people, and perhaps for the first two, the locations of which would answer to a similar practice. Hence Cyprian, who held the office of *referendarius* before that of *comes sacrarum largitionum* (*var*. 8.21.4f.), must have levelled his accusation against Albinus prior to 1 September 524. It is true that in about 527 Cassiodorus could describe Cyprian as old ('senescis quidem corpore', *var*. 8.21.2), as opposed to the blooming youth he had enjoyed when serving Theoderic against the Bulgars, as referendarius and as *comes sacrarum largitionum* ('florida iuventus', *var*. 8.21.5), but the youthfulness need not apply to all this period, and an author who could see Clovis as a 'rex iuvenis' in about 507 (*var*. 3.2.2) cannot be pressed closely on youth. Further, that Cassiodorus wrote concerning problems arising during the consulship of Maximus in 523 (*var*. 5.42) while *magister officiorum*, the office held by Boethius at the time Cyprian accused him (Anon. Vales. 85), suggests that Boethius' fall from office occurred before the end of this year.

Secondly, the expression 'rediens igitur rex Ravennam' (Anon. Vales. 88) may seem awkward in the sequence of events provided by the manuscripts, but the use of 'igitur' in a sense less strong than in classical Latin is typical of the author,[3] and while Theoderic's return to Ravenna would have been understandable if he had received a report that a woman had given birth to four dragons there (84), it makes perfect sense in the sequence which occurs in the manuscripts: Theoderic, who had been at Pavia (87), returned to Ravenna, whither he later summoned Pope John (88) and Symmachus (92). That I believe that the execution of Boethius should be located later than the Anonymous suggests (87) does not alter the fact that the sequence provided by the manuscripts makes perfect sense and does not of itself suggest a need for rearrangement.

Finally, our author holds that Italy enjoyed good government for 30 years under Theoderic, although he was aware that Theoderic had ruled for 33 years (59). Assuming that 30 is not meant as an approximation

[3] Adams, *Vulgar Latin Chronicle*, 80 on *igitur* as an 'introductory connective'.

of 33 and that the author's material was not hopelessly out of control, he must have had some definite turning-point in mind. Of the possible stages in his narrative, 'ex eo enim invenit diabolus locum' (83) is too early, for it is followed by a reference to a comet which seems to have been visible in 519 (84; see above, Ch. 7), whereas the other possibility, 'post haec coepit adversus Romanos rex subinde fremere' (85), introduces the attacks on Albinus and Boethius, which we have suggested occurred in 523, precisely 30 years after Theoderic's coming to power in 493. I therefore believe we would be well advised to adhere to this date.

BIBLIOGRAPHY

Primary Sources

Acta synhodorum habitarum Romae, ed. T. Mommsen (*MGH AA* 12).

AGATHIAS, *Historiarum libri quinque*, ed. R. Keydell (Berlin, 1967).

AGNELLUS qui et ANDREAS, *Codex pontificalis ecclesiae Ravennatis*, ed. A. Testi-Rasponi (2 vols.; RIS 2/3).

AMMIANUS MARCELLINUS, ed. J. C. Rolfe (3 vols.; London, 1950–2).

ANONYMOUS VALESIANUS, ed. T. Mommsen (*MGH AA* 9); ed. R. Cessi (RIS 24/4); ed. V. Velkov and M. Moreau (Leipzig, 1968); ed. and trans. J. C. Rolfe (London, 1952), within vol. 3 of his edn. of Ammianus Marcellinus.

ARATOR, *De actibus apostolorum*, ed. A. P. McKinlay (CSEL 72).

Auctarium Prosperi Hauniensis, ed. T. Mommsen (*MGH AA* 9).

AVITUS OF VIENNE, *Opera*, ed. R. Peiper (*MGH AA* 6).

BLOCKLEY, R. C., ed. and trans., *The Fragmentary Classicizing Historians of the Later Roman Empire* (Liverpool, 1981).

BOETHIUS, *In categorias Aristotelis* (PL 64).

——*De institutione arithmetica* . . ., ed. G. Friedlein (Leipzig, 1867).

——*De interpretatione*, ed. C. Meisner (Leipzig, 1877–80).

——*Philosophiae consolatio*, ed. L. Bieler (CCSL 94); ed. L. Obertello (Milan, 1979); (with *Opuscoli teologi*).

——*Theological Tractates*, ed. and trans. H. F. Stewart, E. K. Rand, and S. J. Tester (London, 1973).

CAESARIUS OF ARLES, *Opera*, ed. G. Morin (2 vols.; Mardesous, 1937, 1942).

CASSIODORUS, *De anima*, ed. J. W. Halporn (CCSL 96).

——*Chronica*, ed. T. Mommsen (*MGH AA* 11).

——*Historia ecclesiastica tripartita*, ed. W. Jacob and R. Hanslik (CSEL 71).

——*Institutiones*, ed. R. A. B. Mynors (Oxford, 1937).

——*Variae, Ordo generis Cassiodororum*, ed. T. Mommsen (*MGH AA* 12); ed. Å. J. Fridh (CCSL 96).

Chronica Caesaraugustana, ed. T. Mommsen (*MGH AA* 11).

Chronica Gallica, ed. T. Mommsen (*MGH AA* 9).

Chronicon Pascale, ed. L. Dindorf (Bonn, 1832).

Codex Theodosianus, ed. T. Mommsen and P. Meyer (3 vols.; Berlin, 1905).

Codicis Euricioni fragmenta (*MGH* Leg. 1/1).

Collectio Avellana, ed. O. Guenther (CSEL 35).

Concilia Africae, ed. C. Munier (CCSL 259).
Concilia Galliae, 1, ed. C. Munier (CCSL 148); 2, ed. C. de Clercq (CCSL 148A).
Concilios Visigóticos e Hispano-Romanos, ed. J. Vives (Barcelona, 1963).
CONSTANTINE PORPHYROGENITUS, *Le Livre des cérémonies*, ed. A. Vogt (Paris, 1935ff.).
Constitutio Silvestri (PL 8.829–40).
Consularia Italica, ed. T. Mommsen (*MGH AA* 9).
CORIPPUS, *In laudem Iustini Augusti minoris*, ed. A. Cameron (London, 1976).
Corpus iuris civilis, ed. P. Krueger, R. Schoell (3 vols.; Berlin, 1912–22).
CYRIL OF SCYTHOPOLIS, *Vita Sabae*, ed. E. Schwartz (Texte und Untersuchungen, 49).
DIONYSIUS EXIGUUS, *Collectio decretorum pontificum Romanorum* (PL 67.229–316).
——*La Vie Latine de saint Pachome*, ed. H. von Craneburgh (Brussels, 1969).
——*Liber de paschate, epistola de ratione paschae* (PL 67.483–520).
DOBSCHUTZ, E. VON, ed., *De libris recipiendis et non recipiendis* (Leipzig, 1912).
DRACONTIUS, *Opera*, ed. F. Vollmer (*MGH AA* 14).
Edictum Theodorici regis, ed. F. Bluhme (*MGH* Leg. 5).
ENNODIUS, *Opera*, ed. F. Vogel (*MGH AA* 7); ed. W. Hartel (CSEL 6).
Epistolae Areletenses genuinae, ed. W. Gundlach (*MGH Ep.* 3).
EUGIPPIUS, *Opera*, ed. P. Knoell (CSEL 9).
EVAGRIUS, *Historia ecclesiastica*, ed. J. Bidez and L. Parmentier (London, 1898).
Fasti Vindobonenses priores, ed. T. Mommsen (*MGH AA* 9).
FERRANDUS, *Vita Fulgentii* (PL 65).
FIEBIGER, O., and SCHMIDT, L., *Inschriftensammlung zur Geschichte der Ostgermanen* (Vienna, 1917; supplements, 1939, 1944).
FLODOARD, *Historia ecclesiae Remensis*, ed. J. Heller and G. Waitz (*MGH SS* 13).
FREDEGARIUS SCHOLASTICUS, *Chronica*, ed. B. Krusch (*MGH SRM* 2).
FULGENTIUS, *Opera*, ed. J. Fraipont (CCSL 91, 91A).
GELASIUS, *Lettre contre les Lupercales*, ed. G. Pomares (SC 65).
Gesta de Xysti purgatione (PLS 3.1249–52).
Gesta Liberii (PL 8.1387–93).
Gesta Polychronii (PLS 3.1252–5).
GREGORY of NYSSA, *De conditione seu opificio hominis*, trans. Dionysius Exiguus (PL 67.347–407).
GREGORY of ROME, *Dialogues*, ed. A. de Vogüé and P. Antin (3 vols.; SC 251, 260, 265, 1978–80).

GREGORY OF ROME, *Epistularum registrum*, ed. P. Ewald and L. M. Hartmann (*MGH Ep.* 1 f.).
——*Moralia in Iob*, ed. M. Adriaen (CCSL 143).
GREGORY OF TOURS, *Opera*, ed. B. Krusch *et al.* (*MGH SRM* 1).
HYDATIUS, *Continuatio chronicorum Hieronymianorum*, ed. T. Mommsen (*MGH AA* 11).
ISIDORE OF SEVILLE, *De viris illustribus*, ed. C. C. Merino (Theses et studia philologica salmanticensia, 12; Salamanca, 1964).
——*Etymologiae*, ed. W. M. Lindsay (Oxford, 1911).
——*Historia Gothorum*, ed. T. Mommsen (*MGH AA* 11).
——*Institutionum disciplinae*, ed. P. Paschal, *Traditio* 13 (1957).
JOHN OF BICLARUM, *Chronicon*, ed. J. Campos (Madrid, 1960).
JOHN THE DEACON, *Epistola ad Senarium*, ed. A. Wilmont (Studi e testi, 59; 1933).
JOHN THE LYDIAN (John Lydus), *On Powers*, ed. and trans. A. C. Bandy (Philadelphia, 1983).
JOHN MALALAS, *Chronographia*, ed. L. Dindorf (Bonn, 1831).
JOHN OF NIKIU, *The Chronicle of John of Nikiu*, trans. R. H. Charles (London, 1916).
JORDANES, *Getica, Romana*, ed. T. Mommsen (*MGH AA* 5).
JULIAN OF TOLEDO, *Historia Wambae*, ed. J. N. Hillgarth (CCSL 115).
Laterculus regum VVandalorum et Alanorum, ed. T. Mommsen (*MGH AA* 13).
LIBERATUS OF CARTHAGE, *Breviarium*, ed. E. Schwartz (ACO 2/5.98–144).
Liber pontificalis, ed. L. Duchesne and C. Vogel (Rome, 1886–92; 1957).
Lives and Legends of the Georgian Saints, trans. D. M. Lang (London, 1976).
LÖWENFELD, S., ed., *Epistulae pontificum Romanorium ineditae* (Leipzig, 1885).
MAASSEN, F., ed., *Geschichte der Quellen und der Literatur des canonischen Rechts in Abendlande bis zum Ausgang des Mittelalters* (Leipzig, 1870).
MARCELLINUS *comes*, *Chronicon*, ed. T. Mommsen (*MGH AA* 11).
MARINI, G., *I papiri diplomatici* (Rome, 1805).
MARIUS OF AVENTICUM, *Chronica*, ed. T. Mommsen (*MGH AA* 11).
MAXENTIUS, *Opuscula*, ed. Fr. Glorie (CCSL 85A).
MICHAEL THE SYRIAN, *Chronique de Michel le Syrien*, ed. and trans. J. B. Chabot (3 vols.; Paris, 1899–1904).
MÜLLER, C., ed., *Fragmenta historicorum Graecorum*, vols. 4, 5 (Paris, 1851, 1870).
Notitia dignitatum, ed. O. Seeck (Frankfurt, 1876).
Paschale campanum, ed. T. Mommsen (*MGH AA* 9).
Passio Sigismundi regis, ed. B. Krusch (*MGH SRM* 2).
PAUL THE DEACON, *Historia Langobardorum*, ed. L. Bethmann and G. Waitz (*MGH SRLI*).

——*Historia Romana*, ed. H. Droysen (*MGH AA* 2).
PETER ABELARD, *Theologia Christiana*, ed. E. M. Buytaert (CCCM 12).
PHILOSTORGIUS, *Historia ecclesiastica*, ed. J. Bidez and F. Winkelmann (Berlin, 1982).
Poetae Latini minores, ed. E. Baehrens (5 vols.; Leipzig, 1879–83).
PROCOPIUS, *Opera*, ed. and trans. H. B. Dewing (London, 1914–40); ed. J. Haury (Leipzig, 1905–13).
PROSPER TIRO, *Chronicon*, ed. T. Mommsen (*MGH AA* 9).
SCHWARTZ, E., ed., *Publizistische Sammlungen zum Acacianischen Schisma* (Abhandlungen der Bayerischen Akademie der Wissenschaften, Phil.-hist. Abteilung, Heft, 10; Munich, 1934).
SEVERUS, *Liber contra impium grammaticum*, trans. J. Lebon (Corpus Scriptorum Christianorum orientalium, 102 (Scriptores Syri, 51).
SIDONIUS APOLLINARIS, *Poems and Letters*, ed. W. B. Anderson (2 vols.; London, 1936, 1965).
Synodi Sinuessanae gestae (PL 6.11–20).
THEODORE LECTOR, *Epitome historiae ecclesiasticae*, ed. G. C. Hanson (Berlin, 1971).
THEOPHANES, *Chronographia*, ed. C. de Boor (2 vols.; Leipzig, 1883).
THIEL, A., ed., *Epistulae Romanorum pontificum* (Braunsberg, 1868).
TJÄDER, J.-O., *Die nichtliterarischen Lateinischen Papyri Italiens aus der Zeit 445–700*, 1 (Lund, 1955); 2 (Stockholm, 1982).
VENANTIUS FORTUNATUS, *De virtutibus sancti Hilarii*, ed. B. Krusch (*MGH SRM* 4/2).
VICTOR OF TUNNUNA, *Chronica*, ed. T. Mommsen (*MGH AA* 11).
VICTOR OF VITA, *Historia persecutionis Africanae provinciae*, ed. C. Halm (*MGH AA* 3/1).
Vita Caesarii episcopi Arelatensis, ed. B. Krusch (*MGH SRM* 3).
Vita Hilari (AASS Mai 3).
Vita Remigii, ed. B. Krusch (*MGH SRM* 3).
ZACHARIAH OF MYTILENE, *Historia ecclesiastica*, trans. F. J. Hamilton and E. W. Brooks (London, 1899).
ZOSIMUS, *Histoire nouvelle*, ed. and trans. F. Paschoud (Paris, 1971–89).

Select List of Secondary Works

ÅBERG, N., *Die Goten und Langobarden in Italien* (Uppsala, 1923).
ADAMS, J. N., *The Text and Language of a Vulgar Latin Chronicle (Anonymous Valesianus II)* (London, 1976).
ALFOLDI, M. R., 'Il medaglione d'oro di Teodorico', *Rivista italiana di numismatica*, 80 (1978), 133–42.
ALFONSI, L., 'Ennodio letterato (nel XV centenario della nascita)', *Studi romani*, 23 (1975), 303–10.

Antoniadis-Bibicou, H., '*Études d'histoire maritime de Byzance, à propos du 'thème des caravisiens*' (Paris, 1966).

Atti della settimana di studi su Flavio Magno Aurelio Cassiodoro (1983) (1986).

Bach, E., 'Théodoric, Romaine ou barbare?', *Byzantion*, 25–7 (1955–7), 413–20.

Bachrach, B. S., *Early Medieval Jewish Policy in Western Europe* (Minneapolis, 1977).

——'Procopius and the Chronology of Clovis' Reign', *Viator*, 1 (1970), 20–31.

Baldwin, B., 'Illiterate Emperors', *Historia*, 38 (1989), 124–6.

Bark, W., 'The Legend of Boethius' Martyrdom', *Speculum*, 21 (1946), 312–17.

——'Theoderic vs. Boethius: Vindication and Apology', *American Historical Review*, 49 (1944), 410–26.

Barnes, T. D., *Constantine and Eusebius* (Cambridge, Mass., 1983).

——'*Patricii* under Valentinian III', *Phoenix*, 29 (1975), 155–70.

Barnish, S. J. B., 'Maximian, Cassiodorus, Boethius, Theodahad: Literature, Philosophy and Politics in Ostrogothic Italy', *Nottingham Medieval Studies*, 34 (1990), 16–32.

——'Pigs, Plebians and *Potentes*: Rome's Economic Hinterland, *c.*530–600 A.D.', *Papers of the British School at Rome*, 42 (1987), 157–85.

——'Taxation, Land and Barbarian Settlement in the Western Empire', *Papers of the British School at Rome*, 54 (1986), 170–95.

——'The *Anonymous Valesianus II* as a Source for the Last Years of Theoderic', *Latomus*, 42 (1983), 572–96.

——'The Genesis and Composition of Cassiodorus' "Gothic history"', *Latomus*, 43 (1984), 336–61.

——'Transformation and Survival in the Western Senatorial Aristocracy, *c.* AD 400–700', *Papers of the British School at Rome*, 56 (1988), 120–55.

Barrett, H. M., *Boethius: Some Aspects of his Times and Work* (Cambridge, 1940).

Bartoli, A., 'Lavori nella sede del senato romano al tempo di Teoderico', *Bullettino della Commissione archeologica comunale di Roma*, 73 (1949–50), 77–90.

Behr, B., *Das alemannische Herzogtum bis 750* (Berne, 1975).

Bernareggi, E., 'Il medaglione d'oro di Teoderico', *Rivista italiana di numismatica*, 71 (1969), 89–106.

Berschin, W., *Griechisch-Lateinisches Mittelalter* (Berne, 1980).

Bertolini, O., *Roma di fronte a Bisanzio e ai Longobardi* (Bologna, 1941).

Beumann, H., 'Die Historiographie des Mittelalters als Quelle für die Ideengeschichte des Königtums', *Historische Zeitschrift*, 80 (1955), 449–88.

BEYERLE, F., 'Süddeutschland in der politischen Konzeption Theoderichs des Grossen', in *Grundfragen der Alemannischen Geschichte* (Vorträge und Forschungen, 1; Lindau, 1955), 65–81.

BIERBRAUER, V., *Die ostgotischen Grab- und Schatzfunde in Italien* (Spoleto, 1975).

——'Frühgeschichtliche Akkulturationprozesse in den Germanischen Staaten am Mittelmeer (Westgoten, Ostgoten, Langobarden) aus der Sicht des Archäologen', in *Atti dei 6° congresso internazionale di studi sull'alto medioevo*, 1978 (Spoleto, 1980), 89–105.

——'Frühmittelalterliche Castra in Östlichen und Mittleren Alpengebiet: Germanische Wehranlagen oder Romanische Siedlungen?', *Archäologisches Korrespondenzblatt*, 15 (1985), 497–513.

——'Zur ostgotische Geschichte in Italien', *Studi medievali*, 3rd ser., 14 (1973), 1–37.

BINDING, C., *Das Burgundisch-Romanische Königreich (von 443 bis 532 n. Chr.)*, 1 (Leipzig, 1868).

BLOCH, H., 'Ein datierter Ziegelstempel Theoderichs des Grossen', *Mitteilungen des Deutschen Archäeologischen Instituts: Römische Abt.* 66 (1959), 196–203.

BLOCKLEY, R. C., 'Roman–Barbarian Marriages in the Late Empire', *Florilegium*, 4 (1982), 63–79.

BLUMENKRANZ, B., *Juifs et Chrétiens dans le monde occidentale 430–1096* (Paris, 1960).

BODMER, J.-P., *Der Krieger der Merowingerzeit und seine Welt* (Zurich, 1957).

BOGNETTI, G. P., 'Teodorico di Verona e Verona longobardica, capitale de regno', in *Studi giuridici in onore de Mario Cavalieri* (Padua, 1960), 1–39.

BÓNA, I., 'Severiana', *Acta antiqua Academiae scientiarum Hungaricae*, 21 (1973), 281–338.

BONFANTE, G., *Latini e Germani in Italia*, 3rd edn. (Genoa, 1965).

BONNET, M., *Le Latin de Grégoire de Tours* (Paris, 1890).

BORMANN, F., 'Motivi tucididei in Procopio', *Atene e Roma*, 19 (1974), 138–50.

BOSL, K., *Gesellschaftsgeschichte Italiens im Mittelalter* (Stuttgart, 1983).

BRANDT, S., 'Entstehungszeit und zeitliche Folge der Werke von Boethius', *Philologus*, 62 (1903), 141–54.

BRATÓZ, R., *Severinus von Noricum und seine Zeit* (Österreichische Akademie der Wissenschaften, Phil./Hist. Klasse, Denkschriften Bd. 165; Vienna, 1983).

BRION, M., *Théodoric roi des Ostrogoths 454–526* (Paris, 1935).

BROWN, T. S., *Gentlemen and Officers: Imperial Administration and Aristocratic Power in Byzantine Italy A.D. 554–800* (Rome, 1984).

BUCHOWIECKI, W., *Handbuch der Kirchen Roms* (3 vols.; Vienna, 1967–74).

Büdinger, M., 'Eugipius [*sic*]: Eine Untersuchung', *Sitzungsberichte der Kaiserlichen Akademie der Wissenschaften (Vienna)* (Phil.-hist. Klasse, 91; 1878), 793–814.

Bullough, D. A., 'Germanic Italy: The Ostrogothic and Lombard Kingdoms', in D. Talbot Rice, ed., *The Dark Ages* (London, 1965), 157–74.

Burns, T. S., *A History of the Ostrogoths* (Bloomington, Ind., 1984).

——*The Ostrogoths: Kingship and Society* (Wiesbaden, 1980).

——'Calculating Ostrogothic Population', *Acta antiqua Academiae Scientarum Hungaricae*, 26 (1978), 457–63.

——'Ennodius and the Ostrogothic Settlement', *Classical Folia*, 32 (1978), 153–68.

Bury, J. B., *History of the Later Roman Empire* (2 vols.; London, 1923).

Cameron, Alan, *Circus Factions* (Oxford, 1976).

——'Boethius' Father's Name', *Zeitschrift für Papyrologie und Epigraphik*, 44 (1981), 181–3.

——'The Date of Priscian's *De laude Anastasii*', *Greek, Roman and Byzantine Studies*, 15 (1974), 313–16.

——'Junior Consuls', *Zeitschrift für Papyrologie und Epigraphik*, 56 (1984), 159–72.

——and Schauer, D., 'The Last Consul: Basilius and his Diptych', *Journal of Roman Studies*, 72 (1982), 126–45.

Cameron, Averil, *Procopius and the Sixth Century* (London, 1985).

Cantarelli, L., 'Il patrizio Liberio e l'imperatore Giustiniano', in *Studi romani e bizantini* (Rome, 1915), 289–303.

Capelle, W., *Die Germanen der Völkerwanderung* (Stuttgart, 1940).

Capizzi, C., *L'imperatore Anastasio I (491–518)* (Rome, 1969).

Cappuyns, M., 'Boèce', *Dictionnaire d'histoire et de géographie ecclésiastiques*, 9 (1937), 348–80.

——'Cassiodore', *Dictionnaire d'histoire et de géographie ecclésiastiques*, 11 (1949), 1349–408.

Caspar, E., *Geschichte des Papsttums*, 2 (Tübingen, 1933).

Castritius, H., 'Korruption im Ostgotischen Italien', in W. Schuller, ed., *Korruption in Altertum* (Munich, 1982), 215–38.

Cavallo, G., 'La circolazione libraria nell'età di Giustiniano', in *L'imperatore Giustiniano: Storia e mito* (Milan, 1978), 201–36.

——'La cultura a Ravenna tra corte e chiesa', in *Le sedi della cultura nell' Emilia Romagna: L'alto medioevo* (Milan, 1983), 29–51.

Cecchelli, C., 'L'Arianesimo e le chiese ariane d'Italia', in *Le chiese nei regni dell'Europa occidentale e i loro rapporti con Roma sino all'800* (Settimane di studio, 3; Spoleto, 1960), 742–74.

——'"Spazio cristiano" e monumenti eretici', in *Atti del VI Congresso nazionale di archeologia cristiana* (Florence, 1983), 287–96.

Cesa, M., *Ennodio: Vita del beatissimo Epifanio vescovo della chiesa Pavese* (Como, 1988).

——'Hospitalitas o altre "techniques of accomodation"? A proposito di un libro recente', *Archivio storico italiano*, 140 (1982), 539–52.

CESSI, R., 'La vita di papa Giovanni I', *Archivio muratoriano*, 19/20 (1917), 463–87.

——'Theodericus inlitteratus', in *Miscellanea di studi critici in onore di Vincenzo Crescini* (Cividale, 1927), 221–36.

——'Un vescovo pugliese del sec. VI', *Atti del reale instituto Veneto di scienze, lettere ed arti*, 73 (1914), 1141–57.

CHADWICK, H., *Boethius: The Consolations of Music, Logic, Theology, and Philosophy* (Oxford, 1981).

——'Theta on Philosophy's Dress in Boethius', *Medium Aevum*, 49 (1980), 175–9.

CHARANIS, P., *Church and State in the Later Roman Empire* (Madison, 1939).

CHASTAGNOL, A., *La Préfecture urbaine à Rome sous le bas-empire* (Paris, 1960).

——*Le Sénat romain sous le règne d'Odoacre* (Bonn, 1966).

CHATILLON, F., 'Dracontiana', *Revue du moyen âge latin*, 8 (1952), 177–212.

CHAUVOT, A., 'Observations sur la date de *l'Éloge d'Anastase* de Priscien de Césarée', *Latomus*, 36 (1977), 539–50.

CHRISTIE, N., and GIBSON, S., 'The City Walls of Ravenna', *Papers of the British School at Rome*, 56 (1988), 156–97.

CHRYSOS, E. K., 'Die Amaler-Herrschaft in Italien und das Imperium Romanum: Der Vertragsentwurf des Jahres 535', *Byzantion*, 51 (1981), 430–74.

——'The Title *ΒΑΣΙΛΕΥΣ*' in Early Byzantine International Relations', *Dumbarton Oaks Papers*, 32 (1978), 29–75.

CIPOLLA, C., 'Della supposta fusione degli Italiani con Germani nei primi secoli del medioevo', *Reniconti della reale accademia dei Lincei*, 5th ser. 9 (1900), 329–60, 369–422.

——'Ricerche intorno all' "Anonymus Valesianus II"', *Bollettino dell'istituto storico italiano*, 11 (1892), 7–98.

CLASSEN, P., 'Der erste Römerzug in der Weltgeschichte', in H. Beumann, ed., *Historische Forschungen für Walter Schlesinger* (Cologne, 1974), 325–47.

CLAUDE, D., *Adel, Kirche und Königtum im Westgotenreich* (Sigmaringen, 1971).

——*Der Handel im westlichen Mittelmeer während des Frühmittelalters* (Göttingen, 1985).

——*Geschichte der Westgoten* (Stuttgart, 1970).

——'Gentile und territoriale Staatsideen im Westgotenreich', *Frühmittelalterliche Studien*, 6 (1972), 1–38.

——'Millenarius und Thiuphadus', *Zeitschrift der Savigny-Stiftung für Rechtsgeschichte, Germanische Abt.* 88 (1971), 181–90.

CLAUDE, D., 'Universale und partikulare Züge in der Politik Theoderichs', *Francia*, 6 (1978), 19–58.

——'Zur Geschichte der frühmittelalterlichen Königsschatz', *Early Medieval Studies*, 7 (*Antikvariskt arkiv*, 54; 1973), 5–24.

——'Zur Königserhebung Theoderichs des Grossen', in K. Hauck and H. Mordek, eds., *Geschichtschreibung und Geistigen Leben in Mittelalter: Festschrift für Heinz Löwe* (Cologne, 1978), 1–13.

CLAUSS, M., *Der Magister officiorum in der Spätantike (4.–6. Jahrhundert)* (Munich, 1980).

CLAVADETSCHER, O. R., 'Churrätien im Übergang von der Spätantike zum Mittelalter nach den Schriftquellen', *Vorträge und Forschungen*, 25 (1979), 159–78.

CONTI, P. M., *'Devotio' e 'viri devoti' in Italia da Diocleziano ai Carolinigi* (Padua, 1971).

COOK, G. M., *The Life of Saint Epiphanius by Ennodius: A Translation with an Introduction and Commentary* (Washington, DC, 1942).

COSTER, C. H., *Late Roman Studies* (Cambridge, Mass., 1968).

——*The Iudicium quinquevirale* (Cambridge, Mass., 1935).

——and PATCH, H. R., 'Procopius and Boethius', *Speculum*, 23 (1948), 284–7.

COURCELLE, P., *Histoire littéraire des grandes invasions germaniques*, 3rd edn. (Paris, 1964).

——*Late Latin Writers and their Greek Sources* (Cambridge, Mass., 1969).

COURTOIS, C., *Les Vandales et l'Afrique* (Paris, 1955).

CROKE, B., 'A.D. 467: The Manufacture of a Turning Point', *Chiron*, 13 (1983), 81–119.

——'Mundo the Gepid: From Freebooter to Roman General', *Chiron*, 12 (1982), 125–35.

CURTIUS, E., *Europäische Literatur und Lateinisches Mittelalter*, 3rd edn. (Berne, 1961).

CUSCITO, G., 'Testimonianze archeologico monumentali sul più antico culto dei santi nella "Venetia et Histria"', *Aquileia nostra*, 45/6 (1974/5), 631–68.

DALEY, B. E., 'Boethius' Theological Tracts and Early Byzantine Scholasticism', *Mediaeval Studies*, 46 (1984), 158–91.

DEÉR, J., 'Byzanz und die Herrschaftszeichen des Abendlandes', *Byzantinische Zeitschrift*, 50 (1957), 405–36.

DE FRANCISCI, P., 'Per la storia del Senato Romano e della curia nei secoli V e VI', *Atti della Pontifica accademia Romana di archeologica rendiconti*, 22 (1946/7), 275–317.

DEGANI, M., *Il tesoro Romano-barbarico di Reggio Emilia* (Florence, 1959).

DEGRASSI, A., *I fasti consolari dell'impero Romano* (Rome, 1952).

DEICHMANN, F. W., *Ravenna Haupstadt des spätantiken Abendlandes* (Wiesbaden, 1958–76).

——'La corte dei re Goti a Ravenna', *XXVII Corso di cultura sull'arte Ravennate e bizantina* (1980), 41–53.

DELBRUECK, R., *Die Consulardiptychen und verwandte Denkmäler* (Berlin, 1926–9).

DELLA CORTE, F., 'Sui presunti rapporti fra Boezio e Bisanzio', *Rivista di studi Bizantini e Neohellenici*, pamphlet 1 (1965), 185–8.

DELLA VALLE, G., 'Moenia', *Rendiconti della Accademia di archeologia lettere e belle arti* (Naples), 33 (1958), 167–76.

——'Teoderico e Roma', *Rendiconti della Accademia di archeologia lettere e belle arti*, 34 (1959), 119–76.

DEMOUGEOT, E., *La Formation de l'Europe et les invasions barbares*, 2 (Paris, 1979).

——'Bedeutet das Jahr 476 das Ende des Römischen Reiches im Okzident?', *Klio*, 60 (1978), 371–81.

——'Une lettre de l'empereur Honorius sur *l'hospitium* des soldats', *Revue historique de droit français et étranger*, 4th ser. 34 (1956), 25–49.

DIESNER, H.-J., 'Die Auswirkungen der Religionspolitik Thrasamunds und Hilderichs auf Ostgoten und Byzantiner', *Sitzungsberichte der Sächsischen Akademie der Wissenschaften zu Leipzig, phil.-hist. Klasse*, 113.3 (Berlin, 1967).

DITTEN, H., 'Zu Prokops Nachrichten über die deutschen Stämme', *Byzantinoslavica*, 36 (1971), 1–24, 184–91.

DREW, K. F., 'Law, German: Early Germanic Codes', in J. R. Strayer, ed., *Dictionary of the Middle Ages*, 7 (1986), 468–75.

DUBOIS, A., *La Latinité d'Ennodius* (Paris, 1903).

DUCHESNE, L., *L'Église au VIe siècle* (Paris, 1925).

DUMOULIN, M., 'Le Gouvernement de Théodoric et la domination des Ostrogoths in Italie d'après les œuvres d'Ennodius', *Revue historique*, 78 (1902), 1–7, 241–65; 79 (1903), 1–22.

DUVAL, N., 'Que savons-nous du palais de Théodoric à Ravenne?', *Mélanges d'archéologie et d'histoire*, 72 (1960), 337–71.

DUVAL, Y.-M., 'Les Lupercales de Constantinople aux Lupercales de Rome', *Revue des études latines*, 55 (1977), 222–70.

DYGGVE, E., *Ravennatum palatium sacrum* (Copenhagen, 1941).

EFFENBERGER, A., *Frühchristliche Kunst und Kultur* (Leipzig, 1986).

ENSSLIN, W., 'Beweise der Römverbundenheit in Theoderichs des Grossen Aussen- und Innenpolitik', in *I Goti in occidente problemi* (Settimane di studio, 3; Spoleto, 1956).

——*Theoderich der Grosse*, 2nd edn. (Munich, 1959).

——'Der erste bekannte Erlass des Königs Theoderich', *Rheinisches Museum für Philologie*, 92 (1944), 266–80.

——'Nachmals zu der Ehrung Clodowechs durch Kaiser Anastasius', *Historisches Jahrbuch*, 56 (1936), 499–507.

ENSSLIN, W., 'Papste Johannes I. als Gesandter Theoderichs des Grossen bei Kaiser Justinos I.', *Byzantinische Zeitschrift*, 44 (1951), 127–34.
——'Rex Theodericus inlitteratus?', *Historisches Jahrbuch*, 60 (1940), 391–6.
ERCOLANI COCCHI, E., 'Osservazioni sull'origine del tipo monetale Ostrogoto "Felix Ravenna"', *Studi romagnoli*, 31 (1980), 21–44.
ERRINGTON, M., 'Malchos von Philadelphia, Kaiser Zenon und die zwei Theoderiche', *Museum Helveticum*, 40 (1983), 82–110.
EWIG, E., 'Die Fränkischen Teilungen und Teilreiche (511–613)', in *Akademie der Wissenschaften und der Literatur: Abhandlungen der Geistes- und Sozialwissenschaftlichen Klasse*, 9 (Wiesbaden, 1952).
FELLETTI-MAJ, B. M., 'Una carta di Ravenna romana e bizantina', *Rendiconti della pontifica accademia romana di archeologia*, 41 (1968–9), 85–120.
FICARRA, R., 'Fonti letterarie e motivi topici nel panegirico a Teodorico di magno Felice Ennodio', in *Scritti in onore di Salvatore Pugliatti*, 5 (Milan, 1978), 233–54.
FINGERLIN, G., GARBSCH, J., and WERNER, J., 'Die Ausgrabungen in langobardischen Kastell Ibligo-Invillino (Friaul)', *Germania*, 46 (1968), 73–110.
FLAMANT, J., *Macrobe et le néo-Platonisme latin à la fin du IV*[e] *siècle* (Leiden, 1977).
FO, A., 'L'Appendix Maximiani', *Romanobarbarica*, 8 (1984–5), 151–230.
FONTAINE, J., *Naissance de la poésie dans l'occident chrétien* (Paris, 1981).
——'Ennodius', *Reallexikon für Antike und Christentum*, 5 (1962), 398–421.
FRANCOVICH, G. DE, *Il palatium di Teoderico e la cosidetta 'Architettura di Potenza'* (Rome, 1970).
FREND, W. H. C., *The Rise of the Monophysite Movement* (Cambridge, 1972).
FRIDH, Å., *Terminologie et formules dans les Variae de Cassiodore* (Stockholm, 1956).
FUCHS, S., *Kunst der Ostgotenzeit* (Berlin, 1944).
——'Bildnisse und Denkmaler aus der Ostgotenzeit', *Die Antike*, 19 (1943), 109–53.
FUHRMANN, M., and GRUBER, J., eds., *Boethius* (Darmstadt, 1984).
GAMILLSCHEG, E., *Romania Germanica*, 2 (Berlin, 1935).
GARCIA IGLESIAS, L., 'El intermedio ostrogodo en Hispania (507–549 d.C.)', *Hispania antiqua*, 5 (1975), 89–120.
GAUDENZI, A., *Sui rapporti tra l'Italia e l'impero d'oriente fra gli anni 476 e 554 D.C.* (Bologna, 1888).
GAUPP, E. T., *Die germanischen Ansiedlungen und Landtheilungen in den Provinzen des römischen Westreiches* (Breslau, 1844).

GEARY, P. J., *Before France and Germany: The Creation and Transformation of the Merovingian World* (New York, 1988).

GIARDINA, A., ed., *Società romana e impero tardoantico*, 3. *Le merci gli insediamenti* (Rome, 1986).

GIBSON, M., ed., *Boethius: His Life, Thought and Influence* (Oxford, 1981).

GIESECKE, H.-E., *Die Ostgermanen und der Arianismus* (Leipzig, 1939).

GIUNTA, F., *Jordanes e la cultura dell'alto medioevo* (Palermo, 1952).

GOFFART, W., *Barbarians and Romans* (Princeton, 1980).

——*The Narrators of Barbarian History* (Princeton, 1988).

GOUBERT, P., 'Autour du voyage à Byzance du pape saint Jean I. (523–526)', *Orientalia christiana periodica*, 24 (1958), 339–52.

GRAHN-HOEK, H., *Die Fränkische Obersicht im 6. Jahrhundert* (Sigmaringen, 1976).

GRIERSON, P., 'The Date of Theoderic's Gold Medallion', *Hikuin*, 11 (1985), 19–26.

——'The Tombs and Obits of the Byzantine Emperors (337–1042)', *Dumbarton Oaks Papers*, 16 (1962), 1–63.

——and BLACKBURN, M., *Medieval European Coinage*, 1 (Cambridge, 1986).

GRILLMEIER, A., and BACHT, H., eds., *Das Konzil von Chalkedon*, 1 (Würzburg, 1951).

GRUBER, J., *Kommentar zu Boethius De consolatione philosophiae* (Berlin/New York, 1978).

GRUNDMANN, H., 'Litteratus-illitteratus', *Archiv für Kulturgeschichte*, 40 (1958), 1–65.

GUILLOU, A., *Régionalisme et indépendance dans l'empire byzantine au VIIe siècle* (Rome, 1969).

GUREVICH, A., *Medieval Popular Culture: Problems of Belief and Perception* (Cambridge, 1988).

HAENSSLER, F., *Byzanz und Byzantiner: Ihr Bild im Spiegel der Uberlieferung der Germanischen Reiche im früheren Mittelalter* (Berne, 1960).

HAHN, W., *Moneta imperii Byzantini*, 1. *Von Anastasius I. bis Iustinianus I. (491–565)*; 2. *Von Iustinius II. bis Phocas (565–610)* (with supplement to vol. 1) (Vienna, 1973–5).

HALLER, J., *Das Papsttum: Idee und Wirklichkeit*, 1 (Basle, 1951).

HAMMOND BRAMMEL, C. P., 'Products of Fifth-Century Scriptoria Preserving Conventions Used by Rufinus of Aquileia', *Journal of Theological Studies*, n.s. 30 (1979), 430–62.

HANNESTAD, K., *L'Évolution des ressources agricoles de l'Italie du IVe au VIe siècle de notre ère* (Copenhagen, 1962).

——'Les Forces militaires d'après la Guerre gothique de Procope', *Classica et mediaevalia*, 21 (1960), 136–83.

HARTMANN, L. M., *Geschichte Italiens im Mittelalter*, 1. *Das Italienische Königreich* (Leipzig, 1897).

Hartung, W., *Süddeutschland in der frühen Merowingerzeit* (Wiesbaden, 1983).

Hauck, K., 'Von einer spätantiken Randkultur zum karolingischen Europa', *Frühmittelalterliche Studien*, 1 (1967), 3–93.

Heather, P., 'Cassiodorus and the Rise of the Amals: Genealogy and the Goths under Hun Domination', *Journal of Roman Studies*, 79 (1989), 103–28.

Heidenreich, R., and Johannes, H., *Das Grabmal Theoderichs zu Ravenna* (Wiesbaden, 1971).

Heinzle, J., and Metzner, E. E., 'Dietrich v. Bern', *Lexikon des Mittel Alters*, 3 (1986), 1016–21.

Helbling, H., *Goten und Wandalen* (Zurich, 1954).

Helm, R., 'Priscianus', PW 22.2328–46.

Hemmerdinger, B., 'Les Lettres latines à Constantinople jusqu'à Justinien', *Byzantinische Forschungen*, 1 (1966), 174–8.

Hendy, M. F., *Studies in the Byzantine Monetary Economy c.300–1450* (Cambridge, 1985).

——'From Public to Private: The Western Barbarian Coinages as a Mirror of the Disintegration of Late Roman State Structures', *Viator*, 19 (1988), 29–78.

Herz, P., *Studien zur römischen Wirtschaftsgesetzgebung (Historia Einzelschriften* 55; Stuttgart, 1988).

Hessen, O. von, Kurze, W., and Mastrelli, C. A., *Il tesoro di Galognano* (Florence, 1977).

Heuberger, R., 'Das Ostgotische Rätien', *Klio*, 30 (1937), 77–109.

——'Ein angebliches Edikt Theoderichs des Grossen vom Jahre 505 aus dem Castrum Maiense über dem Laureinerberg', in *Festschrift Karl Piveč* (Innsbruck, 1966), 201–3.

Hodges, R., and Whitehouse, D., *Mohammed, Charlemagne and the Origins of Europe* (London, 1983).

Hodgkin, T., *Italy and her Invaders*, 3, 2nd edn. (Oxford, 1896).

——*The Letters of Cassiodorus* (London, 1886).

——*Theodoric the Goth* (New York and London, 1891).

Holfer, O., 'Der Sakralcharacter des germanischen Königtums', *Vörtrage und Forschungen*, 3 (1954), 75–104.

——'Theoderich der Grosse und sein Bild in der Sage', *Anzeiger der Österreichischen Akademie der Wissenschaften, phil.-hist. Klasse*, 111 (1974), 349–72.

Hope, D. M., *The Leonine Sacramentary* (Oxford, 1971).

Iandiorio, L., 'Le lettere siciliane di Cassiodoro', *Orpheus*, 24/5 (1977/8), 171–86.

Jahn, O., 'Über die Subscriptionen in den Handschriften römischer Classiker', *Berichte über die Verhandlungen der königlich-sächsischen Gesellschaft der Wissenschaften zu Leipzig, phil.-hist. Klasse*, 3 (1851), 327–72.

JAHN, W., 'Felicitas est secuta Italiam: Bemerkungen zur Lage der römischen Bevölkerung im 6. Jahrhundert in Italien', *Klio*, 71 (1989), 410–13.

JASCHKE, K.-U., and WENSKUS, R., eds., *Festschrift für Helmut Beumann* (Sigmaringen, 1977).

JIMENEZ GARNICA, A. M., *Origenes y desarrollo del reino Visigodo de Tolosa (a. 418–507)* (Valladolid, 1983).

JOHNSON, M. J., 'Toward a History of Theoderic's Building Program', *Dumbarton Oaks Papers*, 42 (1988), 73–96.

JONES, A. H. M., *The Later Roman Empire 284–602: A Social, Economic and Administrative Survey* (Oxford, 1964).

——'The Constitutional Position of Odoacer and Theoderic', *Journal of Roman Studies*, 52 (1962), 126–30.

——MARTINDALE, J., and MORRIS, J., *The Prosopography of the Later Roman Empire*, I. AD *260–395* (Cambridge, 1971).

JONES, C. W., 'The Victorian and Dionysiac Paschal Tables in the West', *Speculum*, 9 (1934), 408–21.

KAHANE, A., THREIPLAND, L. M., and WARD-PERKINS, J., 'The Ager Veientanus, North and East of Rome', *Papers of the British School at Rome*, 36 (n.s. 23) (1968), 1–218.

KAMPERS, G., 'Anmerkungen zum lateinisch-gotischen Ravennater Papyrus von 551', *Historisches Jahrbuch*, 101 (1981), 141–51.

KASTER, R. A., *Guardians of Language: The Grammarian and Society in Late Antiquity* (Berkeley, 1988).

KENT, J. P. C., 'Julius Nepos and the Fall of the Western Empire', in *Corolla memoriae Erich Swoboda dedicata* (Böhlau, 1966), 146–50.

KING, P. D., *Law and Society in the Visigothic Kingdom* (Cambridge, 1972).

KISS, A., 'Ein Versuch die Funde und das Siedlungsgebiet der Ostgoten in Pannonien zwischen 456 und 471 zu bestimmen', *Acta archaeologica*, 31 (1979), 329–39.

KLINGSHIRN, W., 'Charity and Power: Caesarius of Arles and the Ransoming of Captives in Sub-Roman Gaul', *Journal of Roman Studies*, 75 (1985), 183–203.

KORSUNSKIJ, A. R., 'K Diskusii ob "Edikte Theodericha"', in *Europa v sredie veka: Ekonomika, politika, kul'tura (Fest. C. D. Skazkin)* (Moscow, 1972), 16–31.

KRAUTHEIMER, R., *et al.*, *Corpus Basilicarum Christianarum Romae* (5 vols.; Rome, 1937–77).

——*Rome: Profile of a City, 312–1308* (Princeton, NJ, 1980).

——*Three Christian Capitals* (Berkeley, 1983).

KRAUTSCHICK, S., *Cassiodor und die Politik seiner Zeit* (Bonn, 1983).

——'Bemerkungen zur PLRE II', *Historia*, 35 (1986), 121–4.

——'Zwei Aspekte des Jahres 476', *Historia*, 35 (1986), 344–71.

KRÜGER, B., ed., *Die Germanen* (2 vols.; Berlin, 1983).

KRÜGER, K. H., *Königsgrabkirchen der Franken, Angelsachsen und Langobarden bis zur Mitte des 8. Jahrhunderts* (Munich, 1971).

KRUSCH, B., 'Die Einführung des griechischen Paschalritus in Abendlande', *Neues Archiv*, 9 (1883), 100–69.

——'Ein Bericht der päpstlichen Kanzlei an Papst Johannes I. von 526', in A. Brackmann, ed., *Papsttum und Kaisertum: Festschrift Paul Kehr* (Munich, 1926), 48–58.

LAMMA, P., *Oriente e occidente nell'alto medioevo* (Padua, 1968).

——*Teoderico* (Brescia, 1950).

LAPEYRE, G.-G., *Saint Fulgence de Ruspe* (Paris, 1929).

LECCE, M., 'La vita economica dell'Italia durante la dominazione dei Goti nelle "Variae" di Cassiodoro', *Economia e storia*, 3 (1956), 354–408.

LEONARDI, C., MINIO-PALUELLO, L., PIZZANI, U., and COURCELLE, P., 'Boezio', *Dizionario biografico degli Italiani*, 11 (1969), 142–65.

LÉVÊQUE, P., 'Le Palais de Théoderic-le-Grand, à Galeata', *Revue archéologique*, 6th ser. 28 (1947), 58–61.

LEVILLAIN, L., 'La Crise des années 507–508 et les rivalités d'influence en Gaule de 508 à 514', in *Mélanges offerts à M. Nicholas Iorga* (Paris, 1933), 537–67.

LEWIS, C. S., *The Discarded Image* (Cambridge, 1964).

LIPPOLD, A., 'Chlodovechus', PW supplementary vol. 13.139–74.

LLEWELLYN, P. A. B., *Rome in the Dark Ages* (London, 1971).

——'Le indicazioni numeriche del Liber Pontificalis relativemente alle ordinazioni del V secolo', *Rivista di storia della chiesa in Italia*, 29 (1975), 439–43.

——'The Roman Church during the Laurentian Schism: Priests and Senators', *Church History*, 45 (1976), 417–27.

——'The Roman Clergy during the Laurentian Schism (498–506): A Preliminary Analysis', *Ancient Society*, 8 (1977), 245–75.

LONCAO, E., *Fondazione del regno di Odoacre* (Scansano, 1907).

LOT, F., 'Du régime de l'hospitalité', *Revue belge de philologie et d'histoire*, 7 (1928), 975–1011.

——'La Conversion de Clovis', *Revue belge de philologie et d'histoire*, 17 (1938), 63–9.

LOTTER, F., *Severinus von Noricum: Legende und historische Wirklichkeit* (Stuttgart, 1976).

LÖWE, H., *Von Cassiodor zu Dante* (Berlin, 1973).

——'Theoderich der Grosse und Papst Johann I.', *Historisches Jahrbuch*, 72 (1953), 83–100.

——'Theoderichs Gepidensieg im Winter 488/489', in *Historische Forschungen und Probleme: Festschrift Peter Rassow* (Wiesbaden, 1961), 1–16.

——'Von Theoderich dem Grossen zu Karl dem Grossen', *Deutsches Archiv für Erforschung des Mittelalters*, 9 (1952), 353–401.

——review of W. Ensslin, *Theoderich der Grosse* (1st edn.), *Historische Zeitschrift*, 170 (1950), 566–74.

LUISELLI, R., 'La società dell'Italia romano-gotica', in *Atti del 7° Congresso internazionale di studi sull'alto medioevo, 1980* (Spoleto, 1982), 49–116.

——'Sul De summa temporum di Iordanes', *Romanobarbarica*, 1 (1976), 83–133.

LUKMAN, N. K., *Skjoldunge und Skilfinge* (Copenhagen, 1943).

LUMPE, A., 'Die konziliengeschichtliche Bedeutung des Ennodius', *Annuarium historiae conciliorum*, 1 (1969), 15–36.

——'Ennodiana', *Byzantinische Forschungen*, 1 (1966), 200–10.

MACCORMACK, S. G., *Art and Ceremony in Late Antiquity* (Berkeley, Calif., 1981).

MCCORMICK, M., *Eternal Victory: Triumphal Rulership in Late Antiquity, Byzantium, and the Early Medieval West* (Cambridge, 1986).

——'Odoacer, Emperor Zeno and the Rugian Victory Legation', *Byzantion*, 47 (1977), 212–22.

MAENCHEN-HELFEN, O. J., *The World of the Huns* (Berkeley, Calif., 1973).

MAGI, L., *La sede romana nella corrispondenza degli imperatori e patriarchi bizantini (VI–VII sec.)* (Rome, 1972).

Magistra barbaritas: I barbari in Italia (Milan, 1984).

MALORNY, A., *Saint Césaire évêque d'Arles* (Paris, 1894).

MARTINDALE, J. R., *The Prosopography of the Later Roman Empire*, 2. *AD 395–527* (Cambridge, 1980).

MATHISEN, R. W., 'Patricians as Diplomats in Late Antiquity', *Byzantinische Zeitschrift*, 79 (1986), 35–49.

MATHWICH, J., 'De Boethi morte', *Eunomia*, 4 (1960), 26–37.

MATTHEWS, J., *Western Aristocracies and Imperial Court, AD 364–425* (Oxford, 1975).

MEIXNER, I., 'Three Unknown Coins of King Theoderic', *Numizmaticke vijesti*, 15 (1968), 53–5.

MENENDEZ PIDAL, R., 'Los Godos y el origen de la epopeya española', in *I Goti in occidente problemi* (Settimane di studio, 3; 1956), 285–322.

MIEROW, C. C., *The Gothic History of Jordanes* (Princeton, NJ, 1915).

MOELLER, C., 'Le Chalcédonisme et le néo-chalcédonisme en Orient de 451 à la fin du VIe SIÈCLE', IN A. GRILLMEIER AND H. BACHT, EDS., *Das Konzil von Chalkedon*, 1 (Würzburg, 1954), 637–720.

MOISL, H., 'Anglo-Saxon Royal Genealogies and Germanic Oral Tradition', *Journal of Medieval History*, 7 (1981), 215–48.

MOMIGLIANO, A., 'Cassiodoro', *Dizionario biografico degli Italiani*, 21 (1978), 494–504.

MOMIGLIANO, A., 'Cassiodorus and Italian Culture of his Time', *Proceedings of the British Academy*, 41 (1955), 207–45 (Also in *Studies in historiography* (London, 1966), 181–210).

——'Gli Anicii e la storiografia latina del VI sec. D.C.', in his *Secondo contributo alla storia degli studi classici* (Rome, 1960), 231–53 (orig. pub. *Rendiconti accademia dei Lincei*, 8th ser. 11 (1956), 279–97).

——'La caduta senza rumore di un impero nel 476 D.C.', *Annali della scuola normale superiore di Pisa*, 3rd ser. 3 (1973), 394–418.

——'Un appunto di I. Casaubon dalle "Variae" di Cassiodoro', in *Tra Latino e volgare: Per Carlo Dionisotti* (*Medioevo e umanesimo*, 17f.; Padua, 1977), 615–17.

MOMMSEN, T., *Gesammelte Schriften*, 6 (Berlin, 1910).

——*Römisches Staatsrecht*, 1 (Leipzig, 1887).

MOORHEAD, J., 'Boethius and Romans in Ostrogothic Service', *Historia*, 27 (1978), 604–12.

——'Clovis' Motives for Becoming a Catholic Christian', *Journal of Religious History*, 13 (1985), 329–39.

——'Culture and Power among the Ostrogoths', *Klio*, 68 (1986), 112–22.

——'Italian Loyalties during Justinian's Gothic War', *Byzantion*, 53 (1983), 575–96.

——'*Libertas* and *Nomen Romanum* in Ostrogothic Italy', *Latomus*, 46 (1987), 161–8.

——'The Decii under Theoderic', *Historia*, 33 (1984), 107–15.

——'The Last Years of Theoderic', *Historia*, 32 (1981), 106–20.

——'Theoderic, Zeno and Odovacer', *Byzantinische Zeitschrift*, 77 (1984), 261–6.

MORIN, G., 'L'Origine du symbole d'Athanase', *Revue bénédictine*, 44 (1932), 207–19.

MOROSI, R., 'I *saiones*, speciali agenti di polizia presso i Goti', *Athenaeum*, 59 (1981), 150–65.

MORTON, C., 'Marius of Avenches, the *Excerpta Valesiana*, and the Death of Boethius', *Traditio*, 38 (1982), 107–36.

MÜLLER, W., ed., *Zur Geschichte der Alemannen* (Darmstadt, 1975).

MUSSET, L., *The Germanic Invasions* (Eng. trans.; London, 1975).

NAGL, A., 'Odoacer', PW 17.1888–96.

——'Theoderich der Grosse', PW 2/10.1745–71.

NAVARRA, L., 'Contributo storico di Ennodio', *Augustinianum*, 14 (1974), 315–42.

——'Le componenti letterarie e concettuali delle "Dictiones" di Ennodio', *Augustinianum*, 12 (1972), 465–78.

NÉDONCELLE, M., 'Les Variations de Boèce sur la personne', *Revue des sciences religieuses*, 29 (1955), 201–38.

NEHLSEN, H., *Sklavenrecht zwischen Antike und Mittelalter* (Göttingen, 1972).

NETZER, N., 'Redating the Consular Ivory of Orestes', *Burlington Magazine*, 125 (1983), 265–71.
NORDENFALK, C., *Die spätantike Kanontafeln* (Göteburg, 1938).
NORDSTROM, C.-O., *Ravennastudien* (Stockholm, 1953).
OBERTELLO, L., *Severino Boezio* (2 vols.; Genoa, 1974).
——ed., *Congresso internazionale di studi Boeziani Atti* (Rome, 1981).
O'DONNELL, J. J., *Cassiodorus* (Berkeley, Calif., 1979).
——'Liberius the Patrician', *Traditio*, 37 (1981), 31–72.
OPELT, I., and SPEYER, W., 'Barbar', *Jahrbuch für Antike und Christentum*, 10 (1967), 251–90.
PANCIERA, S., 'Inscrizione senatorie di Roma e dintorni', *Tituli*, 4 (1982), 591–678.
PALMIERI, S., 'Reminiscenze gotiche nelle fonti napoletane d'età ducale', *Koinonia*, 6 (1982), 61–72.
PARADISI, B., 'Critica e mito dell'editto teodericiano', *Bollettino dell'Istituto di dirotto romano*, 68 (1965), 1–47.
PATCH, H. R., 'The Beginnings of the Legend of Boethius', *Speculum*, 22 (1947), 443–5.
PEPE, G., *Le Moyen Âge barbare en Italie* (Paris, 1956).
PERONI, A., *Oreficerie e metalli lavorati tardoantichi e altomedievali del territorio di Pavia* (Spoleto, 1967).
PERRIN, O., *Les Burgondes* (Neuchâtel, 1968).
PFEILSCHIFTER, G., *Der Ostgotenkönig Theoderich der Grosse und die katholische Kirche* (Münster, 1896).
PFERSCHY, B., 'Das Problem der Getreidepreise unter Theoderich', in *Siedlung Macht und Wirtschaft: Festschrift Fritz Posch* (Graz, 1981), 481–6.
PICOTTI, G. B., 'Il "patricius" nell'ultima età imperiale e nei primi regni barbarici d'Italia', *Archivio storico italiano*, 7th ser. 9 (1928), 3–80.
——'Il senato Romano e il processo di Boezio', *Archivio storico italiano*, 7th ser. 15 (1931), 205–28.
——'I sinodi romani nella scisma laurenziano', in *Studi storici in onore di Gioacchino Volpe*, 3 (Florence, 1958), 743–86.
——'Osservazioni su alcuni punti della politica religiosa di Teoderico', in *I Goti in occidente problemi* (Settimane di studio, 3; Spoleto, 1956), 173–226.
——'Sulle relazioni fra re Odoacre e il senato e la chiesa di Roma', *Rivista storica italiana*, 5th ser. 4 (1939), 363–86.
PIETRELLA, E., 'La figura del santo-vescovo nella "Vita Epifani" di Ennodio di Pavia', *Augustinianum*, 24 (1984), 213–26.
PIETRI, C., *Roma Christiana: Recherches sur l'église de Rome, son organisation, sa politique, son idéologie de Miltiade à Sixte III (311–440)* (Rome, 1976).
——'Aristocratie et société cléricale dans l'Italie chrétienne au temps

d'Odoacre et de Théodoric', *Mélanges d'archéologie et d'histoire*, 93 (1981), 417–67.

PIETRI, C., 'Evergétisme et richesses ecclésiastiques dans l'Italie du IVe à la fin du Ve s.: L'Example romain', *Ktema*, 3 (1978), 317–37.

——'Les Aristocraties de Ravenne (V–VI s.)', *Studi romagnoli*, 34 (1983), 643–73.

——'Le Sénat, le peuple chrétien et les partis du cirque à Rome sous le pape Symmaque (498–514)', *Mélanges d'archéologie et d'histoire*, 78 (1966), 123–39.

PIZZANI, U., 'Boezio "consulente tecnico" al servizio dei re barbarici', *Romanobarbarica*, 3 (1978), 189–242.

PLATNER, S. B., and ASHBY, T., *A Topographical Dictionary of Ancient Rome* (Oxford, 1929).

PONTIERI, E., *Le invasioni barbariche e l'Italia del V e VI secolo* (Naples, 1960).

RAND, E. K., *The Founders of the Middle Ages* (Cambridge, Mass., 1928).

RASI, P., 'Sulla paternità del c.d. Edictum Theodorici Regis', *Archivio giuridico*, 145 (1953), 105–62.

RECCHIA, V., 'San Benedetto e la politica religiosa', *Romanobarbarica*, 7 (1982–3), 201–52.

REISS, E., *Boethius* (Boston, 1982).

REYDELLET, M., *La Royauté dans la littérature latine de Sidoine Apollinaire à Isidore de Seville* (Paris, 1981).

RICHARDS, J., *The Popes and the Papacy in the Early Middle Ages 476–752* (London, 1979).

RICHÉ, P., *Education and Culture in the Barbarian West* (Eng. trans.; Columbia, SC, 1976).

RIGHINI, V., 'I bolli laterzi di Teodorico e l'attività edilizia teodericana in Ravenna', *XXXIII Corso di cultura sull'arte ravennate e bizantina* (Ravenna, 1986), 371–98.

RIJK, L. M. DE, 'On the Chronology of Boethius' Works on Logic', *Vivarium*, 2 (1964), 1–49, 125–62.

RODA, S., 'Alcune ipotesi sulla prima edizione dell'epistolario di Simmaco', *La parola del passato*, 34 (1979), 31–54.

ROMANO, G., and SOLMI, A., *Le dominazioni barbariche in Italia (395–888)* (Milan, 1940).

RÖSCH, G., *Onoma Basileus* (Vienna, 1978).

ROSENFELD, H., 'Ost- und Westgoten', *Die Welt als Geschichte*, 17 (1957), 245–58.

ROUCHE, M., *L'Aquitaine des Wisigoths aux Arabes 418–781* (Paris, 1979).

ROUGÉ, J., 'Quelques aspects de la navigation en Méditerranée au Ve siècle et dans la première moitié du VIe siècle', *Cahiers d'histoire*, 6 (1961), 129–54.

ROUSSEAU, P., 'The Death of Boethius: The Charge of "Maleficium"', *Studi medievali*, 20 (1979), 871–89.
RUBIN, B., *Daz Zeitalter Iustinians I* (Berlin, 1960).
RUGGINI, L. C., *Economia e società nell 'Italia annonaria'* (Milan, 1961).
——'Ebrei e Orientali nell'Italia settentrionale fra il IV e il VI secolo d. Cr.', *Studia et documenta historiae et iuris*, 25 (1959), 186–308.
——'La Sicilia tra Roma e Bisanzio', *Storia della Sicilia*, 3 (Naples, (?) 1980), 1–96.
——'Vicende rurali dell'Italia antica dall'età tetrarchia ai Longobardi', *Rivista storica italiana*, 76 (1964), 261–86.
RUSSELL, J. C., *Late Ancient and Medieval Population* (Philadelphia, 1958).
SAITTA, B., '"Religionem imperare non possumus": Motivi e momenti della politica di Teoderico il Grande', *Quaderni catanese*, 8 (1986), 63–8.
SALAMON, M., 'Priscianus und sein Schülerkreis in Konstantinopel', *Philologus*, 123 (1979), 91–6.
SANDE, S., 'Zur Porträtplastik des sechsten nachchristlichen Jahrhunderts', *Institutum romanum Norwegiae acta*, 6 (1975), 65–106.
SANTIFALLER, L., 'Die Urkunde des Königs Odovakar vom Jahre 489', *Mitteilungen des Österreichische Geschichtsforschung*, 60 (1952), 1–30.
ŠAŠEL, J., 'Antiqui barbari', *Vorträge und Forschungen*, 25 (1979), 125–39.
SCARDIGLI, P., *Die Goten Sprache und Kultur* (Munich, 1973).
SCHÄFERDIEK, K., 'Ein neues Bild der Geschichte Chlodwigs?', *Zeitschrift für Kirchengeschichte*, 84 (1973), 270–7.
——'Germanenmission', *Reallexikon für Antike und Christentum*, 10 (1978), 492–548.
SCHAFFRAN, E., 'Zur Nordgrenze des Ostgotischen Reiches in Kärnten', *Jahresheft des Österreichischen Archäologischen Instituts in Wien*, 42 (1955), 111–30.
SCHANZ, M., HOSIUS, C., and KRUEGER, G., *Geschichte der römischen Literatur*, 4/2 (Munich, 1920).
SHANZER, D., 'Ennodius, Boethius and the Date and Interpretation of Maximanus' *Elegia III*', *Rivista di filòlogia e di instruzione classica*, 3 (1983), 183–95.
SCHMIDT, B., 'Theoderich der Grosse und die damaszierten Schwerter der Thuringer', *Ausgrabungen und Funde*, 14 (1969), 38–40.
SCHMIDT, L., *Die Ostgermanen* (Munich, 1934).
——'Theoderich, römischer Patricius und König der Goten', *Zeitschrift für schweizerische Geschichte*, 19 (1939), 404–14.
SCHMIEDT, G., 'Città scomparse e di nuovo formazione in Italia', in *Topografia urbana e vita cittadina nell'alto medioevo in occidente Spoleto* (Settimane di studio, 21; 1974), 503–607.
SCHNEIDER, F., *Die Reichsverwaltung in Toscana von der Gründung des Langobardenreich bis zum Ausgang der Staufer (568–1268)* (Rome, 1914).

SCHNEIDER, F., *Rom und Romgedanke im Mittelalter* (Munich, 1926).

SCHÖNFELD, M., *Wörterbuch der Altgermanischen Personen- und Völkernamen* (Heidelberg, 1911).

SCHOTT, C. I., 'Der Stand der Leges-Forschung', *Frühmittelalterliche Studien*, 13 (1979), 29–55.

SCHRAMM, P. E., *Herrschaftszeichen und Staatssymbolik*, 1 (Stuttgart, 1954).

SCHURR, V., *Die Trinitätslehre des Boethius im Lichte der 'Skythischen Kontroversen'* (Paderborn, 1935).

SEPPELT, F. X., *Der Aufstieg des Papsttums* (Leipzig, 1931).

SESTAN, E., *Stato e nazione nell'alto medioevo* (Naples, 1952).

SIMEONI, L., 'Note Teodericiane', *Accademia delle scienze dell'Istituto di Bologna: Classe di scienze morali memorie*, 4th ser. 8 (1945–8), 149–98.

SIMONETTI, M., 'Arianesimo Latino', *Studi medievali*, 3rd ser. 8 (1967), 663–744.

SINNIGEN, W. G., 'Administrative Shifts of Competence under Theoderic', *Traditio*, 21 (1965), 456–67.

——'Comes consistoriani in Ostrogothic Italy', *Classica et mediaevalia*, 24 (1963), 158–65.

SORACI, R., *Aspetti di storia economica italiana nell'età di Cassiodoro* (Catania, 1974).

——*Ricerche sui conubia tra Romani e Germani nei secoli IV–VI* (Catania, 1974).

SÖRRIES, R., *Die Bilder der Orthodoxen im Kampf gegen den Arianismus* (Frankfurt, 1983).

STAAB, F., 'Ostrogothic Geographers at the Court of Theoderic the Great: A Study of Some Sources of the Anonymous Cosmographer of Ravenna', *Viator*, 7 (1976), 27–64.

STAUBACH, N., 'Germanisches Königtum und lateinischen Literatur vom fünften bis zum siebten Jahrhundert', *Frühmittelalterliche Studien*, 17 (1983), 1–54.

STAUFFENBERG, A. S., GRAF VON, 'Theoderich der Grosse und seine römische Sendung', in *Würzburger Festgabe Heinrich Bulle (Würzburger Studien zur Altertumswissenschaft*, 13; Stuttgart, 1938), 115–29.

STEIN, E., *Histoire du bas-empire*, 2 (Paris/Bruges, 1949).

——'Beiträge zu Geschichte von Ravenna in spätrömischer und byzantinischer Zeit', *Klio*, 16 (1920), 40–71.

STEINACKER, H., 'Die römische Kirche und die griechischen Sprachkenntnisse des Frühmittelalters', *Mitteilungen des Instituts für Österreichische Geschichtsforschung*, 62 (1954), 28–66.

STEVENS, S. T., 'The Circle of Bishop Fulgentius', *Traditio*, 38 (1982), 327–41.

STEWART, H. F., *Boethius: An Essay* (Edinburgh, 1891).

Storia di Brescia, 1 (Brescia, 1963).

Storia di Milano, 2 (Milan, 1954).
Storia d'Italia, 2 (Turin, 1974).
STROHEKER, K. F., *Euric König der Westgoten* (Stuttgart, 1937).
——*Germanentum und Spätantike* (Zurich, 1965).
——'Zu den ersten Begegnungen der Germanen mit dem spätantiken Bildungsgedanken', in *Silvae Festschrift Ernest Zinn* (Tübingen, 1970), 233–43.
SUERBAUM, W., *Von Antiken zum frühmittelalterlichen Staatsbegriff*, 3rd edn. (Munster, 1977).
SUNDWALL, J., *Abhandlungen zur Geschichte des ausgehenden Römertums* (Helsinki, 1919).
TABARRONI, G., 'La cupola monolitica del mausoleo di Teodorico', *Felix Ravenna*, 105/6 (1973), 119–42.
TAMASSIA, N., 'Sulla seconda parte dell'anonimo Valesiano', *Archivio storico italiano*, 71 (1913), 3–22.
TEILLET, S., *Des Goths à la nation gothique* (Paris, 1984).
TESSIER, G., *Le Baptême de Clovis* (Paris, 1964).
THOMPSON, E. A., *Romans and Barbarians* (Madison, Wis., 1982).
——'Christianity and the Northern Barbarians', in A. Momigliano, ed., *The Conflict between Paganism and Christianity in the Fourth Century* (London, 1963), 56–78 (orig. pub. *Nottingham Medieval Studies*, 1 (1957)).
——'The Visigoths from Fritigern to Euric', *Historia*, 12 (1963), 105–26.
THOMSEN, R., *The Italic Regions from Augustus to the Lombard Invasion* (Copenhagen, 1947).
THORDEMAN, B., 'Il cosidetto palazzo di Teoderico a Ravenna un palazzo reale Longobardo?', *Opuscula Romana*, 10 (1974), 23–40.
TJÄDER, J.-O., 'Der Codex argenteus in Uppsala und der Buchmeister Viliaric in Ravenna', in V. E. Hagberg, ed., *Studia Gotica* (Stockholm, 1972), 144–64.
TOYNBEE, J. M. C., 'Roman Medallions: Their Scope and Purpose', *Numismatic Chronicle*, 6th ser. 4 (1944), 27–44.
TRÄNKLE, H., 'Ist die *Philosophiae Consolatio* des Boethius zum vorgesehenen Abschluss gelangt?', *Vigiliae Christianae*, 31 (1977), 148–56.
——'Philologische Bemerkungen zum Boethiusprozess', in *Romanitas et Christianitas: Studia I. H. Waszink oblata* (Amsterdam, 1973), 329–39.
TRONCARELLI, F., *Tradizioni perdute: La 'Consolatio philosophiae' nell'alto medioevo* (Padua, 1981).
UDAL'TSOVA, Z. V., *Italiia i Vizantiia v VI Veke* (Moscow, 1959).
ULLMANN, W., *Gelasius I (492–496)* (Stuttgart, 1981).
USENER, H., 'Das Verhältnis des Römischen Senats zur Kirche in der Ostgothenzeit', in *Commentationes philologiae in honorem Theodori Mommseni* (Berlin, 1877), 759–67.

Vaccari, P., 'Concetto ed ordinamento dello stato in Italia sotto il governo dei Goti', in *Goti in occidente problemi* (Settimane di studio, 3; Spoleto, 1956), 585–94.

Vanderspoel, J., 'Cassiodorus as Patricius and Ex-patricio', *Historia*, 39 (1990), 499–503.

Van de Vyver, A., 'La Unique Victoire contre les Alamans et la conversion de Clovis', *Revue belge de philologie et d'histoire*, 17 (1938), 793–813.

——'La Victoire contre les Alamans et la conversion de Clovis', *Revue belge de philologie et d'histoire*, 15 (1936), 859–914; 16 (1937), 35–94.

van Vytfanghe, M., 'Les Avatars contemporains de l'"hagiologie"', *Francia*, 5 (1977), 639–71.

Várady, L., *Epochenwechsel um 476: Odoaker. Theoderich d. gr. und die Unnwandlungen* (Budapest, 1984).

Vasiliev, A. A., *Justin the First* (Cambridge, Mass., 1950).

Verona in età Gotica e Longobarda (Verona, 1982).

Vetter, G., *Die Ostgoten und Theoderich* (Stuttgart, 1938).

Viden, G., *The Roman Chancery Tradition: Studies in the Language of Codex Theodosianus and Cassiodorus' 'Variae'* (Göteborg, 1984).

Vieillard-Troiekouroff, M., *Les Monuments religieux de la Gaule d'après les œuvres de Grégoire de Tours* (Paris, 1976).

Vismara, G., *Edictum Theoderici* (*Ius Romanum medii aevi*, pts. 1, 2.b.aa.α; Milan, 1967).

——'Romani e Goti di fronte al dirotto nel regno Ostrogoto', in *I Goti in occidente problemi* (Settimane di studio, 3; Spoleto, 1956), 409–63.

Vogel, C., 'Le Liber Pontificalis dans l'édition de Louis Duchesne: État de la question', in *Monseigneur Duchesne et son temps* (Collection de l'École française de Rome, 23; Rome, 1975), 99–127.

Vogel, C. J. de, 'Boethiana I', *Vivarium*, 9 (1971), 49–66.

Volbach, W. F., *Elfenbeiarbeiten der Spätantike und des frühen Mittelalters* (Mainz, 1952).

Wagner, N., *Getica* (Berlin, 1967).

Wallace-Hadrill, J. M., *Early Germanic Kingship in England and on the Continent* (Oxford, 1971).

——*The Barbarian West 400–1000*, 3rd edn. (London, 1967).

——*The Long-Haired Kings and Other Studies in Frankish History* (London, 1962).

Ward-Perkins, B., *From Classical Antiquity to the Middle Ages: Urban Public Building in Northern and Central Italy* AD *300–850* (Oxford, 1984).

——'The Decline and Abandonment of a Roman Town', in H. Blake, T. Potter, and D. Whitehouse, eds., *Papers in Italian Archaeology*, 2 (Oxford, 1978), 313–21.

Wartburg, W. von, *Die Entstehung der romanischen Völker* (Halle, 1939).

WEBER, S., 'Die Leges Barbarorum aus germanischer, römischer und byzantinischer Sicht: Ein Beitrag zu ihrer historischen Analyse', in V. Vavrinek, ed., *From Late Antiquity to Early Byzantium* (Prague, 1985), 167–71.
——'Zur Ansiedlung der Germanen nach den Leges Barbarorum', *Zeitschrift für Archäologie*, 19 (1985), 207–11.
WEISS, R., *Clodwigs Tauf: Reims 508* (Berne, 1971).
WENSKUS, R., *Stammesbildung und Verfassung* (Cologne, 1961).
WERNER, J., *Die Langobarden in Pannonien* (Munich, 1962).
——'Der Grabfund von Taurapilis, Rayon Utna (Litauen) un die Verbindung der Balten zum Reich Theoderichs', in G. Kossack and J. Reichstein, eds., *Archäologische Beiträge zur Chronologie der Völkerwanderungziet* (Bonn, 1977), 87–92.
WERNER, K. F., *Histoire de France*, 1. *Les Origines (avant l'an mil)* (Paris, 1984).
WES, M. A., *Das Ende des Kaisertums im Westen des römischen Reichs* (The Hague, 1967).
WHITTAKER, C. R., 'Agri deserti' in M. I. Finley, ed., *Studies in Roman Property* (Cambridge, 1976), 137–65.
WICKHAM, C., *Early Medieval Italy: Central Power and Local Society 400–1000* (London, 1981).
WIRTH, G., 'Zur Frage der foederierten Staaten in den späteren römischen Kaiserzeit', *Historia*, 16 (1967), 231–51.
WOLFRAM, H., *History of the Goths* (Berkeley, Calif., 1988).
——*Intitulatio*, 1. *Lateinische Königs- und Fürstentitel bis zum Ende des 8. Jahrhunderts* (Graz, 1967).
——'Die Aufnahme germanischer Völker ins Romerreiche: Aspekte und Konsequenzen', in *Popoli e paesi nella cultura altomedievale* (Settimane di studio, 29; Spoleto, 1983), 87–117.
——'Gotisches Königtum und römisches Kaisertum von Theodosius dem Grossen bis Justinian I.' *Frühmittelalterliche Studien*, 13 (1979), 1–28.
——'Zur Ansiedlung reichsangehöriger Föderaten', *Mitteilungen des Instituts für Österreichische Geschichtsforschung*, 91 (1983), 5–35.
—— and DAIM, F., *Die Fölker an der mittleren und unteren Donau im fünfsten und sechsten Jahrhundert* (Vienna, 1980).
—— and SCHWARCZ, A., ed., *Die Bayern und ihre Nachbarn*, 1 (Vienna, 1985).
WOOD, I. N., 'Gregory of Tours and Clovis', *Revue belge de philologie et d'histoire*, 63 (1985), 249–72.
WORMALD, P., 'The Decline of the Western Empire and the Survival of its Aristocracy', *Journal of Roman Studies*, 66 (1976), 217–26.
WOZNIAK, F. E., 'East Rome, Ravenna and Western Illyricum', *Historia*, 30 (1981), 351–82.
WREDE, F., *Über die Sprache der Ostgoten in Italien* (Strasburg, 1891).

WROTH, W., *Catalogue of the Coins of the Vandals, Ostrogoths and Lombards in the British Museum* (London, 1911).

ZECCHINI, G., 'I "Gesta de Xysti purgatione" e le fazioni aristocratiche a Roma alla metà del V secolo', *Rivista di storia della chiesa in Italia*, 34 (1984), 60–74.

——'Il 476 nella storiografia tardoantica', *Aevum*, 59 (1985), 3–23.

ZEILLER, J., *Les Origines chrétiennes dans les provinces danubiennes de l'empire romain* (Paris, 1918).

ZEISS, H., 'Die Nordgrenze des Ostgotenreiches', *Germania*, 12 (1928), 25–34.

ZIMMERMANN, F. X., 'Der Grabstein der ostgotischen Königstochter Amalafrida Theodenanda in Genazzano bei Rom', in *Festschrift für Rudolf Egger*, 2 (Klagenfurt, 1953), 330–55.

ZIMMERMANN, O. J., *The Late Latin Vocabulary of the 'Variae' of Cassiodorus* (Washington, DC, 1944).

ZÖLLNER, E., *Geschichte der Franken bis zur Mitte des sechsten Jahrhunderts* (Munich, 1970).

INDEX

PHYSICALISM